Springer Series in
ADVANCED MICROELECTRONICS   19

Springer Series in
ADVANCED MICROELECTRONICS

Series Editors: K. Itoh   T. Lee   T. Sakurai   W. M. C. Sansen   D. Schmitt-Landsiedel

The Springer Series in Advanced Microelectronics provides systematic information on all the topics relevant for the design, processing, and manufacturing of microelectronic devices. The books, each prepared by leading researchers or engineers in their fields, cover the basic and advanced aspects of topics such as wafer processing, materials, device design, device technologies, circuit design, VLSI implementation, and subsystem technology. The series forms a bridge between physics and engineering and the volumes will appeal to practicing engineers as well as research scientists.

Volumes 1–17 are listed at the end of the book.

V.A. Perevostchikov   V.D. Skoupov

Gettering Defects in Semiconductors

With 70 Figures

 Springer

Professor Victor A. Perevostchikov
Dr. Vladimir D. Skoupov
Nizhny Novgorod State University
after N.I. Lobatchevsky
Gagarin ave. 23
603950 Nizhny Novgorod, Russia

Translator:
Dr. Victor Gloumov
RGGU Russian State University
for Humanities
Nizhny Novgorod Branch
Oktyabrskaya Str. 25
603005 Nizhny Novgorod, Russia

Series Editors:
Dr. Kiyoo Itoh
Hitachi Ltd., Central Research Laboratory, 1-280 Higashi-Koigakubo
Kokubunji-shi, Tokyo 185-8601, Japan

Professor Thomas Lee
Stanford University, Department of Electrical Engineering, 420 Via Palou Mall, CIS-205
Stanford, CA 94305-4070, USA

Professor Takayasu Sakurai
Center for Collaborative Research, University of Tokyo, 7-22-1 Roppongi
Minato-ku, Tokyo 106-8558, Japan

Professor Willy M. C. Sansen
Katholieke Universiteit Leuven, ESAT-MICAS, Kasteelpark Arenberg 10
3001 Leuven, Belgium

Professor Doris Schmitt-Landsiedel
Technische Universität München, Lehrstuhl für Technische Elektronik
Theresienstrasse 90, Gebäude N3, 80290 München, Germany

ISSN 1437-0387

ISBN-10 3-540-26244-X Springer Berlin Heidelberg New York
ISBN-13 978-3-540-26244-2 Springer Berlin Heidelberg New York

Library of Congress Control Number: 2005927733

Springer is a part of Springer Science+Business Media.

springeronline.com

© Springer Berlin Heidelberg 2005
Printed in The Netherlands

Camera-ready by the Author and SPI, India
Cover concept by eStudio Calmar Steinen using a background picture from Photo Studio "SONO".
Courtesy of Mr. Yukio Sono, 3-18-4 Uchi-Kanda, Chiyoda-ku, Tokyo
Cover design: *design & production* GmbH, Heidelberg

Printed on acid-free paper   SPIN: 10973776   62/3100/SPI - 5 4 3 2 1 0

Preface

The level achieved by semiconductor micro– and nanoelectronics and its orientation upon high integration level and functional complexity of devices closely relate to the problems of precise regulation of impurities in the original materials and semifinished items. These problems also cover a development of new and perfection of traditional technologies. The below problems also fall in this list: producing especially pure substances; growing monocrystals of semiconductors and dielectrics, their by-stage treatment up to a ready product with preassigned electrophysical and functional properties; developing the technologies which could drop the probability of defect formation and unregulated pollution by unwanted impurities of active and passive elements in device structures, in each of these operations.

These problems can be solved in several ways. From one side, by improving the technological equipment; its automatizing; and by a use of soft (not severe) treating conditions; by diagnosing the material parameters in the *in situ* regime – in order to timely correct the modification of the material properties, etc. From the other side, these problems may be solved through developing and introducing new techniques into technological processes; for example, for preventing or significantly decreasing – in each operation – the negative changes in the impurity and defect composition. A totality of such techniques or each of them separately is an essence of gettering.

In contrast to basic, well studied and tested, technological operations (such as growing the crystals, preparing a wafer, epitaxy, oxidation, diffusion, ionic implantation, lithography, etc), gettering techniques are usually mentioned casually and their significance in controlling a structure and properties of semiconducting compositions is thought to be additional and secondary. However, as micro– and nanoelectronics are farther and farther advancing into the area of submicronic and nanotechnologies there increasingly appears a more precise understanding of the negative effects of even slight unregulated spectral variations and defect concentrations upon the electrophysical and functional characteristics of devices, upon their aging and degrading processes under the action of environmental destabilizing factors.

The channels of unregulated changes in the structure and in the properties of device layer components can be suppressed only via correctly choosing and exploiting some particular gettering technique or techniques – and this is impossible to be done without a good knowledge of them.

vi

This book touches the problems of gettering defects in semiconducting materials and structures; and a need in this book is determined by its ability to help you to fulfil, at least, the following three basic purposes:
  – to systematize the experience so far gained in exploiting various gettering techniques in micro– and nanoelectronics;
  – to substantiate new directions in the fundamental and experimental research of this field, certainly perspective for professionals and young researchers and specialists;
  – to fill a gap in the contemporary educational literature on the semiconducting material theory, since the students usually have no constant access to foreign and national periodic publications and patents suggested in this book.

The book briefly discusses the basic types and properties of structural defects in semiconductors, the mechanisms of their birth and transformation during basic technological operations in the manufacture of semiconducting devices and ICs. In it, great attention is paid to classifying and describing the specific gettering techniques, to their specificities both in a general technological process and regimes and conditions of their application. A separate chapter is dedicated to low–temperature gettering techniques which are of high practical significance in the manufacture of devices with active elements of submicronic and nanometer topological sizes.

The authors tried to discuss not only standard already exploited gettering techniques but also to describe the contemporary tendencies in gettering technologies, according to the publications of Russian and foreign researchers.

This book is suggested both to engineers-technologists interested in semiconducting materials theory and also to students, MS holders, post-graduate students being trained in solid–state micro– and nanoelectronics. In this connection, the book has a relatively large list of references that may help students and specialists to find more easily their place among contemporary technologies of gettering.

Our great gratitude for assistance and consultations is expressed to professors N.N. Gerasimenko, G.A. Maximov, B.N. Mordkovitch, D.I. Tetel'baum; to A.N. Mikhailov, a post-graduate student of the physical faculty, NNGU; to PhD V.L. Levshinova; to A.N. Kiselev, a post–graduate student – for cooperation in preparating the Russian manuscript; to leading specialists M.B. Kraeva, V.M. Boubnova, L.M. Semenova and to other colleagues from the physical faculty and the Research Institute of Physics and Engineering, the Nizhny Novgorod State University after N.I. Lobatchevsky.

We are much obliged to Nastya Gloumova and post–graduate student Artyem Morozov for their great effort in preparing numerous figures and an electronic version of the manuscript.

Our special thanks to the translator, PhD Victor Gloumov, who has thoroughly got into the matter and successfully converted our ideas in a clear–cut fashion.

This research was conducted under the financial support of the Federal special program "Governmental support to the higher education integration and fundamental sciences in 1997-2001"; branch 2.1. Unified educational and scientific centre "Physics and chemistry of solids" of the Nizhny Novgorod state university after N.I. Lobatchevsky and Russian Academy research institutes. (State contract No. A-0047, project No. 0541).

Contents

List of abbreviations

AFM (atomoc–force microscopy)
Ag (silver)
AG (aluminoyttrium garnet)
Al (aluminium)
AP – ammonia – peroxide (solution)
Ar (argon)
As (arsenic)
$a - Si$ (amorphous silicon)
Au (gold)
B (boron)
BESOI (bond etching silicon on insulator)
BW (bonding wafers)
C (carbon)
Ca (calcium)
CCD (charge coupled device)
Cd (cadmium)
Cl (chlorine)
cN (one hundredth of Newton)
Co (cobalt)
conv. u. (conventional units)
Cr (chromium)
Cu (copper)
CZ (the Czochralski technique, Czochralski–grown)
ED clusters (extrinsic defect clusters)
ESFs (epitaxial stacking faults)
EW (elastic waves)
dB (decibel)
F (farad)
F (fluorine)
FDI (fast–diffusing impurity)
Fe (iron)
Ga (gallium)
GaAs (gallium arsenide)
Ge (germanium)
H (hydrogen)
HC (hydrostatically–compressed)

HCl (hydrochloric acid)
HDICs (high-density ICs)
He (helium)
HF (hydrofluoric acid)
IMCs (integrated microcircuits)
In (indium)

J_{leak} (leakage currents)

K (potassium)
Kr (krypton)
lead (Pb)
Li (lithium)
LSI (large-scale integration)
MDS (metal-dielectric-semiconductor)
Mg (magnesium)
Mo (molybdenum)
MOS (metal-oxide-semiconductor)
N (Newton)
N (nitrogen)
Na (sodium)
Nb (niobium)

NH_3 (ammonia)

Ni (nickel)
O (oxygen)
ODC (original defect composition)
ODE (orientation–dependent etching)
OSFs (oxygen-induced stacking faults)
P (phosphorus)
Pb (lead)
PCE (plasmatic-chemical etching)
Pd (palladium)
PDs (point defects)
QDs (quantum dots)
rad (radian)
RTA (rapid thermal annealer)
s (second)
Sb (antimony)
SDB (boron–doped silicon)
Se (selenium)
SEB (silicon of electronic conductivity, boron–doped)
SEP (electronic conductivity, phosphorus–doped silicon)
SES (electronic conductivity, sapphire – doped silicon
SHB (silicon of hole conductivity, boron–doped)
SHF (superhigh frequency)
SHSICs (superhigh–speed integrated circuits)
SiC (silicon carbide)

SiH_4 (silane, silicon hydride)

SiO_2 (silicon dioxide)

Si_3N_4 (silicon nitride)

SIMNI (separation by implantation of nitrogen)

SIMONI (separation by implantation of oxygen and nitrogen)

SIMOX (separation by implantation of oxygen)

SIMS (secondary ionic mass–spectroscopy)

SMOS (silicon–metal–oxide–semiconductor)

SOD (silicon on dielectric)

SOI (silicon on isolator)

SOS (silicon on sapphire)

Ta (tantalum)

Te (tellurium)

$U_{barrier}$ (barrier voltage)

UHV (ultrahigh vacuum)

USI (ultrasonic irradiation)

W (tungsten)

V (vacancy)

V (vanadium)

VATE (vertical anisotropic etching)

ZMR (zone–melting recrystallization)

Zn (zinc)

Zr (zirconium)

Names and abbreviations
of publishing houses and periodicals

Abbreviations and transliteratied names of Russian journals and publishing houses used in the book	In English
Aktivirouemye protzessy tekhnologii mikroelektroniki	Activated processes in the technology of microelectronics (Collection)
Atomnaya energiya	Atomic Energy (Moscow)
Dielektriki i polouprovodniki	Dielectrics and Semiconductors (Moscow)
Doklady Akademii Nauk (DAN)	Reports of Academy of Sciences (Moscow)
Elektronnaya Promyshlennost'	Electronic Industry (Moscow)
Elektronnaya Tekhnika	Electronic Engineering (Moscow)
Energiya	Energy (Moscow)
Energoatomizdat	Atomic Energy Publishing House (Moscow)
Fizika i khimiya obrabotki materialov	Physics and chemistry of material processing (Moscow)
Fizika i Tekhnika Polouprovodnikov (FTP)	Physics and Techniques of Semiconductors (Moscow)
Fizika i Tekhnika Vysokih Davleny (FTVD)	Physics and Techniques of High Pressures (Moscow)
Fizika Metallov in Metallovedeniye (FMM)	Physics of Metals and Science of Metals (Moscow)
Fizika Tverdogo Tela (FTT)	Physics of a solid body (Moscow)
Fizitcheskaya i khimitcheskaya obrabotki materialov	Physical and Chemical Treatments of Materials (Moscow)
Izvestiya Vouzov. Elektronika	Reports of Higher Education Institutes. Electronics (Moscow)
Itogi Nauki i Tekhniki. Pouchki zaryazhennyh tchastitz i tvergoe telo. Raspolozhenie	The Collection "Results of Science and Engineering. Beams of charged particles and a solid body. Location), Moscow, Russian Academy of Sciences
Izvestiya Vouzov, Seriya Materialy Elektronnoi Tekhniki	Reports of Higher Education Institutes. Series: Materials of Electronic Devices (Moscow)
Khimiya	Chemistry (Leningrad)
Kristallographiya	Crystallography (Moscow)
Materialy Elektronnoi Tekhniki	Materials of Electronic Engineering (St. Petersburg)

Metallurgia	Mettalurgy (Mocow)
Metallurgizdat	Mettalurgy Publishing House (Moscow)
Mikroelektronika	Microelectronics (Moscow)
Mir	World (Moscow)
Nauka	Science (Moscow)
Naukova doumka	Scientific Thought (Kiev)
Obzory po elektronnoy tekhnike	Reviews on Electronic Engineering (Moscow)
Optiko–Mekhanitcheskaya Promyshlennost' (OMP)	Opticomechanical Industry (Moscow)
Optoelektronika i Polouprovodnikovaya Tekhnika	Optoelectronics and Semiconducting Engineering (Moscow)
Pis'ma v Zhournal Teoretitcheskoy Fiziki (ZhTF)	Letters to Journal of Theoretical Physics (Moscow)
Peterburgsky zhournal elektroniky	Petersburg Journal of Electronics
Polouprovodnikovaya tekhnika i mikro-elektronika	Semiconducting Engineering and Microelectronics (Moscow)
Poverkchnost'. Fizika, khimiya, mekhanika	Surface. Physics, chemistry, mechanics (Moscow)
Poverkchnost'. X-ray structures and neutron studies	Surface. X-ray structures and neutron studies (Moscow)
Pribory i Teknika Eksperimenta (PTE)	Instruments and Techniques of Experiment (Moscow)
Radio i Svyaz'	Radio and Communication (Moscow)
Sol. St. Technology	Solid State Technology (Moscow)
Vysokotchistye veshestva	Highly Pure Substances (Moscow)
Voprosy mikroelektroniky	Problems of Microelectronics (Moscow)
Zaroubezhnaya Elektronnaya Tekhnika	Foreign Electronic Engineering (Moscow)
Vestnik Nizhegorodskogo Gosudarstvennogo Universiteta: Materiali, prozessi i tekhnologii elektronnoi tekhniki	Reports of Nizhny Novgorod State University: Materials, processes and technologies of electronic devices (Nizhny Novgorod)
Vysshaya shkola	Higher Education School (Moscow)
Zavodskaya Laboratorya. Diagnostika Materialov.	Industrial Laboratory. Diagnostics of Materials (Moscow)
Zhournal Teoretitcheskoy Fiziki (ZhTF)	Journal of Theoretical Physics (Moscow)
Zhournal Eksperimentalnoy i Tekhnitcheskoy Fiziki (ZhETF)	Journal of Experimental and Engineering Physics (Moscow)

Introduction

The presently popular notion "gettering" seemingly first introduced into the terminology of microelectronic technologies by Goetzberger and Shockly [Mil'vidsky (1999)] embraces both particular techniques and also the entire totality of technologies being exploited to improve the performance of semiconducting devices, through eliminating or suppressing the electrical activity of background impurities and extensive structural defects objectively accompanying practically any device manufacturing stage.

Extending defects in semiconductors (dislocations, in particular) were intensively and fruitfully studied in 60s – 80s of the past century. At this stage of research, there have been unveiled the basic properties of "pure" dislocations (i.e. the dislocations without ambients of impurity atoms).

The defect–free linear segments of dislocations in silicon and germanium were found to relate to one-dimensional electronic zones responsible for a number of interesting physical effects, for example, electrodipolar spin resonance (the Rushby effect), spin–dependent recombination on dislocations, and dislocation photoluminescence. Though these processes were of interest, firstly, from the fundamental point of view, soon it has become clear that the dislocations in the active zone of electronic devices affect their performance strongly. The problem of eliminating the dislocations within the active layer of silicon chips has been successfully solved. In reality, man has managed to produce dislocation–free silicon and eliminate process–induced dislocations.

But it turned out so that in some cases the extended defects are simply needed! To know how to incorporate them into silicon in a controllable way has become a necessity. This knowledge is needed by the contemporary technologies of silicon ICs. Contemporary microelectronics is impossible without gettering, i.e. removing harmful impurities (Cu , Au , Fe , Ni) from a working layer of silicon wafers by "sucking" them off into the bulk of the wafer or onto its reverse side.

Dislocations and some other extended defects (for example, SiO_2 precipitates) have been revealed to be wonderful getters. Therefore, they are purposefully generated in the nonworking areas of wafers either by a thoroughly chosen sequence of thermal treatments or by special implantations followed by thermal treatments, or by some other more intricate techniques. Extended defects are thus exploited as a very efficient garbage collector, a collector of parasitical impurities. This kind of gettering is commonly employed by IC designers and producers. The physics of gettering processes is still not clear to the end. For example, still unknown are the acts and reactions leading to a capture of impurities into the nuclei of a dislocation; it is not yet clear what role is played in these processes by dislocation defects and by the energy of the bonds of impurities in the nuclei of dislocations. At present, these problems are still under active study.

Dislocations may also help to solve one more problem of contemporary silicon electronics – manufacturing the silicon light–emitting diodes for optoelectronical communication inside silicon chips. This problem has two most intensively advanced approaches: a) a use of internal emitting transitions of some impurities implanted into silicon (for example, erbium); and b) a use of silicon nanocrystals in a dielectric matrix (for example, "nanoporous" silicon or silicon nanoclusters in SiO_2). Contemporary silicon diodes are still not efficient. Perhaps, the dislocations "correctly dislocated" in silicon will change the situation. The dislocations in silicon are known to produce four luminescent bands. One of these bands, the most promising, lies within the area of $\lambda \approx 1.5\ mcm$. As for the problem of creating dislocations in a strictly definite place, this problem is easily solved by several ways:

• through generating dislocation semiloops encircling the locally–implanted precipitates of some impurities (for example, the same erbium);

• by growing a $Si - SiO_2$ layer on a silicon surface (the differences in the constants for the lattice of these materials give rise to the so–called misfit dislocations in the interface).

It should be emphasized here that earlier main attention of researchers was paid to a study of "pure" dislocations, and the properties of "dirty" dislocations were not studied thoroughly.

Dirty dislocations is an absolutely new and specific object of investigation, since the properties of the impurity atom approaching the dislocation nucleus may drastically differ from the properties of the atom lying in the bulk of the crystal. There has been established an entire branch of science, "defects engineering". In particular, at present Russian researchers are busy with constructing "a gettering simulator" for solar batteries, i.e. a special computer program for simulating and optimizing the gettering processes which can lead to the idea of selforganization.

The gettering techniques make it possible:

• to remove unwanted impurities from the entire bulk or some local regions in a semiconductor;

• suppress the formation of nucleation centres and growth of new crystallographic distortions;

• perform a purposeful spatial redistribution of original (for the particular technological operation) defects in the device composition, in particular, at the expense of reducing there sizes and concentrations.

Gettering is usually combined with a basic technological operation, which modifies the properties and the topology of a semiconducting crystal when the elements of specific discrete devices or ICs are being formed. From the point of view of the technology, they distinguish the below gettering techniques:

• gettering by structurally–damaged layers of the semiconductor itself;

• gettering by the films of the materials whose physical and chemical properties differ from the properties of the material to be gettered;

• gettering by thermally treating in inertial or chemically active ambients;

• gettering by the getterophase drains especially constructed in the bulk of the crystal for mobile point defects (the so-called "internal" gettering).

The primary portion of gettering techniques has been developed and is exploited in the manufacture of Si – based devices. This is so, first, because this material retains still basic in micro- and nanoelectronics; and second, the technology of its growth and purification qualitatively far surpasses the technologies for manufacturing other semiconductors; this allows to exploit more efficiently the gettering processes which provide a minimal probability of uncontrollable transformation of original defect composition in Si crystals. That is why basic attention in this book is paid to gettering the defects in silicon.

In the last years, gettering is also applied to manufacturing the devices on other semiconducting materials and is used in the technology of micro- and nanoelectronics, too. Here, micro- and nanoelectronics play a special role in the scientific and technical development of the competitive and economically–efficient products of new generations. In particular, integration of micro- and nanodevices may determine to a large extent the manufacture of components for micro- and nanoelectronics, quantum computers and the materials for them. In this connection, the specificity of gettering techniques is also considered below.

It must be noted here that sufficiently complete reviews of gettering techniques developed in 70s – 80s and so far exploited were given in [Kock (1981); Fistoul' (1995)].

In our book these reviews are referred to repeatedly, since they cover not only the basic gettering techniques but also the methodological principles which make it easier to understand the qualitative essence of physicochemical processes of removing impurities and defects from semiconducting structures. The latter is significant for analysing the primary mechanisms of new gettering techniques lately developed empirically; these techniques are habitually complex because the number of physical factors in them simultaneously acting on the material is great and the nature of them is complicated. As an example of such techniques there may be the irradiation–, chemically– or electrochemically–stimulated low–temperature gettering by structurally–inhomogeneous homo– or heterophase layers.

After all, a skill of analysing the experimental results and recognizing a dominating process or a group of processes being initiated by a given type of external actions is nothing but a guarantee of trustworthiness in forecasting possible changes in the structure–sensitive properties of the material under study; and, besides, this skill is a single assistance in choosing the most efficient technologies providing the needed operating characteristics and the quality for the item under production.

The content and the structure of this book are first oriented upon making a reader familiar with numerous contemporary gettering techniques differing in the ways of their application. Though, in contrast to traditional narrow–oriented reviews we, except our comments on each technique, try to trace some common regularities in the gettering effects and explain these regularities with a use of contemporary physics of true crystals. Some of our considerations will probably seem disputable and, perhaps, erroneous. Therefore, any constructive comment or complement will be highly appreciated by us.

1 Basic technological processes and defect formation in the components of device structures

There are briefly considered the basic technological processes in contemporary microelectronics: growing the semiconducting crystals (ingots) and manufacturing the wafers from them; depositing the dielectric and oxidized films; diffusive and ionic doping; epitaxy; lithography; chemicodynamical polishing and etching; metallization and the like. Defect formation processes and types of defects are discussed as well.

1.1. General

Semiconducting silicon has been a "number one" material in solid electronics for about 50 years already. During these years, colossal money has been invested into the silicon industry (and into its scientific provision), and the results gained are extremely impressive. Contemporary monocrystalline silicon is the most perfect crystalline material among the vast number of materials ever created by humans or nature. Every year the world manufactures about 9000 tons of highly–perfect dislocation–free monocrystals. A main user of this unique product is microelectronics whose need makes up about 80 percent of the world output of monocrystals [Hannay (1960)].

The silicon microelectronics is entering the XXI-st century with its ultralarge ICs for a dynamical memory ($256\,Mb - 1\,Gb$) and with its microprocessors of timing frequencies $3.0 - 5.0\,GHz$. In 2012, a $64\,Gb$ dynamical memory and $10\,GHz$ microprocessors are supposed to have been developed.

Monocrystalline silicon is also a basic material in the devices for high–current ("powerful") electronics. Each year the nomenclature of high–current devices is widened: powerful diodes and terristors are accompanied with a wide spectrum of powerful transistors and various "powerful" ICs.

Contemporary silicon is also crucial in the development of semiconducting solar engineering. The total electrical energy produced by silicon solar batteries has exceeded the level of thousands of megawatts. For the nearest years it is planned to obtain a multiple increase (as much as ten times) of this power, with the cost of electrical energy being simultaneously reduced to the level which is comparable – or even lower – with the traditional sources of energy (thermal, atomic and that of hydropowerstations). A vast application of solar batteries in power engineering is promising an enormous ecological advantage. The XXI-st century will certainly be an epoch of the massive jump of world power engineering to the exploitation of nontraditional restorable sources of energy. In this very process, the significance of semiconducting solar energy transformers will be extremely high.

Last–year studies have revealed real perspectives for silicon optoelectronics, wide efficient applications of silicon in various sensors, precisional micromechanical systems and in some other newest engineering systems.

A growing complicity of ultralarge ICs is accompanied with a rise of severe demands as to the quality of original wafers (their general and local flatness, a reduction of their surface pollution level, an increase in structural perfection and homogeneity of original monocrystals), with a simultaneous demand for their larger diameters and low costs.

One of the most important objectives for semiconducting electronics is providing a high level of purity and perfection of crystalline structures in semiconducting materials used in the manufacture of discrete devices and integrated microcircuits. These requirements are so high that the most contemporary methods for estimating the concentration of unwanted impurities often turn out to be less sensitive to these impurities than the devices made on this material. Structural defects and side impurities significantly affect the parameters, the percent of output of fit and reliable items.

A sharp rise in assemblage density and a fall in sizes of the working elements in ultralarge ICs determine a need of reducing the working currents and voltages. Under these conditions, an importance of outside disturbances becomes very crucial. Therefore, severe demands to the purity, structural perfection and microhomogeneity in the active zone of the device composition are laid down. Especially heavy are the demands concerning the composition of impurities being able to generate electrical and recombinationally–active centres. In this connection, the problem of purity turns into one of the most global [Kock (1981)]. Here, the purity is implied to be provided throughout all stages of the general cycle in the IC production, starting from the manufacture of superpure monocrystalline silicon. In the most acute way these problems also arise in the wafer technologies and in the production of IC components. The dynamics of demands to the admissible level for surface pollution in the wafers approximately looks like this: $5 \cdot 10^{10} \, cm^{-2}$ (year of 1995), $2.5 \cdot 10^{10} \, cm^{-2}$ (1998), $1 \cdot 10^{10} \, cm^{-2}$ (2000), $5 \cdot 10^{9} \, cm^{-2}$ (2004). Superpure original and minor materials are most crucial in the solution of this problem. We imply here a need in efficient chemically pure technologies and the equipment to manufacture a wide choice of technological container materials, metals, chemical agents where limited residual impurities are on the level of $10^{-9} - 10^{-10}$ percent.

Arranging an absolutely pure manufacture is a hardly applicable problem. It needs a colossal expense. In this connection, a sharp increase in gettering the residual impurities and defects and a construction of efficient gettering ambients are becoming most important. To attack these problems, one needs a reliable control of defects in the system and, first of all, regulating an ensemble of point defects in the crystalline matrix, both at the stages of monocrystal growth and slice preparation and in the course of forming separate IC components themselves.

To solve the problem of purity and increasing quality of device structures, one should widely exploit in a general productive line the low–temperature and fast–running technological processes, especially at such stages as epitaxial growth, plasmochemical irradiation and sedimentation, the $p-n-$junction formation and dielectric isolation, the thermal and photonic anneal. Here, nonthermal techniques for stimulating technological processes deserve some special attention.

If defects are considered as n-dimensional damages in a crystalline lattice, then these defects may be classified as point defects ($n=0$), linear ($n=1$), plane (surface) ($n=2$) and spatial defects ($n=3$). Various n-dimensional defects are able to interact with each other to produce complicated associative defects. The point defects are the foreign impurity atoms (Fig. 1.1) [Fistoul' (1995)] lying interstitially in the matrix lattice of the semiconducting crystal, the vacancies, i.e. the void nodes in the matrix, the interstitial inherent atoms of the matrix, foreign atoms adsorbed on the crystal surface. Planar defects are, in particular, the boundaries of grains in crystal–twins and the boundaries of the crystal itself. Spatial defects are microscopic damages like pores, cracks and foreign phase inclusions. The defects in the form of impurity atoms (depending on the difference in the covalent radii of impurity atoms and the matrix and also depending on the impurity and matrix atoms being different in their electrical negativity) constitute hard substituting or implanting solutions.

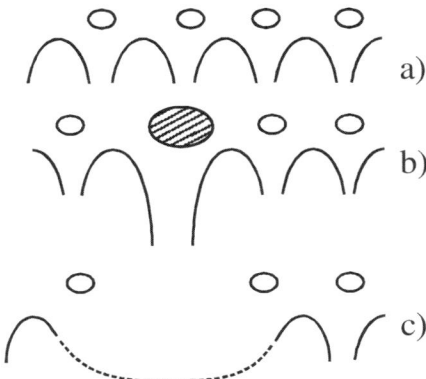

Fig. 1.1. An internal crystalline field **a** in the ideal crystalline lattice, **b** in the lattice with an impurity atom, **c** in the lattice with a vacancy.

Sometimes the defects in silicon are classified in accordance with the way or mechanism of their formation (Fig. 1.2) [Fistoul' (1995)].

I. Defects of inherent disordering
   1.1. Electronic defects
      1.1.1.Free electrons
      1.1.2.Free holes
   1.2. Atomic defects
      1.2.1.Vacancies
         1.1.1.1. Frenkel vacancies
         1.1.1.2. Schottky vacancies
      1.2.2. Interstitial atoms
      1.2.3. Point defects
      1.2.4. Phonons
   1.3. Motion defects
   1.4. Orientation defects
   1.5. Associated defects
      1.5.1. Excitons
      1.5.2. Electron–hole pairs
      1.5.3. Polarons
      1.5.4. Divacancies

2. Impurity disordering
   2.1.Substitutional impurities
   2.2. Interstitial impurities
3. Structural defects
   3.1. Dislocations
   3.2. Small–angle boundaries
   3.3. Crystal boundaries
4. Associated defects
   4.1. Associates
     Atomic
      Vacancies
       Frenkel vacancies
       Schottky vacancies
      Interstitial atoms
      Point defects
      Phonons
     Motion defects
     Orientation defects
     Associated defects
      Excitons
      Electron-hole pairs
      Polarons
      Divacancies
     Substitutional impurities

4.2. Associates
      Atomic
            Vacancies
                Frenkel vacancies
                Schottky vacancies
            Interstitial atoms
            Point defects
            Phonons
        Motion defects
        Orientation defects
        Associated defects
            Excitons
            Electron-hole pairs
            Polarons
            Divacancies
        Substitutional impurities
        Intestitial impurities
        Dislocations
4.3. Associates
      Substitutional impurities
      Interstitial impurities
      Dislocations

Fig. 1.2. Classifying the defects as to the shape of a crystalline lattice disordering.

In this case, mechanical defects imply the defects being induced by elastic–plastic deformation and also during vibrating or striking impacts (slicing; further mechanical treatment: grinding, chemical and mechanical polishing, etc.). Irradiation–induced defects embrace the defects produced by ionic implantation and also under the action of fast electrons, neutrons, $\gamma-$beams, $\alpha-$particles and the like. Thermal defects include the defects emerging during various thermal treatments of crystals. Chemical defects include impurity atoms being introduced into a crystal at its growth stage and also during epitaxy, diffusion and ionic implantation.

    The semiconducting materials being manufactured by the contemporary technology are practically defect-free. However, the exploitation of such materials has revealed a drastic significance of point defects – vacancies, impurity atoms and their clusters. During a technological process, an absence of dislocations, working in a semiconducting material as a drain for point defects, brings a constant rise in the concentration of vacancies and unwanted impurities of nitrogen, potassium, copper, gold, iron, etc. Point defects and their clusters worsen the performance of semiconducting material, cause a degradation of parameters in devices and contribute to a birth of structural defects during thermal treatment. Structural defects, especially decorated with impurities, negatively affect the devices thus decreasing their yield. The diagram (Fig. 1.3) compiled on the basis of studying the silicon transistors and called by the author [Hannay (1960)] "a wheel of failures" shows that practically all technological operations give rise to defects in the material which cause a degradation of transistor parameters.

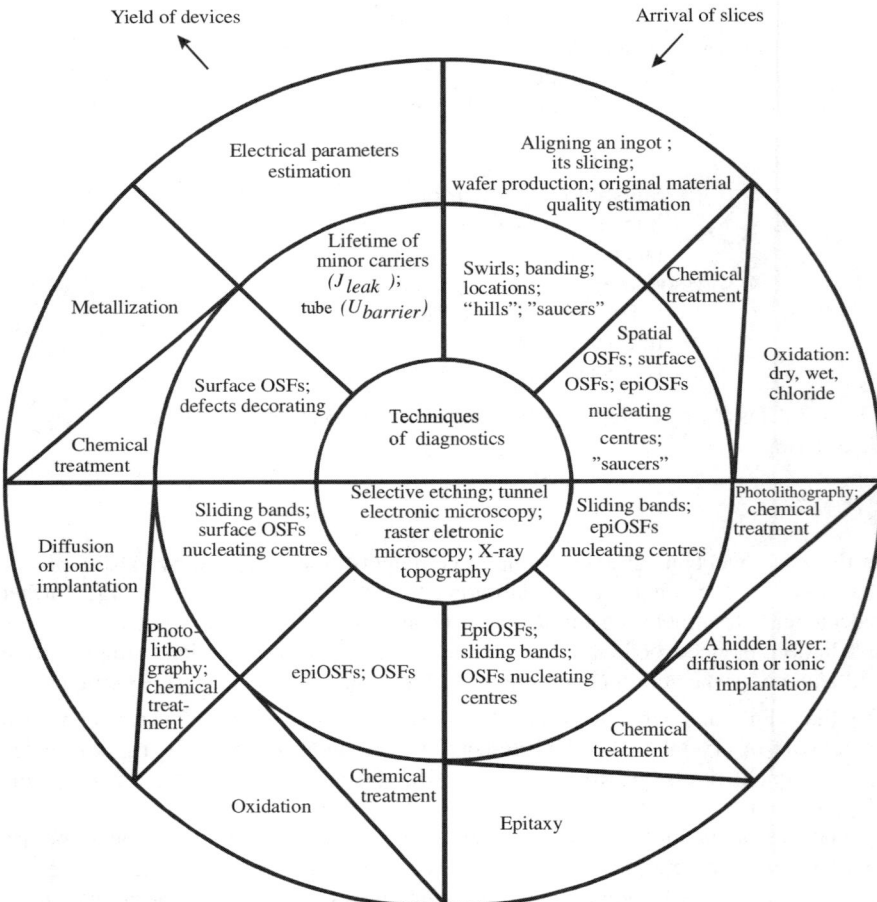

Fig. 1.3. A relation between the basic technological operations for manufacturing a silicon transistor and the basic types of defects.

Contemporary radioelectronic devices and ICs are manufactured with a use of basic technologies, i.e. the technological processes for microelectronic devices

(MED). The techniques of the microelectronic technology are widely applied to manufacturing the instrumentation items, computers, elements and devices for automation, ICs of various integration levels, etc. Contemporary technological processes of the microelectronic technology make use of the latest achievements of physics, chemistry, mathematics, theory of materials, automatization and control.

Microelectronic devices are produced with a use of such types of treatment as mechanical, chemical, thermal, electronic, ionic, plasmic, laser and others. These treatments make it possible in vacuum, regulated gaseous ambient and in usual atmosphere to purposefully produce the wanted mono- and polycrystal or amorphous materials, modify and assign electrophysical parameters for semiconducting, dielectric, resistant, conducting materials and structures built on their basis.

Below is suggested a conventional scheme of basic technological processes in microelectronics being used in manufacturing devices and ICs.

Physicochemical processes in the technology
a) Removing the substance from the hard phase surface
Cutting off
Mechanical cutting and chemicomechanical polishing (removal), abrasive–chemical treatment

Treating by cutting

Abrasive grinding and polishing

Diamond and chemicomechanical polishing

Etching
Chemical and electrochemical etching in solutions and electrolytes

Etching in vapourgaseous mixtures

Evaporating
Vacuum–thermal, plasmic, laser and other types of removals

Etching in vapourgaseous mixtures

Evaporating of subcompounds

Sublimation

b) Depositing the substance onto the hard phase surface
Epitaxy
Oriented build-up of layers during crystalochemical interactions with a wafer

Autoepitaxy

Heteroepitaxy

Hemoepitaxy

Building up

> Nonoriented build-up of films during physicochemical
> interactions with a wafer
> > Polycrystalline films
> > Glass–like and amorphous films

Depositing

> Films depositing during physical interaction (adhesion)
> with a wafer
> > Polycrystalline films
>
> > Glass–like and polymeric films

c) Redistributing impurity atoms / ions in the hard phase volume without

changing geometrical sizes

> **Thermodiffusive processes**
> > Penetrating thermodiffusion–doping
> >
> > Extracting thermodiffusion
> >
> > Exchanging thermodiffusion
>
> **Changing a hard phase composition by thermal or beam action**
> > Forming new compounds in the bulk of hard phase
> >
> > during thermal treatment
>
> **Changing structural components in the bulk, without chemical**
>
> **interaction**
> > Recrystalization and other changes during thermal treatment

Here, the term "a semiconducting structure" implies an ingot of the semiconducting device at some definite stage in the technological process. These structures may be epitaxial, dielectric–semiconductor structures, MDS structures, the structures with diffusive or ionically–implanted layers, and with layers built by metallization, etc.

Main basic technological processes exploited in the manufacture of electronic devices are shown in Fig. 1.4.

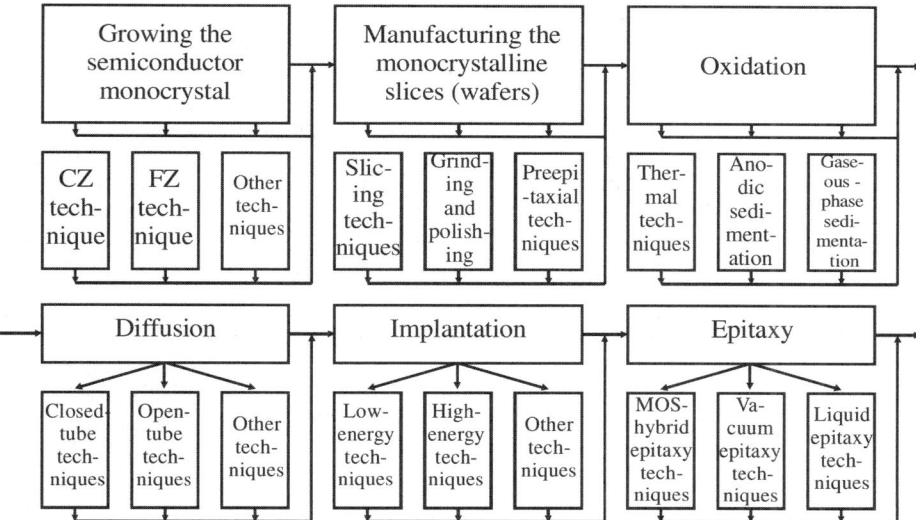

Fig. 1.4. Main basic technological processes to manufacture microelectronic devices.

Silicon, germanium and gallium arsenide are the most important representatives of semiconducting materials most widely employed in contemporary microelectronics. Here, silicon is practically a single material used as a basis for manufacturing ICs of various levels of integration by planar, bipolar or metal-oxide-semiconductor (MOS) technology. A device–manufacturing process implies that the previously prepared silicon slices (matrices) are sequentially subjected to various physical and chemical treatments, in order to establish in the crystal the electrically active and passive elements. Here, the processes used for treating the slices are considered to be the stages of a single productive cycle. Therefore, such productive cycle can be thought to be some set of necessary and obligatory basic technological processes, without which any microelectronic device manufacturing technology becomes impossible. Such technological processes (Table 1.1; Fig. 1.4) include a production (a growing) of crystals, a mechanical treatment of crystals, chemical processes, such as chemical etching, polishing, pickling, diffusion, ionic implantation, epitaxy, lithographic pictures formation, thermal oxidation and metallization.

Table 1.1

Basic technological processes to manufacture
microelectronic devices

Technolo-gical processes	Conditions to perform the processes					
	T (K)	Pressure (Pa)	Techno-logical ambient	Length (hrs)	Mecha-nical pressure (load) kg/mm^2	Depth of structural damages (mcm)
1	**2**	**3**	**4**	**5**	**6**	**7**
a) Manufacturing the monocrystalline materials (silicon, germanium)						
1) CZ growth	Melting temperature	$0.7 \cdot (10^5 - 10^7)$	Vacuum, gaseous (inertial)	24–200	–	Density of dislocations $N_d = 10^2 - 10^5 cm^{-2}$
2) FZ growth	Melting temperature	Same	Same	24–150	–	Density of dislocations $N_d = 10^2 - 10^4 cm^{-2}$
b) Manufacturing the slices of wanted geometrical parameters, quality, structural perfection and physicochemical purity						
Mecha-nical, abrasive–chemical, finishing preplanar and preepitechniques for monocrystal surface treatment						
1) Cutting the mono-crystals	293–298	$0.7 \cdot 10^5$	Standard conditions	1–7	< 0.1	< 30–40
2) Grinding by cohered abrasive	Same	Same	Same	0.5–1.0	< 0.1	< 8–10
Grinding by loose abrasive	Same	Same	Same	1–2	0.1–0.2 (0.8–1.0 MPa)	5–15

1	2	3	4	5	6	7
3) Mechanical polishing by diamond pastes	Same	Same	Same	0.5–1.0	0.1–0.2 (0.5–0.8 MPa)	2–10
4) Chemico-mechanical polishing	Same	Same	Same	1–2	< 0.5–0.6 MPa	< 1.0
5) Chemico-dynamical polishing	Same	Same	Same	0.1–0.3	Not (absent)	Not (absent)
6) Ionic-plasmic or plasmico-chemical etching	< 373	0.6–100	Inertial or active gases	0.5–1.0	Not (absent)	Not (absent)

c) Manufacturing the devices and ICs by constructing in the crystal the electrically–active and passive elements

1	2	3	4	5	6	7
1) *Oxidizing* or- depositing dielectric film – thermal, in the atmosphere of vapour, wet or dry oxygen	1073–1523	Up to $2 \cdot 10^5$	Oxygen, water vapours	1.0–7.0	–	OSFs ≤ 15
– chemical depositing, from gaseous phase	1073–1173	Up to $1 \cdot 10^5$	Oxygen, water vapours	1.0–3.0	–	Same ≥ 10
2) *Diffusion* of donors and acceptor dopants	1073–1273	Up to $1 \cdot 10^5$	Oxygen, com-pounds, vacuum	0.5–1.0	–	SFs ≤ 30
3) *Epitaxy* – chemical depositing, from gaseous phase	773–1473	Up to $0.7 \cdot 10^5$	Hydrogen, oxygen, water vapours	0.5–2.0	–	–

1	2	3	4	5	6	7
– molecular and beam (vacuum) epitaxy	873–1073	$1.3 \cdot 10^{-7}$ $+1 \cdot 10^{-10}$	Vacuum	0.5–3.0	–	–
– liquid phase epitaxy	T_{melt}	Same	Inertial gases, vacuum	1.0–4.0	–	–
4) *Ionic implantation* (ionic doping)	< 373	Up to $0.7 \cdot 10^5$	Vacuum, chemically inertial gases	0.015 – 0.020	–	≤1–2
5) *Metallization* – electronic and beam evaporation (thermovacuum deposition of thin metal films)	T_{melt}	$10^{-11} - 1 \cdot 10^3$	Vacuum	Same	–	–
– magnetronic and plasmic techniques	< 373	$10^{-1} - 10^1$	Argon or vacuum	Same	–	–
– electrochemical techniques	293	$0.7 \cdot 10^5$	Standard conditions	≤1.0	–	–
6) *Lithography* – photolithography	293	Same	Same	Same	–	–
– electronic lithography	293	Same	Same	Same	–	–
–X-ray lithography	293	Same	Same	≤1.0	–	–

It is expedient now to briefly consider the basic technological processes responsible for manufacturing semiconducting devices and ICs.

1.2
General information on the technology of growing crystals and their mechanical treatment

In industry, monocrystals are generally grown by two techniques: by withdrawing, with rotation, from the melt a crystalline seeding (the Czochralski technique) and by float–zone melting technique (Fig. 1.5) [Hannay (1960); Sally and Fol'kovitch (1970); Ravi (1981)].

The Czochralski–grown crystals (the CZ crystals) are produced (Fig. 1.5) by placing polycrystalline silicon into a high–frequency quartz crucible being heated up to the silicon melting temperature ($\approx 1693\ K$). Into the melt there is dropped a seeding crystal of the necessary orientation. The seeding crystal and the melt are being rotated in the opposite directions and through pulling the seeding crystal at some definite rate one obtains the ingots of monocrystalline silicon. This process is arranged in the atmosphere of inertial gas. The purity of the grown monocrystals is determined by the degree of their contamination by the impurity atoms penetrating into the melt during thermal dissociation of quartz crucibles. In this material, basic uncontrollable impurities are oxygen and carbon. In the course of thermal treatment, the presence of these impurities gives rise to microdefects and oxygen donor centres. For example, as a result of the effects exerted by the electrically active oxygen–held aggregates, a growing of monocrystals of specific resistance more than $25\ Ohm \cdot cm$ seems problematic. Besides, unevenly distributed impurities (such distribution occurred during a crystal growing process) affect the deformation of wafers made of slices.

A monocrystal growing is one of the most responsible stages in manufacturing device structures. The slices being cut from a monocrystal are used either to immediately form ICs and discrete devices on them or are used as wafers in manufacturing thin–film epitaxial structures. The CZ technique retains to be basic in the manufacture of Si monocrystals. The technology of manufacturing dislocation–free Si monocrystals is being developed along the way of increasing a diameter of ingots and simultaneously imposing severe demands on the perfection of a crystalline structure and the homogeneous distribution of electrophysical characteristics in the entire bulk of the material.

Meanwhile, the most substantial impact on the working characteristics of ultralarge ICs is exerted just by the growth microdefects located in the slices [Ravi (1981); Park and Ushio, et al. (1994)]. In this connection, with a reduction of topological sizes of the IC working elements there appears a severe demand to the admissible number of submicronic defects to be held on a slice surface.

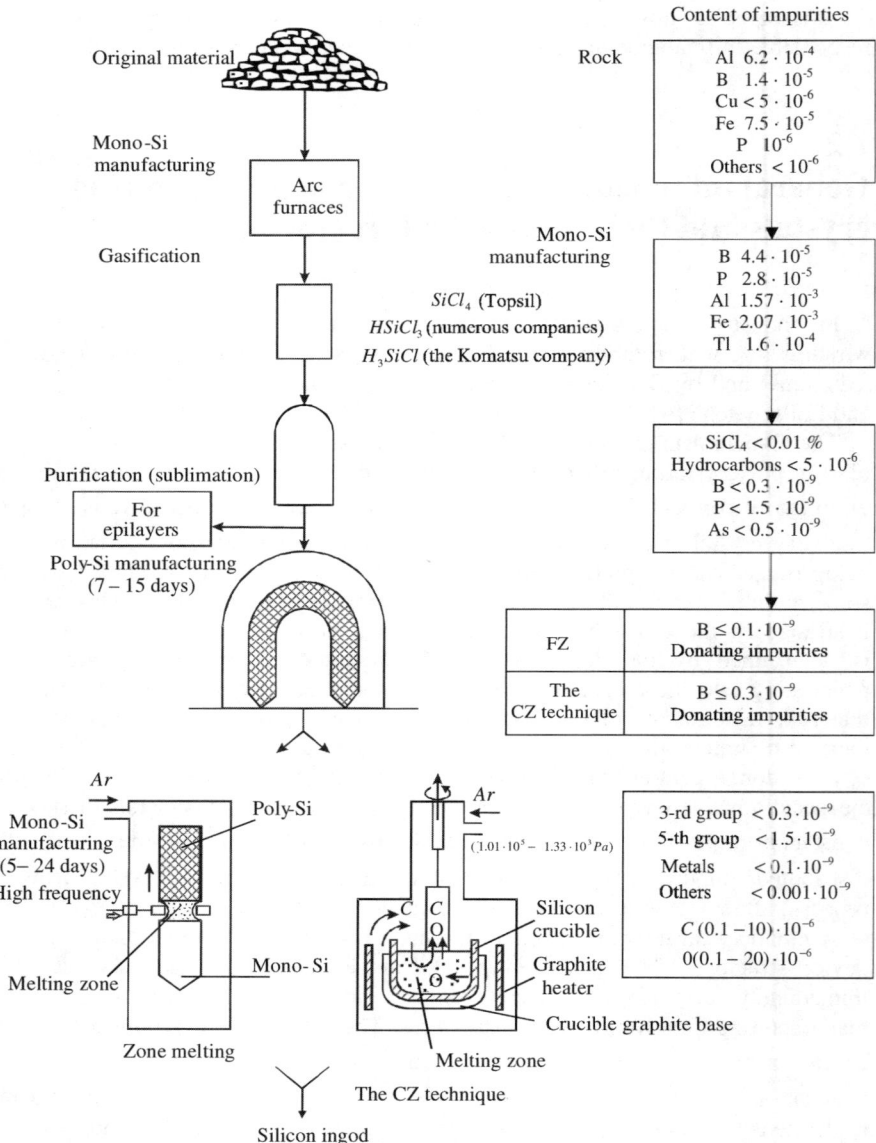

Fig. 1.5. The technological process for growing Si monocrystals and their mechanical treatment.

Intrinsic point (IP) defects – vacancies, interstitial atoms and oxygen – play a basic role in the birth of growth microdefects (for example, see [Park and Rozgonyi (1996); Takeno and Kato, et al. (1996); Puzanov and Eidenson (1992);

Puzanov and Eidenson (1997); Mil'vidsky M.G. (1998); Dornberger and Ammon (1996); Voronkov and Falster (1998)]. Under real growth conditions, already at sufficiently small distances from the crystallization front a monocrystal gets considerably oversaturated with IP defects; these oversaturations are accounted by the balanced concentrations sharply becoming thermally dependent in the Si crystalline lattice. The emerging excessive unbalanced IP defects are annihilating in the drains. As these drains there serve a lateral surface in the ingot and larger defects (mostly dislocations) present in the bulk of the ingot. However, the dislocation–free monocrystals does not contain such efficient internal drains, and the ingot's lateral surface – because of purely diffusive constraints – is not able to fully kill the oversaturations. As a result of this, a crystal in its bulk has oversaturated hard solutions of IP defects which, during a postcrystallization cooling, will dissolve to form specific aggregates of submicronic sizes – usually called microdefects. Another source of growth microdefects is oxygen; in the monocrystals, being grown, its concentration is also sufficient to form an oversaturated hard solution during the postcrystalization cooling. In spite of the concentration of oxygen in crystals largely exceeding the concentration of IP defects, the latters are crucial for a defect formation process. This is so because of the aggregation force being provided not by an absolute concentration of point defects (impurity defects or intrinsic) but rather by an oversaturation of the associated hard solution; this oversaturation is sufficiently higher, namely, for IP defects, due to a sharp fall of their balanced concentrations during a fall of temperature.

The enthalpies of forming interstitial atoms and vacancies in the Si crystalline lattice are comparatively equal. The result of this is that within a sufficiently wide thermal range (the crystallization temperature falling into this range) the balanced concentrations of these defects become comparable, and their recombination plays a decisive role in the microdefect formation process. As shown in [Vanhellemont (1999)], during a quick recombination of interstitial atoms and vacancies the major parameter regulating the nature, sizes and a composition of growth microdefects in the growing crystal is the ratio of the growth rate (V) to the axial temperature gradient in the ingot near the crystallization front: $(G) - \xi = \dfrac{V}{G}$.

At some critical value (ξ_t) of this parameter and with a limited content of oxygen in crystals, no growth microdefects must not be generated. At $\xi > \xi_t$ there emerge interstitial microdefects in crystals; these defects represent themselves the aggregates of interstitial atoms and implanted dislocation loops. Such crystals are usually called interstitial. At $\xi < \xi_1$, there emerge the vacancy pores and oxygen–vacancy aggregates. Such crystals are referred to as vacancy–type crystals. The most negative effects on the parameters of superhigh–density ICs are exerted by the interstitial loops and atoms arising when the crystals are being grown under conditions significantly deviated from ξ_t. In this case, the aggregates of interstitial atoms affect the transistor characteristics immediately; they increase the leakage currents through the $p - n -$ junction, and the vacancy pores worsen, firstly, the quality of a thin layer in the subgate dielectric.

Nevertheless, the CZ technique produces perfect and definitely oriented monocrystals having a minimal number of defects in the form of small–angle boundaries, twins and cracks.

Structural defects emerge at different growth stages, the basic of which being the following.

Seeding

To obtain a low number of dislocations in the growing monocrystal, their number in the seeding crystal should be minimal. For this, surface damages are removed by chemical polishing or chemical etching of the seeding crystal. A cross-section in the seeding crystal must be minimal. At the instant when a cold seeding crystal is getting into contact with a surface of the melt the dislocations will emerge and will be reproduced in the seeding crystal. Therefore, prior to being submerged into the melt the seeding crystal is heated above the melt to high temperatures.

Approaching a necessary diameter

The monocrystal must approach a needed diameter with a small angle of growth, this will drop thermal strains and in this way the density of dislocations. For example, with a growth angle of the monocrystal being equal to $60°$, the density of dislocations in it will be $N_D \approx 10^5 \, cm^{-2}$, and at $10°$ it will be $N_D \approx 10^3 \, cm^{-2}$.

Growing a monocrystal cylindrical portion

The inhomogeneous distribution of temperatures along the length and the cross-section in the monocrystal results in its nonuniform cooling. Thermal strains induce dislocations, too. Thermal strains and thermal gradients cause a travell of dislocations along the sliding planes. In the lattices of diamond (silicon) or gallium arsenide such planes are the $(111) -$ planes. Growing the monocrystals in the directions different from the $<111> -$ direction gives rise to twins. The growth orientation most susceptible to twinning is the $(001) -$ plane. The radial distribution of dislocation densities in ingots is characterized by the maximums available in a lateral surface and in the central part of the cross–section (with triple or six-fold symmetry expressed in the $<111> -$ growth direction and with four–fold for the $<001> -$ direction). The dislocation density is observed to be monotonously increasing along the ingot length, especially in the terminal part of the monocrystal.

More high–ohmic and highly–pure materials are produced by the float–zone (FZ) technique. In this technique (Fig. 1.5), a melted zone of polycrystalline Si

rod is displaced from one end of the reactor (carrying a seeding crystal) to the other. A chozen area is being melted by inductive heating. When the monocrystal is grown in vacuum or inertial gas, impurities are pressed back to an ingot terminal or evaporated from the monocrystal. With some double passes through the reactor's working zone it becomes possible to produce a high–ohmic crystal; it has minor long–life carriers and a small content of oxygen because in this technique the growing semiconductor is not touching the heater's walls (the inductor coil) and other components of the unit. In contrast to the CZ–grown monocrystals being doped with polycrystalline silicon, in the FZ technique the dopants are mostly incorporated into the gas–carrier (argon, helium, etc) or the impurity is incorporated during manufacturing the original semicrystalline rods. Of extraordinary importance in crystals is a homogeneity in the distribution of their impurities; this homogeneity depends on the coefficient K_{ef} of impurity distribution (in other words, the segregation coefficient) between hard and liquid phases. At $K_{ef} > 1$, the impurities are being accumulated in the beginning of the ingot, whereas at $K_{ef} < 1$ they are being driven to the end of the ingot.

There exists an interrelation between the impurity distribution inhomogeneity in crystals and the density of microdefects. There are three types of microdefects emerging in dislocation–free crystals being grown by both techniques. In the cross-section, these defects are often distributed in a spiral–like or flaky manner; that is why they are called swirl defects [Hannay (1960); Sally and Fol'kovitch (1970); Mil'vidsky (1999); Vanhellemont (1999)].

The spiral–like distribution of microdefects betrays a heterogeneous nature of their birth and declares its relation to periodic changes in the growth rate of the crystallization from the melt. Such microdefects produce not only impurity inhomogeneities. Their presence largely affects various technological operations in the device manufacturing line. Therefore, microdefects and other structural distortions are able to cause a degradation of characteristics in semiconducting devices and they can turn into oxygen–induced stacking faults (OSFs) during oxidation and epitaxial growth of silicon films [Kock (1981); Matlock (1977)].

Swirl–defects vary in sizes, structure and conditions of their birth. The largest of them, A − defects, are prismatic interstitial dislocation loops and their aggregates. The sizes of these microdefects depend on the growth rate, the degree of crystal perfection and make up $1 - 5\,mcm$. During a selective chemical etching they give rise to larger pits or etching hillocks. Other defects, B − defects, are either the clusters of interstitial atoms or small inclusions of other phase particles. In both cases, B − defects do not exceed $0.05 - 0.1\,mcm$ in size. They come to being in the shape of shallow pits or etching hillocks. At last, the smallest of the defects, C − defects, appear mostly in the monocrystals grown in silicon crucibles; so far their structure has been studied slightly. The microdefect density is mostly affected by the crystal growth rate.

At the growing rate $v \approx 1.6 \cdot 10^{-3} \ cm/s$ the concentration of $A-$ and $B-$ defects makes up $10^6 - 10^7 \ cm^3$. A rise of the growth rate ($v > 3.3 \cdot 10^{-3} \ cm/\sec$) for the cubicle crystals and $v > 6.7 \cdot 10^{-3} \ cm/\sec$ for the cubicle–free crystals create a possibility to suppress the $A-$ defect formation. To eliminate the $B-$ defects, the growth rate $v \geq 8.3 \cdot 10^{-3} \ cm/s$ is needed. However, the rate $v > 10^{-2} \ cm/s$ brings about new microdefects called $D-$ defects [Karban and Koi (1982)]. These structural distortions, in contrast to swirl-defects, are distributed in crystals uniformly, with the concentration of $\approx 10^9 \ cm^3$, they are vacancy clusters.

Impurity–induced defects

Oxygen penetrating into CZ monocrystal Si ingots accounts to a large extent for material defectiveness (oxygen donors). Its presence determines the emergence and behaviour of the following defects: OSFs; oxygen precipitates, including those decorated by metals; dislocations and dislocation loops. The concentration of O_2 in the CZ ingots is $2 \cdot 10^{17} - 2 \cdot 10^{18} \ cm^3$. The concentration of interstitial O_2 can be estimated through a saturation line at the wave number $1106 \ cm^{-1}$ in the IR-area of the spectrum.

Oxygen donors are generally induced by the thermal treatment within the range $673 - 723 \ K$; their habitual concentration being equal to $10^{14} - 10^{16} \ cm^3$ brings changes in the crystal specific resistance. Therefore, in order to eliminate the oxygen donors and make the ingot's specific resistance similar to the nominal (according to technical requirements), the ingots are annealed at $873 - 923 \ K$ for $0.5 - 1.5 \ hrs$, with a subsequent fast cooling used. A low–temperature anneal has been experimentally detected to make thermodonors electrically active and have an abnormally high concentration. Interstitial oxygen seemingly occupies its position along the $< 111 > -$ direction between the two silicon atoms. The two oxygen atoms present on both sides of the silicon atom weaken its bonds with the remaining atoms of silicon. The oxygen atoms later penetrating into the cluster can drive off the silicon atom into the interstitial space. The subsequent oxygen atoms incorporated into the cluster slightly affect the central interstitial silicon atom "responsible" for electrical activity. An increase in the number of electrically negative oxygen atoms diminishes a depth in the donor's energy levels and turns a cluster structure into a thread–like one. The oxygen–silicon interstitial aggregate is highly mobile at low temperatures. Oxygen donors vanish at $873 - 923 \ K$. Here, a long thermal treatment at $823 - 1073 \ K$ gives rise to donors but this time their structure differs from that in low–temperature donors. These thermodonors easily encounter in the crystals containing carbon;

their maximal concentration is $\approx 10^{16} cm^3$. A preliminary thermal treatment at $723 - 823\ K$ makes it easier to form thermal donors at $1073\ K$.

Oxygen precipitates

A precipitate is a phase of SiO_X. In the process of its growth, the misfits of volumes in the Si lattice give birth to compressing stresses which are relaxing around a precipitate by pressing out a prismatic dislocation loop. Precipitates relate to a formation of OSFs as well.

The CZ ingots hold two types of original oxygen precipitates: "a growth microprecipitate" of the size $1\ nm$ and "a large growth precipitate" of the size up to $100\ nm$. The first type is homogeneously formed during cooling and during ingot's annealing; the second is nucleated heterogeneously.

Carbon-induced defects

In manufacturing a Si monocrystal, carbon alongside with oxygen is a basic component among the accompanying impurities. Usually its content is $(3-5) \cdot 10^{16}\ cm^{-3}$. Its low segregation coefficient (0.058) makes carbon irregularly distributed along the axis and the diameter in the ingot. This results in a stratified distribution (along the ingot's axis) and a spiral–like distribution (along the ingot's diameter) of carbon. With the solubility threshold increasing locally, carbon will form independent SiC-type precipitates. Earlier we have emphasized the importance of carbon in the $B-$ defect formation in the dislocation–free silicon. Particles of silicon carbide can serve as centres for nucleating other defects.

Besides the defects induced by O_2 and C, there may also emerge the defects induced by boron and phosphorus dopants. The boron distribution coefficient is 0.8 and for phosphorus 0.35. At the low pulling rate and when being pulled in an asymmetric thermal field, a crystal will be partially fused per each revolution; here the concentration of impurities in the hardening part will be a function of instant growth rate. Besides, when being pulled into the melt through a gaseous phase the crystal can obtain the impurities of heavy metals. Their source is the rotating shafts and holders made of molybdenum or stainless steel and connected to a graphite clamp of the crystal's seeding.

There is some definite interrelation between the original Si defects and the electrophysical parameters of crystalline structures and devices. First, the dislocations in the original ingots affect the electrophysical properties of the structures being formed during subsequent technological operations and also the parameters of the devices themselves. The effects of dislocations are usually detected either through a change in the macroscopic properties of original crystals or through local changes in the crystal in the vicinity of some separate dislocations. The effects of dislocations on concentration, mobility and lifetime of charge carriers in original crystals can be detected only at sufficiently high density of dislocations $(N_D > 10^5 cm^{-2})$. The electrically active PD aggregates available in the dislocation ambients can give rise to a spatial charge encircling the defects. For example, in silicon in the place where the $60°$ dislocation is approaching the surface the width of the spatial charge exceeds 10 mcm – this is largely more than expected.

When constructing epitaxial layers, the growth dislocations give rise to misfit dislocations in the "epifilm – wafer" interface and can intergrow into the epitaxial film, etc (for details see Chapters 2 – 3).

When being built up epitaxially, stacking faults (SFs) and swirl-defects will turn into epitaxial stacking faults (ESFs), and during oxidation into OSFs.

From the point of view of their impact on the performance of semiconducting devices, the stacking faults stand near to macroscopically extended dislocations; this is because of stacking faults being constantly related to partial dislocations restricting them. Similarly to the dislocations, the electrical activity of stacking faults is determined by the impurity inclusions decorating them and also by the increased diffusion along the partial dislocation. Accordingly, the stacking faults determine more "soft" volt–ampere characteristics, an emergence of microplasma, a drop in breakthrough voltages of $p - n -$ junctions, a rise of leakage currents in diodes and transistors, a local growth of dark generating current in CCD-matrices.

The above permits to conclude that the Si monocrystals being used to construct in an epitaxy–free way the wafers for epitaxial structures and devices must not have growth dislocations, swirl-defects, carbon and large oxygen precipitates; the content of metal impurities (iron, nickel, copper, gold, etc) in them must be minimal. As was told earlier, the carbon concentration must not exceed $4 \cdot 10^{15} cm^{-3}$. As will be shown later, the technological technique needed for this case will be the gettering technique.

To make a semiconducting material ready for manufacturing the devices and ICs, cylindrical CZ or FZ ingots are cut into slices of some definite thickness; the slices are then treated mechanically. Fig. 1.6 presents a general picture of mechanical techniques for monocrystals and the slices made from them [Ravi (1981)]. It should be noted here that the parameter values suggested in Fig. 1.6 may vary to a large extent depending of the technology exploited. Here, the amount of the semiconducting material to be removed mechanically (by grinding and polishing) will be significantly different for a front side carrying a circuit and a back side.

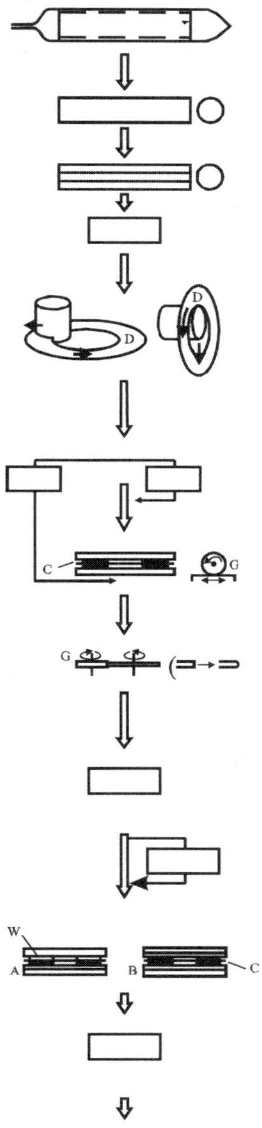

Slices producing and treating technology	Process	Losses of treatment (amount of the material removed)
	Cutting off the ingot terminals	10 – 30 %
	Grinding along the diameter	~ 5 %
	Chamfer grinding	~ 1 %
	Etching	~ 100 mcm
	Slicing (by a usual cutting instrument equipped with an inner cutting edge)	~ 300 400 mcm per a single operation
	Etching (usually not performed) Etching as deep as 50 mcm (if no polishing done)	10 – 50 mcm is removed from both sides
	Polishing or grinding	
	Chamfering	20 – 60 mcm is removed from both sides
	Etching	15 – 80 mcm is removed from both sides of the slice
	Slice back side treating	
	Slice polishing (usually onesided only, sometimes both sided)	A onesided polishing removes 20 – 50 mcm
	Washing, cleaning	Nothing removed
		Total is 65 – 85 %

Fig. 1.6. Silicon slice mechanical treatment.

Upon slicing an ingot, a set of abrasive and chemical operations on the surface of semiconducting slices constitutes a primary block of original operations in the start of the technological line for manufacturing any items of microelectronics, nanomicroelectronics and micromechanics. At present, there are some variants of abrasive and chemical operations being employed depending on specific physical (first of all, mechanical) and chemical properties of crystals [Karban' and Borzakov (1988)]. Each stage of abrasive and chemical operations removes layers of the needed thickness; these layers provide a minimal residual defectiveness in a presurface layer of the slice, prior to a further removing operation. As a result of this, if the operations have been optimized, this block of operations, after a final chemicomechanical or chemicodynamical polishing, produces the wafers of the wanted macro- and microgeometry of the surface and with a minimal (up to interatomic distances as maximum) thickness of the structurally damaged layer. All this sequence of stages in the abrasive and chemical treatment – with the prescribed regimes, conditions for operations and equipment – determines a purposeful and controllable transformation of extrinsic defect composition of the layers in the vicinity of the wafer surface under treatment. Alongside with this, it was experimentally and analytically shown that each of the abrasive or chemical operations results in a uncontrollable change in the extrinsic defect composition, and not only near the surface but in the entire bulk of the crystals [Perevostchikov and Skoupov (1992)]. These effects, in contrast to a purposeful removal to the assigned depth, are better to be called minor effects.

The *abrasive treatment*, starting from the ingot slicing up to a chemicomechanical polishing, gives rise to an injection of unbalanced IP defects into the bulk of the semiconductor from a presurface layer, and induces in it the elastic waves (EW) [Perevostchikov and Skoupov (1992); Bedny and Yershov, et al. (1985); Perevostchikov and Skoupov (1987); Penina and Nazarova, et al. (1988); Perevostchikov and Skoupov (2002)].

The streams of unbalanced IP defects emerge naturally because of dislocations subjected to microplastic deformation and travelling within the local zones where abrasive grains are interacting with the material in the course of removing operations. The traditional literature usually discuss the two mechanisms of IPD formation during plastic deformation [Damask and Dienes (1963)]: a) a mutual annihilation of positive and negative segments in the dislocations with the Bűrgers vector edge component, and this may result in the vacancy chains or intrinsic interstitial atoms, and b) a break-off of atoms from the steps of edge dislocations when the dislocations have a high speed. In abrasive treatment, especially at the first stages – slicing and grinding – a crystal is subjected to a practically striking local stress; under such stress the impulses of mechanical stresses are spreading with a considerable amplitude at speeds close to the longitudinal velocity of sound travelling in the given material; this activates the functioning of the second mechanism [Hirth and Lothe (1970)]. In this case, a rise in the concentration of IP defects is chiefly stimulated evidently by the number of steps directly formed during the local plastic deformation, due to the intersection of edge and swirl defects in the course of their motion.

The concentration of IP defects generated by the first mechanism (i.e. "geometrically") is known to be proportional to plastic deformation, the coefficient of proportionality being $10^{-5} - 10^{-4}$. For silicon in the range of $20 - 30°C$ the deformation during microdamage may reach $3 \cdot 10^{-2}$ [Puzanov and Eidenson (1992)] and the corresponding concentration of unbalanced point defects is $(1 - 2) \cdot 10^{17} \; cm^{-3}$ – that does not largely exceed the IPD concentration at premelting temperatures [Puzanov and Eidenson (1997)]. However, in the event of abrasive treatment followed by quasiimpulsive local heating and by crystal presurface region cooling, the injection of such number of unbalanced IP defects will inevitably transform the original extrinsic defect composition in the crystal down to the depths compatible to a diffusive run of the Frenkel pairs. For example, for silicon the diffusive length of vacancies (according to various estimations) is $3.5 - 22 \; mcm$, and that of intrinsic interstitial atoms is $18 - 20 \; mcm$ [Mil'vidsky (1998); Mil'vidsky (1999)].

Namely, in this range of depths one detects a misfit of microhardness in presurface and bulk layers of silicon slices which underwent various abrasive and chemical treatments [Fistoul' (1995); Voronkov and Falster (1998)]. The dynamical fields of the elastic stresses change the microdefectiveness throughout the entire bulk of the crystals, though most intensively in the presurface region and on the back side of the slice. In the last case, microdefects are transformed due to their interrelations with IP defects being incorporated from the surface by the so-called mechanism of "vacancy pump" [Puzanov and Eidenson (1997)].

That the unbalanced vacancies and interstitial atoms are induced by the abrasive treatment of silicon, gallium arsenide ($GaAs$) and indium antimonide ($InSb$) is proved by X-ray investigations (diffraction and topography), selective etching, and by measurements of microhardness and mass-spectrometry of minor ions [Fistoul' (1995); Tchistyakov and Rainova (1979)]. It has been shown experimentally [Vanhellemont (1999)] that the traditional abrasive treatment (with its subsequent jump from rough abrasive treatment to fine treatment without an intermediate chemicodynamical polishing) gives rise to uncontrolled postoperation accumulation of structural defects increasing the depth of the defective layer. Alongside with this, the fact that IP defects and active elastic waves penetrate anomalously deep may be exploited for regulating an extrinsic defect composition near a front side in the slice, by an abrasive–chemical treatment of its back side [Matlock (1977); Mil'vidsky and Osvensky (1984); Newman (1982)]. For example, the regularities in the long-range (LR) interactions during abrasive treatment are used to purposefully modify the silicon extrinsic defect composition, the unbalanced vacancies and interstitial atoms during the abrasive treatment of silicon, gallium – prior to chemicomechanical and chemicodynamical polishing of slices. Through the X-ray diffractometry, component analysis and metallography and also through measuring the electrical characteristics, stationary and nonstationary microhardness of crystals it has been detected that a diamond single–sided polishing will reduce the mechanical stresses, the thickness of the damaged layer and the density of PD clusters at the reverse side undergoing a finishing treatment. The efficiency of the extrinsic defect composition will be increased by long–range interactions if the processes in the damaged layer are additionally activated by ionic irradiation or by placing the specimens under impulsive hydrostatical compression [Fistoul' (1995); Ravi (1981)]. In the event of long–range effect, the extrinsic defect composition will be transformed more efficiently if the processes in the damaged layer are additionally activated by irradiation or impulsive hydrostatical compression [Perevostchikov and Skoupov (1992); Perevostchikov and Skoupov (2002)]. Irradiation or compression will stimulate a dissolution of impurity ambients around the microdefects and dislocations, and also the unconservative recombination of the latters strengthened by abrasive treatment [Perevostchikov and Skoupov (2002a)]. Similar effect may be achieved by intermediate ultrasonic irradiation of the slices in chemically inactive liquids [Kiselyev and Levshounova, et al. (2002)].

It should be noted here that during the abrasive chemical treatment the uncontrollable minor events, as well as dissolution or growth of microdefects, the sliding or the overcrawling of dislocations, may be largely dependent on the charging state of defects which itself may vary during treatment. These changes are determined by the unbalanced charge carriers being injected from the contacting zone between the semiconductor and the abrasive particles. It is known [Polyakova (1979)] that a shift of minimums of the conductivity zone and a shift of the top in the valency zone may be generally written as a magnitude proportional to a mechanical pressure (P): $\Delta E_g = \alpha_i \cdot P$, where α_i is a numerical constant dependent on the type of semiconductor and the kind of deformation. For silicon, the parameter α_i varies from $1.5 \cdot 10^{-11}$ eV/Pa (in the event of hydrostatical compression) up to $7.4 \cdot 10^{-11}$ eV/Pa if compression is $<100>-$oriented. For any type of deformation, a width of the prohibited zone becomes less, and this results in a growth of the charge carrier concentration. For example, for $n-$material the concentration of the basic and minor charge carriers is described as

$$n_n = N_d + \frac{n_{io}^2}{N_d} \exp\left(-\frac{\Delta E_g}{kT}\right)$$

and

$$p_n = \frac{n_{io}^2}{N_d} \exp\left(-\frac{\Delta E_g}{kT}\right).$$

Here, n_{io} is a concentration of charge carriers in the self–semiconductor; and N_d is a concentration of donors. Similar ratios exist for the $p-$materials.

Taking for silicon the pressure P to be equal to the microhardness $H = 14\,GPa$ during an abrasive action [Karban' and Borzakov (1988); Perevostchikov and Skoupov (1992)], we find that the concentration of minor charge carriers p_n rises as much as $2 \cdot 10^8$ times. Such concentration of minor charge carriers intensifies the recombination processes and this is followed with a local energy release accelerating all the reactions including the microplastic deformation. Besides, the dynamical characteristics of defects will change, in particular, the activating energy of the dislocation travell will drop [Milevscky (1978)].

During abrasive treatment, internal ionization in the material can give rise to minor elastic waves, due to the Coulomb repulsion of nearly lying ions. The amplitude of such elastic waves is determined, in the first approximation of the pair interaction, as

$$P_0 = \frac{q_1\, q_2\, e^2}{4\pi\,\varepsilon\,\varepsilon_0\, r_0\,\Omega}\,,$$

where

$q_{1,2}$ is a multiplicity for ionization of neigbouring ions;

e is an electron charge;

r_O is a distance between ions;

$\Omega \approx \dfrac{4}{3}\pi\, r_O$ is an effective volume of the elastic wave source; and

ε_O, ε are, respectively, a dielectric constant and a relative dielectric permeability of the material.

For silicon, if ions are located in the first coordinate sphere, the estimates give $P_O \geq 1\,GPa$. It is evident that in semiconducting structures, the internal ionization will affect more efficiently the activation and the kinetics of reactions between the components of the extrinsic defect composition, since we have $q_{1,2} > 1$. This is proved by the varying electrophysical properties of $GaAs$, upon the mechanical treatment at large distances from the surface.

In abrasive chemical treatment, especially in abrasive actions with effective excitation frequencies and elastic wave impulse length being, respectively, $40\,kHz$ and $10^{-7} - 10^{-6}\,s$ [Perevostchikov and Skoupov (1992)], the kinetics of creating the extrinsic defect composition, a final spectrum and a spatial distribution of its components will be substantially affected by transient processes that are running during a local contact stress and a material removal by abrasive particles. Here, in addition to basic defect formation processes there are generated dislocation loops in the vicinity of second–phase inclusions – for silicon this mainly occurs near SiO_2 particles; this process creates compressing stresses in the lattice. Besides, there arise new clusters, due to oversaturation with IP defects [Alekhin (1983)], and minor elastic waves induced by recombining vacancies and intrinsic interstitial atoms.

All structural changes in crystals caused by abrasive treatment bring (in accordance with the Le Chatelier – Brown principle) a fall of free energy. However, as soon as the stresses cease (the abrasive treatment is over), the created spectrum and distribution profiles of the extrinsic defect composition get thermodynamically unbalanced and relaxed to a new metastable state; this is betrayed by changes in micro- and macroparameters of slices during their interoperation storage [Perevostchikov and Skoupov (1992)]. From the viewpoint of the technology, this event is important to be considered, for it may increase a damaged layer thickness and bend the slices, especially after the first (rough) treatments of surfaces. How long these semifinished items should be stored between operations – this point has to be solved for each individual case, for a given brand of the semiconducting material and the type of abrasive chemical treatment; the decision should be adopted on the basis of studying the relaxation kinetics of structurally–sensitive properties in crystals. By measuring these properties in the cases of different lengths of storage (at least, for two temperatures), it is possible to determine an efficient activation energy for the relaxation process:

$$U = \frac{k\,T_1\,T_2}{T_2 - T_1}\,\ln\left\{ \frac{t_2(T_1) - t_1(T_1)\ln\dfrac{P_1(T_2)}{P_2(T_2)}}{t_2(T_2) - t_1(T_2)\ln\dfrac{P_1(T_1)}{P_2(T_1)}} \right\},$$

where $P_i(T_i)$ is a numerical value of the detected physical parameter relaxing upon the given operation according to the exponential law $P(t) = P_0 \exp\left(-\dfrac{t}{\tau}\right)$, with the specific relaxation time $\tau_i = \tau_0 \exp\left(\dfrac{U}{kT_i}\right)$ and the constants P_0 and τ_0 .

Through the found value of the energy U , an optimal length of storing the slices between operations may be calculated with a use of the below criterion:

$$\Delta t \leq \tau \cdot \left|\frac{\delta P}{P}\right|,$$

where $\dfrac{\delta P}{P}$ is a relative error for measuring a relaxation parameter.

Thus, a mechanical treatment involves a number of subsequent technological operations. Calibrating the monocrystals makes them strictly cylindrical and of a desired diameter. Ingots are mostly calibrated through round grinding on universal round grinding machines. Upon calibration, a basic and additional marking shears are made on a monocrystal slice, by grinding the slice on a flat grinding machine. The basic shear is needed for wafer orientation during photolithographical and scribing operations. It is set in the $< 110 > -$ direction or at $45°$ with respect to the $< 112 > -$ direction. Additional shears are put to identify the slices of semiconductors of different brands and of various crystallographic orientations.

Upon calibrating and placing shears, there will be a damaged layer $50 - 250\ mcm$ in depth on a monocrystal surface, depending on the semiconductor, treating conditions and regimes.

The experimental data obtained makes it possible to represent [Perevostchikov and Skoupov (1992)] the layers damaged by abrasive treatment as a sequence (Fig. 1.7) of:

- a relief presurface zone of efficient thickness (z_1);

- a zone with cracks and dislocation aggregates (z_2);

- the regions of separate dislocations and dislocation loops induced by unbalanced IPD condensation (z_3);

- the zones of IPD aggregates (z_4 and z_5) with various values of mobility of elementary defects (Fig. 1.7 shows the case $\mu_1 < \mu_2$);

- the zones of elastic deformation (z_6).

The ratios between the thicknesses in the associated layers z_i will certainly be different in each specific case (a wafer material, the type of treatment, etc); at least they will depend on the type of deformation, whether the damage will be brittle or a plastic flow will prevail in this operation. Somehow, in general a structure of the damaged layer will be similar to that shown in Fig. 1.7. This should be taken into account when estimating a thickness of the really damaged layer – both for specifying allowances for a further abrasive or chemical treatment and for analysing possible reasons of spoilage in the technological operations responsible for a direct formation of active regions in devices (oxidation, epitaxy, etc). Such reasons may be the PD aggregates and clusters induced by abrasive treatment and having been not completely removed by the surface finishing operations.

The presence of this damaged layer in the periphery of the wafers can give rise to spalls and in the later high–temperature treatment can induce structural defects that will be spreading throughout a central part of the wafer. To remove the damaged layer, the calibrating and chamfering operations are followed with chemical etching.

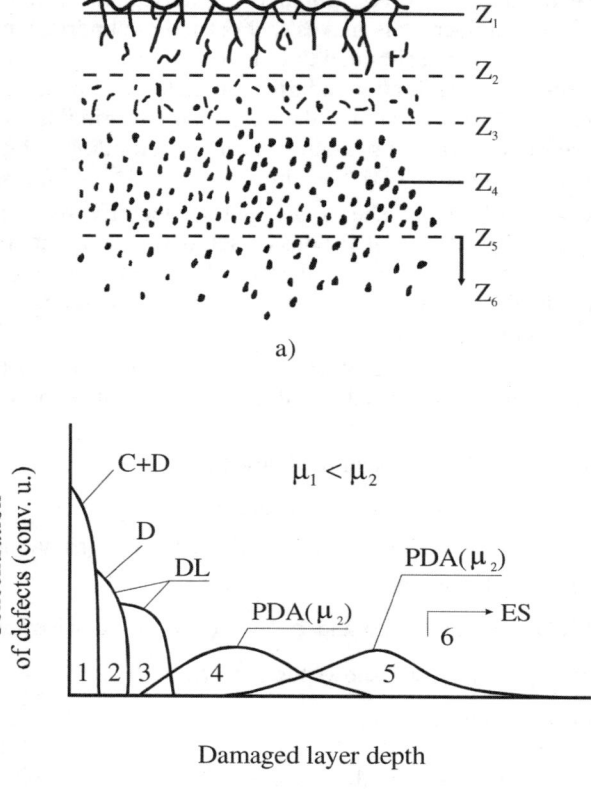

a)

b)

Fig. 1.7. a The structure of the damaged layer and **b** the in-depth concentration distribution profiles for various types of defects: $C + D$ are cracks and dislocations; D, dislocations; DL, dislocation loops; PDA, point defect aggregates and point defect clusters; ES, elastic stresses.

The best of the known technologies of slicing monocrystalline ingots exploits diamond wheels of stainless steel with an internal cutting edge coated with diamond powder [Milevscky (1978); Perevostchikov and Skoupov (2004)]. The cutting speed, the losses of material in the cutting operation and the magnitude of arising residual mechanical distortions are interrelated. Therefore, the regimes and slicing conditions are chosen empirically, jumping from the physicomechanical properties of the semiconducting material under treatment, the diameter and the thickness of the slice, the precision of its geometrical sizes, the surface roughness and the minimal depth of the residual damaged layer. Even with the cutting regimes being the softest and most optimal and the machine tool and the diamond wheel being adjusted precisely, the depth of the damaged layer and the residual stress will retain not less than $20\ mcm$. Therefore, upon slicing the slices obligatorily undergo a chemical etching, to kill structural distortions in the surface. The etching operations eliminate or reduce the effects of microcracks emerging at the stage of deep treatment and increase the mechanical strength of the slices. Here, the density of sliding lines falls sharply if the conditions for the further thermal treatment are not favourable.

However, though the residual mechanical distortions are killed by chemical etching, the slice surface remains not yet smooth and microscopically flat; it should be so in order to be used for manufacturing the devices. The basic techniques to kill defects, reduce a depth of the structurally damaged layer, make the slice sides flat parallel and strictly fitting the declared sizes are slice grinding and polishing.

Depending on the way and nature of the abrasive used, a grinding process may be a cohered abrasive grinding (performed by an abrasive or diamond tool) and a loose abrasive grinding (done by abrasive suspensions).

A cohered abrasive grinding is a high–precision and efficient treatment of a semiconductor slice surface by diamond grinding wheels. Here, a depth of the residual damaged layer will depend on the grains in the grinding wheel, its rotation frequency and the rate of material removal.

When grinding semiconducting slices by a loose abrasive, the averaged grinding parameters (i.e. a specific pressure onto the slices, the cutting rate and orientation, the number of passes for separate abrasive grains) are approximately similar (as the entire square of the slice is being cut by a large array of grains in a nonoriented fashion); this provides a homogeneous structure for the surface. Treating the slices by a loose abrasive makes their surface smoothly dull and eliminate the orientation marks.

To provide a highly–efficient mechanical treatment of monocrystal slices (if the presurface deformation is of small depth), polishing and finishing are recommended to be performed along a closed contour of the least defect producing directions. Here, the surfaces of slices in the $(111)-$plane must be treated along a closed six-edged contour of the $<110>-$directions. If a presurface layer of larger depth is needed to be available (for example, in order to introduce into the slice a getter from a back side or to rise a process rate), then grinding along a $[112]-$sided hexagon is necessary. For the $(001)-$oriented slices, one can choose the two families of optimal directions forming a square. A use of the oriented grinding makes it possible to reduce by $15-30$ percent the depth of the damaged layer, in contrast to an arbitrary (nonoriented) treatment.

Polishing

Polishing provides a minimal depth of the damaged layer and a minimal roughness at a semiconducting slice surface. For this, a semiconductor layer has to be removed as much as two-three times of the depth of the layer damaged by the previous grinding operation. According to the nature of the effects the polishing substances exert on the slice surface, a polishing operation is distinguished as mechanical, chemicomechanical and chemical. The first is used in the preliminary polishing operation.

Mechanical polishing by thinly dispersing (submicronic) diamond pastes does not differ in its nature from a loose abrasive grinding and represents itself a microcutting with a large array of grains. Therefore, this process creates the surface with the microscratches of the sizes similar to the graininess of the polishing diamond powder. These microscratches establish on a slice surface a so-called diamond background representing itself a grid of chaotic micromarks being identified if the slice is illuminated by slanting beams. Therefore, a final polishing (a finishing or a superfinishing) is made chemicomechanically.

A slice is simultaneously acted upon by a chemically active (etching) ambient and mechanical particles or by a soft polishing. The particles remove chemical reaction products from the microprotuberances on the surface and this accelerates the etching reactions in these points of the surface. This evacuating action is also increased if fresh portions of etcher are brought into the microprotuberances, due to mutual movements of the slices and the grinding wheel.

A chemicomechanical polishing is performed both by abrasive and abrasive-free compositions (the solutions of sodium hypochlorite, brom in methanol, brom in phosphorus acid, and others). Alkali or acid solutions with oxidants are used as a chemically active ambient for preparing suspensions, and amorphous high-dispersive powders of silicon oxides, zirconium, aluminium with particles of $0.04 - 0.12 \; mcm$ are used as a hard phase. Such polishing creates the semiconducting slices with a practically defect-free structure of the surface, with the roughness of $\leq 3 \cdot 10^{-2} \; mcm$ and a residual surface concentration of adsorbed impurities equal to $\leq 1 \cdot 10^{14} - 10^{15} \; atoms / cm^2$.

A well-arranged mechanical treatment implies no defects and distortions left on a front (planar) side by the previous treatment. A damaged, or so-called deformed, layer, a layer of the finishing (chemicomechanical polishing) operation, must not exceed $5 - 10 \; nm$. The distortions produced by accidental defects (micromarks, microknock-outs) are significantly larger in depth; the quantity of these distortions is limited by technical regulations.

However, chemicomechanical polishing may result in a sharp rise of pollutants in the presurface layer. For example, due to a high content of phosphorus in the polishing agents (a blue and brown chamois hold $10^{-3} - 10^{-2}$ percent of phosphorus), a phosphorus impurity in a thin presurface Si layer is present in the form of aggregates which, during a later oxidation, change the oxide composition and give rise to the inclusions of phosphor-silicate glass.

In thermally-treated high-ohmic p - slices, phosphorus pollutants in the presurface layer can induce parasitic $p - n$ - junctions. The traces of phosphorus in this layer are sharply reduced or fully cancelled by auxiliary oxidation and through a removal of oxide in the hydrofluoric acid.

Physicochemical treatment of semiconducting wafers also involves a preliminary, an interoperational and a finishing clearing of the surface from organic and inorganic pollutants. The wafers are cleaned in a liquid or vapor–gaseous ambients, this cleaning process being intensified by ultrasonic and low–temperature plasmic irradiation, thermal heating in vacuum, ionic bombardment, etc.

To have a wafer surface prepared completely, a preepitaxial or a preplanar treatment is used. At this technological stage, a wafer surface, immediately prior to its epitaxial build-up, undergoes a chemical polishing, an etching by gaseous ambients, an ionic–plasmic etching or etching in melts.

In any case, a basic technological process for manufacturing and treating the surfaces in semiconducting wafers must meet the technical requirements specifying such admissible structural imperfections as dislocations, second–phase discharges, PD clusters, OSFs and the like. A common requirement to a structural perfection of monocrystalline wafers implies the following: the ingots treated by multistage mechanical operations must hold the residual defects not exceeding the defectiveness level in the original semiconducting monocrystal.

In the course of further basic technological operations (oxidation, dielectric film deposition, epitaxy, diffusion), the structural distortions and microheterogeneity of impurities and defects in the produced wafers drastically affect the formation, interaction and rearrangement of dislocations and other structural defects.

Hunting the defects in silicon and checking the quality of mechanical treatment of semiconducting wafers involve the following: detecting the distortions and defects via chemical etching; microscopy in visible light and IR-irradiation; X-ray-diffractional irradiation and electronic microscopy; analysing the distortions by checking the mechanical properties in specimens; by–layer polishing with a further metallographical study; measuring on angle and spherical metallographical specimens; measuring a depth of flat–bottomed wells, etc.

A selective surface quality check upon chemicomechanical polishing is exercised by a selective etching. Latent defects are detected via a high–temperature oxidation at $383\ K$. Here, mechanical defects are turning to OSFs easily detectable by selective etching.

1.3
Technological processes of dielectric film deposition and semiconductor oxidation

The deformations in wafers arising in the IC manufacture may be classified in the following way:

- elastic and plastic deformations being induced by highly–concentrated impurity diffusion and determined by a difference in sizes of impurity atoms and matrix;
- elastic deformations determined by the inequality existent between the coefficient for a semiconductor thermal expansion and the coefficient for the oxides, nitrides or metals being deposited onto the semiconductor;
- elastic and plastic deformations induced by thermal stresses arising during thermal treatment in the course of oxidation or diffusion, as a result of thermal gradients in the structure.

The deformations in the last category are usually characterized by high values and to a large extent determine the yield of the ready products.

In manufacturing the devices and ICs by the planar technology with a use of polished monocrystalline Si wafers, the necessary stage is oxidation or dielectric film deposition. Silicon devices employ thin amorphous SiO_2 films, Si_3O_4 films, two–layer structures $(SiO_2)+(Si_3O_4)$ of phosphosilicate and boronsilicate glasses. In the last years the films of TiO_2, Ti_2O_3, etc find a wide application.

The SiO_2 films and dielectric films are used as a masking cover in local epitaxy, for local diffusion and local gaseous or chemical etching, for protecting and passivating a semiconductor surface; they are also used in the capacity of subgate dielectric in MOS–based devices, to dielectrically isolate the IC active and passive elements. The SiO_2 doped films are used as a source of diffusion. Amorphous films are produced by various techniques: by a thermal oxidation in the atmosphere of dry or moist oxygen, by pyrolysis, cathodic, plasmic and magnetronic deposition, etc.

Dielectric films should meet the following general requirements:
- good adhesion to a semiconductor and simultaneously good adhesion during depositing a photoresist and metal onto the film;
- mechanical strength;
- high continuity and homogeneity of layers;
- high chemical resistance against selective etchants;
- low permeability for impurity ions;
- good masking properties when boron, phosphorus and arsenic are diffused;
- good dielectric properties (high electric strength, small loss tangent of a dielectric; high dielectric permittivity – in a number of cases);
- providing good properties in the interface (small magnitude and stability of charge, small density of surface states on the silicon surface, etc).

The defects arising in oxidized silicon may be conventionally divided into two categories. The first is determined by the thermomechanical stresses and can give rise to dislocations; the second is determined by chemical reasons contributing to oxidative–restoring reactions on the wafer surfaces, and can induce OSFs.

Besides, in the amorphous dielectrics there arise the crystalline modifications SiO_2, Si_3N_4 and Si_2ON_2 which reveal themselves in the amorphous films in the form of defects. Alongside with this, these films have the through–capillaries (pores) and "dead–end" pores.

Crystalline modifications of SiO_2, Si_3N_4, Si_2ON_2. Crystalline modifications of SiO_2 in amorphous films are revealed as defects due to changes in a specific volume; these changes bring about a local cracking of films. Si_3N_4 has two crystalline modifications: $\alpha - Si_3N_4$ and $\beta - Si_3N_4$. There is also a modification of silicon oxynitride, Si_2ON_2.

The through–capillaries and "dead–end pores" in dielectric films
These defects in silicon are mainly induced at the expense of mechanical defects, the defects in epitaxial layers, by surface pollutants brought about by a nonqualitative wash and dust specks coming from etchants, dissolvents and air.

In the silicon wafers being treated thermally, the regions with local pollutants serve as crystallization centers causing a local cracking. Particles of ion–exchange resins in poorly–filtrated deionized water contain up to some percent of alkoli metals and other impurities favouring crystallization. Pores may be defined as microscopic defects being local distortions of oxide continuity.

Besides cracking, the through and nonthrough ("dead–end") pores emerge because of vacancies being generated and aggregated in SiO_2 during relaxation of mechanical stresses in the SiO_2 interface.

Thermomechanical stresses and dislocations induced by a nonuniform heating and cooling of wafers are reduced by various techniques. When conducting a thermal oxidation, the wafers are sequentially and vertically placed in a special ьoat–like cassette. When placing the cassette into an oven or unloading from it, a central part of each wafer is shielded by neighbouring wafers. In this case, there appears some thermal gradient from the edges of wafers to the centre. In the first approximation a parallel row of wafers looks similar to a cylindrical crystal during radial heating and cooling; this similarity is increasing as the wafers are being mated closer and closer.

Thermal stresses in wafers can be calculated as

$$\sigma = \alpha E \cdot \frac{\Delta T}{t - v},$$

where $\alpha = 4 \cdot 10^{-6}\ K^{-1}$ and $E = 130 - 170\ GPa$. Here, if $\Delta T \approx 50 - 100\ K$, then $\sigma = 40 - 80\ MPa$; this is sufficient for plastic deformation to run at $\geq 1273\ K$. Plastic deformation starts in the places of local aggregates of stresses (spalls in the periphery, local inclusions, etc). Therefore, for the large–diameter wafers the demands of structure homogeneity, a quality of wafer surface mechanical polishing and an absence of bends in wafers increase severely.

Distribution of sliding lines and bands of dislocations has a triple symmetry in $\{111\}$–wafers and a four–fold symmetry in $\{100\}$–wafers. As a rule, the dislocation bands emerge in the periphery. They give birth to mechanical stresses in the wafers, which (if a wafer polishing is two–sided) are detected via the damage–proof technique of X-ray topograms or the damaging technique of selective etching. Dislocation lines and bands can be killed or reduced if a cassette with wafers is slided into or slided out of the oven, and radiation is shielded by metallic foil and if a chamfer is done in the periphery of wafers, etc.

Thermal stresses can also arise on the boundary with dielectric coats. If a SiO_2 layer, as thick as d_2, is deposited at temperature T_0 onto a silicon wafer of thickness d_1, then the SiO_2 layers during their cooling by ΔT will have the compressing stresses:

$$\sigma_2 = (\alpha_{Si} - \alpha_{SiO_2}) \cdot \frac{\Delta T\, E_2}{1 - v_2} \,,$$

where α_{Si}, α_{SiO_2} are the coefficients of linear thermal expansion; E_2 is the Young modulus; and v_2 is the Poisson coefficient for SiO_2.

In a silicon wafer, near the interface of phases, there arise the extension stresses:

$$\sigma_1 = -\frac{4\sigma_2 d_2}{d_1} = -4d_2 \left(\alpha_{Si} - \alpha_{SiO_2}\right) \frac{\Delta T\, E_2}{d_1 (1 - v_2)} \,.$$

As a rule, within a silicon plasticity interval (at $T > 1273\ K$) at the oxide thickness being equal to $d \approx 1\,mcm$, the arising stresses will be of the values insufficient for generating dislocations.

Dislocations induced by local oxidation

When oxidation is done locally or when in the thick oxide there appear windows for diffusion, the stresses and their aggregates near the window boundaries will be redistributed. The stresses may exceed a flow limit and this leads to a formation of a dislocation oriented grid.

A thickness of Si_3N_4 films affects a dislocation density significantly. A critical thickness of such films is from 20 to $50\ nm$, for different processes of local oxidation.

The recommendations on reducing the dislocations and other defects during local oxidation usually imply the following:

- incorporate a buffer polysilicon film, a three–layer mask and a "framed" mask;
- reduce mechanical stresses in the masking Si_3N_4 films; this will depend on the manufacturing technology of Si_3N_4 and its stechiometric composition;
- Si_3N_4 is to be ionically implanted;
- depositing the Si_3N_4 films with a germanium sublayer;
- the gaseous mixture is to be added with nitrogen–containing gases (methane, ethane, propane, etc), a volume portion being equal to $1-5$ percent.

Besides the stresses determined by a difference of thermal expansion coefficients for silicon and oxide, there also arise the stresses induced by structural misfits between a silicon lattice and an ordered structure of SiO_2 in the vicinity of the phase interface.

The difference between the specific volumes of the ordered SiO_2 immediately contacting with Si and the amorphous SiO_2 also induces mechanical stresses in SiO_2 and Si. All this leads to defects in the $Si - SiO_2$ interface.

In dielectrically–coated wafers at $1273\ K$ no dislocations occur and at lower temperatures the wafer bends elastically. The reverse bending radius is

$$\frac{1}{\rho} = -6 \cdot (\alpha_1 - \alpha_2) \cdot \frac{E_2\, d_2}{E_1\, d^2{}_1}$$ (the Poisson coefficients for the wafer and the

coating are assumed equal). Here, with the SiO_2 or Si_3N_4 layer being thick, significant mechanical stresses can emerge affecting a number of parameters in devices.

Oxygen–induced stacking faults (OSFs)

OSFs are born along the traces of scratches left by mechanical processes and on swirl-defects as well. The number of OSFs grows with an increase of oxidation time. This type of defects is mainly induced by a stream of interstitial atoms migrating from the boundary of the growing oxide and heterogeneously depositing onto the defects in the presurface region of silicon.

Large–size OSFs emerge near the interface surface, and they become less in size when they start to be deeper from the surface.

Anisotropy of the sizes of OSFs increases with a fall of oxidation temperature, i.e. at low temperatures the OSFs "strive for growing" along the surface. A repeated oxidation of wafers may cause (besides a growth of OSFs already available) some additional OSFs near the earlier existing faults.

OSFs are assumed to be born rather not on $A-$ but on $B-$defects. If a growing OSF encounters a $B-$cluster, then its growth is accelerated, i.e. the OSF is "absorbing" this $B-$defect.

Long thermal treatment at high temperatures gives birth to OSFs near the SiO_2 discharges in the ingots, with oxygen not higher than 10^{18} cm^{-3}. With oxygen precipitates growing, silicon atoms are driven off from the precipitate–surrounding matrix, defuse into the bulk and can cause OSFs. Besides, a two–stage thermal treatment, i.e. a repeated heating and cooling of the crystal, can also give rise to OSFs in the cases when a single high–temperature treatment produces no noticeable effect.

Annealing in vacuum at 1373 K can drop the size of OSFs, though this kind of annealing is technologically not so good because of erosion in the surface. With a surface orientation being deviated by $3-5$ percent from the $\{100\}$-plane, no nucleation of OSFs is observed.

In FZ silicon (with a small content of oxygen), the OSF density in $\{100\}$-wafers is greatly less than in $\{111\}$-wafers, under similar oxidative conditions. The small–size OSFs or if they are slightly deviated from the surface can be cancelled by multiple oxidations (an oxide layer being removed after each oxidation) and also by low–temperature oxidation which hampers a growth of OSFs.

A basic way to prevent OSFs and remove them is a chloride oxidation at temperatures not higher than 1423 K. Such oxidation also reduces sodium ions and charge instability in the oxide. Implanting a chlorine–holding impurity into an oxidative ambient on the $Si - SiO_2$ interface creates excessive vacancies which are recombined in the presurface region with excessive interstitial atoms. As a result of this, the Si excessive interstitial atoms are annihilated and OSFs suppressed.

Oxidizing at low ($1173-1223$ K) or very high ($1473-1543$ K) temperatures also kills OSFs. Here, the silicon's property is exploited, i.e. during a heating stage the balance between a formation of interstitial atoms and vacancies is biased to vacancies, due to a higher formation energy an interstitial atom obtains in contrast to a vacancy.

To reduce to minimum the possibility of oxidation– and diffusion–induced dislocations, one tries to perform oxidation and diffusion at low temperatures, and also to thoroughly regulate the heating rates and cooling cycles. When the round wafers, held in the wafer–holders vertically and supported along the perimeter in several points, are being put into a hot zone of the oxidizing or diffusing oven or being removed from it, radial thermal changes in wafers can induce thermomechanical strains in them. At this, if wafers are loaded into or removed in a fast fashion, the thermal strains may be sufficiently large to cause a plastic deformation in wafers. A fast cooling in the initial period of extracting the wafers from a hot zone in the oven is controlled by the emitting heat transfer. A drop of temperature allows to reduce radial and tangent strains in wafers, and in this way to prevent their plastic deformation (i.e. to prevent a birth and motion of dislocations). However, low temperatures also reduce an oxidation rate, i.e. a productivity of the process drops, so this technique usually finds no application. A more popular technique is regulating the heating and cooling rates, for maximum strains are induced by sharp thermal changes in wafers. A slow wafer heating and cooling during a thermal treatment reduce a probability of thermomechanically–induced dislocations.

Here, the heating and cooling rates needed depend on the oxidation temperature, a diameter of the wafer and its thickness. The wafers of large–diameter crystals ($>150\,mm$) are usually thicker (up to $500\,mcm$ and more), in order to minimize the probability of formation and sliding of dislocations under thermal strains.

Dislocations in wafers can be also produced by mechanical strikes (friction) or pushes during a load-in or a load-out of wafers from the oxidation or diffusion oven. Similar events are usually eliminated by special precision tools used for wafer load-in and load-out from the diffusion–oxidation systems, providing a minimal friction between the wafer–holders and reaction chambers.

As was emphasized above, a distribution of dislocations induced by thermal strains in silicon usually has a triple symmetry in the $\{111\}$-oriented wafers and a quadruple symmetry in the $\{100\}$-wafers. Consequently, they lie within the $\{111\}$-planes and usually have a higher density on wafer edges as compared against the centre; this represents a distribution of strains in wafers and also the fact that dislocations are nucleated on wafer edges in an easier way.

It was said above that 2-dimentional defects (OSFs) are most popular defects in silicon oxidation. Even if a dislocation–free material is used as original and no mechanical distortions in silicon surface are available, then, nevertheless, silicon oxidation at high temperatures (usually at $1173 - 1523\,K$) in dry or wet oxygen or in water vapours gives rise to OSFs in presurface regions of wafers. This is conditioned by inevitably present microdefects and second–phase discharges which serve as nuclei for generating the above OSFs. In order to have OSFs generated, it is necessary to remove (in the case of a vacancy–type defect) or incorporate (if the defect is interstitial) an atomic "extrasemiplane" between closely packed $\{111\}$-planes. Stacking faults in silicon lie within the $\{111\}$-planes, intersect the wafer surface along the $<110>$ -directions and as a rule are interstitial. Thus, defects are induced because of a birth and condensation of excessive interstitial atoms in silicon. During a silicon oxidation, oxygen atoms are assumed to be transported through the layer of the growing oxide and then they interact with the silicon atoms on the SiO_2 interface. Having approached an interphase boundary, an oxygen atom can occupy a vacancy or get into an occupied lattice site, thus driving off a silicon atom and creating an interstitial atom. Besides, during oxidation, diffusing oxygen causes a growth of SiO_2 precipitates; this growth creates excessive interstitial atoms on the boundary between the discharge and the matrix.

A growth rate of stacking faults depends on temperatures: a low–temperature oxidation produces less stacking faults, in contrast to high–temperature oxidation (with the oxidation time being similar). In the wafers with a repeated oxidation, additional defects may arise in the immediate vicinity of earlier defects. An increase in the layer thickness, time and oxidation temperature leads to a growth of defect sizes, namely, a length of the defect, i.e. a ratio of the length of the defect to its width will increase, and this, in its turn, governs the anisotropy of the stacking fault shape. Swirl-defects in silicon exert larger effects on the formation of OSFs. It was found that during wafer oxidation the aggregates of point growth defects, having a swirl-distribution, work as nuclei for stacking faults. The PD aggregates constructing a swirl–picture are known to be identified in the shape of unlimited flat–bottomed etching wells. As the PD aggregates in silicon are distributed throughout its entire bulk, an increase in oxidation time brings a rise in the number of stacking faults being formed on these aggregates. Besides, whereas the stacking faults emerging in the mechanically–damaged places on the surface are practically of the same size, the stacking faults arising on the PD aggregates are of deferent density and size and emerge both on a wafer surface and in the bulk as well.

Though oxidation is thought to be a basic source of stacking fault concentration, a substantial role in creating stacking faults is also played by impurity atoms, especially, such as oxygen, hydrogen, and fast–diffusing elements of groups I and VIII also.

1.4
Diffusive processes in semiconductors

The $p - n -$ junctions are formed by diffusing impurity (dissolved) atoms being introduced into the lattice to change its electrophysical properties. Impurities in the lattice travell by sequentially jumping in three directions. If there is a gradient of the concentration of intrinsic or extrinsic atoms, the point defects (an atom lying interstitially or in a vacancy) affect their diffusion. The basic diffusive mechanisms involve the following:

• **the vacancy mechanism**, i.e. an extrinsic atom is diffusing to the place of vacancy and then diffusing again to the nearest or fit vacancy;

• **the interstitial mechanism**, i.e. an extrinsic atom is diffusing interstitially;

• **the relay mechanism**, i.e. an extrinsic atom is diffusing from the interstitial position into the interstitial site driving off the silicon atom into the interstitial site;

• **the dissociative mechanism**, i.e. an impurity is simultaneously diffusing across vacancies and interstitial sites. Here, the extrinsic atoms can move being in various charge states – this causes some mutually related diffusive streams between which there occurs a constant exchange of particles;

- the $E-$ **centre mechanism**, i.e. an impurity is diffusing together with a vacancy; here, $E-$ centres can assume different charge states ($\bar{E}, E^0, E^+$) corresponding to different concentrations and different diffusion rates for the associated "vacancy – impurity" pairs;
- the **mechanism of dislocation tubes**, i.e. a mechanism of the accelerated diffusion caused by an increased number of vacancies and interstitial sites and by a change in interplane distances in the vicinity of the dislocation nucleus;
- the **mechanism of diffusion along grain boundaries**, i.e. an accelerated diffusion determined by an increased number of defects on grain boundaries.

The first Fick law for the one-dimensional diffusion is

$$J = -D \frac{[\partial C(x,t)]}{\partial x} \, ,$$

where J is a diffusive flow $(cm^2 \cdot c^{-1})$; C is a concentration of the dissolved substance (cm^{-3}); and D is a diffusion coefficient $(cm^2 \cdot c^{-1})$.

The second Fick law is

$$\frac{\partial C(x,t)}{\partial t} = \frac{\partial}{\partial x} \left(D \, \frac{\partial C(x,t)}{\partial x} \right)$$

$$\frac{\partial C}{\partial t} = D \, \frac{\partial^2 C(x,t)}{\partial x^2} \, .$$

The last equation is called the Fick simple differential equation which is valid for the constant diffusion coefficient (at the given temperature).

To have this second–order partial differential equation solved in partial derivatives, one should know the initial and two boundary conditions. For a small impurity concentration, the dependence of the diffusion coefficient on temperature is described as

$$D = D_D \exp\left(-\frac{E}{kT}\right) \, ,$$

where D_D is a diffusion coefficient ($cm^2 \cdot c^{-1}$), and E is a process activating energy (eV).

The values of D_D and E for some elements are given in Table 1.2.

Table 1.2

The diffusion coefficients for some silicon impurities

Element	$D_0 \, (cm^2 \cdot c^{-1})$	$E \, (eV)$	Range of temperatures (K)
Boron	0.76	3.46	1173 – 1473
Phosphorous	3.85	3.66	1173 – 1473
Arsenic	23 – 24	4.1 – 4.08	1173 – 1473
Antimony	0.214	3.65	1373 – 1473
Hydrogen	$9.4 \cdot 10^{-3}$	0.48	1240 – 1480
Lithium	$2.5 \cdot 10^{-3}$	0.66	298 – 1623
Aluminium	4.8	3.35	1323 – 1653
Germanium	$6,26 \cdot 10^5$	5.28	1423 – 1623
Iron	$6,2 \cdot 10^{-3}$	0.86	1373 – 1533
Gold	$1.1 \cdot 10^{-3}$	1.12	1073 – 1473
Oxygen	$7 \cdot 10^{-2}$	2.44	973 – 1513
Copper	$4 \cdot 10^{-2}$	1.0	1073 – 1373
	$4.7 \cdot 10^{-3}$	0.43	673 – 973
Sodium	$1.6 \cdot 10^{-3}$	0.76	1073 – 1373
Tin	32	4.25	1323– 1473
Carbon	0.33	2.92	1343 – 1673

The most of impurities, except the third– and fourth–group impurities, diffuse according to interstitial, relay or dissociative mechanisms.

Let us specify some peculiarities of the diffusive process.

Boron is a single acceptor impurity of the limiting solubility being sufficiently large for creating an efficient emitter in transistors and other applications needing highly–concentrated impurities. Its maximum concentration at temperatures from 1173 to $1473 \, K$ varies from 10^{20} to $(2-3) \cdot 10^{20} \, cm^{-3}$. At $1473 \, K$ the diffusion coefficient will, depending upon the concentration, change from $2 \cdot 10^{-12}$ to $(0.8-1.0) \cdot 10^{-11} \, cm^2 \cdot c^{-1}$. The boron diffusion coefficient for a high–concentration region within the range of temperatures $1323 - 623 \, K$ can be calculated as $D = 16 \exp(-3.69 \, kT)$.

The diffusion of boron is assumed to run according to the vacancy mechanism. Oxidation accelerates boron diffusion, i.e. the depth of the $p-n-$junction increases under the growing oxide. Diffusion is also supposed to be accelerated, alongside with the action of the vacancy mechanism, by the stream of interstitial atoms from the growing oxide.

Arsenic at $1373 - 1523\ K$ has a high limiting concentration $(0.8 - 2) \cdot 10^{21}\ cm^{-3}$ and a slightly expressed retrogressive solubility. Within the range of $1073 - 1373\ K$, the maximum concentration of arsenic makes up $1.6 \cdot 10^{20}$ ($1073\ K$), $2 \cdot 10^{20}$ ($1173\ K$), $4 \cdot 10^{20}$ ($1273\ K$) and $8 \cdot 10^{20}$ ($11373\ K$) cm^{-3}. The small diffusion coefficient of arsenic being concentration–dependent relates to the purely vacancy diffusion mechanism. The dependence of the diffusion coefficient on concentration is expressed as

$$D(n) = KD(n_i) \frac{n}{n_i} \ ,$$

where $K = 2 - 4$. At $T\ (K)$ the concentration n_i may be found from the ratio:

$$n^2_i = 1.5 \cdot 10^{33} T^3 \exp\left(-\frac{E_g}{kT}\right) \ ,$$

where $E_g = 1.21 - 7.1 \cdot 10^{-10} n_i^{1/2}\ (eV)$.

At $1473\ K$ the diffusion coefficient is not affected by oxidation practically, whereas at $1273\ K$ it increases as much as 2-3 times. The acceleration mechanism here is similar to that boron has.

Phosphorus is mainly applied to forming the basic regions of bipolar $p - n - p -$ transistors, the sources and drains of MOS-transistors and to gettering the fast–diffusing impurities. In the range of $1373 - 1473\ K$ the maximum concentration of phosphorus is equal to $10^{21}\ cm^{-3}$. At high surface concentrations ($10^{21}\ cm^{-3}$), a significant portion (≈ 60 percent) of phosphorus in a presurface region is fixed and inactive at low temperatures. The phosphorus diffusion coefficient, as well as that of arsenic and boron, grows during oxidation but the effects of the phosphorus diffusion acceleration at low temperatures are greater than those of arsenic. At $1173\ K$ the diffusion coefficient increases about 8 times, and this supports the idea of duality of the diffusive mechanism.

In the $n - p - n -$ transistor manufacturing, when an emitting zone is constructed by phosphorus diffusion, boron diffusion is often observed to rise in the base under the emitter. This effect is called in a different way: a deep–effect, a push–effect (a effect of emitter ejection), a cooperative diffusion, an abnormal diffusion, etc.

A push–effect may be explained by the boron diffusive layer being oversaturated with point defects (with vacancies or Si interstitial atoms). As was shown experimentally, these are the Si interstitial atoms that are generated mainly than vacancies. When the emitters in $n - p - n -$ transistors are constructed by arsenic diffusion, the deep–effect does not appear usually. Though, in low–temperature treatment ($773 - 1293\ K$) the arsenic solubility falls sharply, and arsenic, forming precipitates ($SiAs$), will generate dislocation loops and excessive vacancies.

With a sequence of diffusion of boron and phosphorus being changed, a boron concentration profile will undergo the largest changes. The diffusing phosphorus gives rise to a counter–stream of vacancies which captures boron vacancies.

Phosphorus is diffusing into the opposite side of wafers continuously if the external gettering provides maximum concentration. Since a radius of the phosphorus atom (as well as that of boron) is less than a silicon atom radius, silicon will obtain elastic mechanical stresses. If these stresses exceed a stress in the plastic flow of silicon at the diffusing temperature, then there will arise dislocations in the bulk of the diffusive layer.

The diffusion–induced dislocations usually form a flat grid being parallel to a diffusion front; this grid lies at the distance $(0.3 - 0.6) \cdot X_j$, where X_j is a depth of the $p - n -$ junction (or a depth of the profile during the diffusion into a wafer of the same conductivity). All the dislocations are edge dislocations, with the Bűrgers vector $\vec{b} = \dfrac{a}{2} < 110 >$.

A diffusive layer can be considered to be a layer with a changed parameter of lattice. In particular, in the course of boron and phosphorus diffusion a lattice constant diminishes, and for arsenic diffusion it retains nearly unchangeable.

As nuclei for dislocations there may serve the local defects: precipitates, localized defects in the shape of small wells of $\approx 0.2\ mcm$ on a polished surface. A grid of misfit dislocations may arise because of a sliding of the half–fixed interstitial or vacancy–type dislocations. According to this mechanism, the impurity should not be obligatorily located in the replacement positions, i.e., for example, phosphorus is able to create both compressing stresses and extending stresses.

During a local diffusion of boron, the dislocation grids are formed at $C_{subst} \geq (2 - 3) \cdot 10^{19}\ cm^{13}$, whereas during a continuous diffusion at $C_{subst} \geq 1 \cdot 10^{20}\ cm^{13}$. The diffusion density in the diffusive windows exceeds the dislocation density during the continuous diffusion as much as 10 times.

In general, diffusion is anisotropic. In the case of the silicon lattice being cubic, a lattice symmetry makes the diffusion isotropic. Parameters of a diffusive layer are determined by a type of dopant and its concentration. A reproducible creation of surface concentrations of impurities lower than $10^{18}\ cm^{-3}$ is technologically complicated, especially retaining high surface homogeneity in large–size wafers. A reproducible and controllable surface concentration of impurity in a semiconducting wafer is obtained with a use of the following:

- depositing a diffusant onto the wafers in the course of diffusion (an external source). Here, it is necessary to regulate the content of impurity in the ambient encircling the semiconducting wafer during diffusion;
- depositing a diffusant onto the silicon wafers before diffusion. In this case, the amount of the impurity being deposited onto the wafers before the diffusion has to be regulated and a homogeneity in the diffusive surface has to be provided.

The basic demand to a diffusive system is transporting a diffusant to a wafer surface and performing a diffusion at the needed temperature for some definite time. External sources for diffusants can be gaseous, liquid or solid. The impurity coatings from which the impurity is being diffused are doped oxides, predoped semicrystalline or amorphous silicon and silicides of high–melting metals (molybdenum $/Mo/$, titanium $/Ti/$, tungsten $/W/$), photoresistors–diffusants and high–molecular impurity–carrying polymers.

Basic diffusive techniques are: diffusing in a sealed-off ampoule (a closed tube); diffusing in vacuum; diffusing in a closed volume (a box technique); open–tube techniques; impulsive diffusion techniques and an irradiation–stimulated diffusion.

A dopant is being usually diffused into a wafer upon its oxidation; hence, the oxidation–induced defects, such as dislocations and stacking faults, can alter their type under the action of the diffusant. The stresses accompanying the diffusion can lead to an annihilation of stacking faults, i.e. the oxidation–implanted stacking faults in the crystal turn to full dislocations which, being interactive, create a dislocation grid. It was emphasized above that oxidation affects the concentration of interstitial atoms in crystal crucially; thus developing a motive force for creating the stacking faults and dislocations; but diffusion, as it turned out, regulates these phenomena as well. The diffusing atoms of impurities are occupying vacancies in the silicon lattice or driving off the silicon atoms from interstitial sites in the way very much similar to oxygen oxidation; this makes silicon drastically oversaturated interstitially. Since the extrinsic atoms are travelling with intrinsic point defects (vacancies and interstitial atoms of silicon) participating in this process, the diffusive process will cause a change in the balanced concentration of point defects. Incorporating foreign atoms into a crystal lattice gives rise to strains inside the crystal and as a result changes lattice parameters. Besides, diffusion is also accompanied with such phenomenon as discharge (precipitate) generation, usually in the form of impurity–silicon compounds. Such precipitates are capable of creating large strains giving a birth to new defects. In impurity diffusion, a thermal treatment affects a birth of defects in the same manner as in oxidation, because in both cases the wafers are heated up to high temperatures. Thermomechanical stresses induced in wafers by fast heating and cooling can bring dislocations as well.

The role of earlier existent defects and defect nuclei becomes more important in a multiple diffusive process; for example, in manufacturing transistors when after creating a base region one creates an emitter region, these two diffusive processes being accompanied with oxidation.

Among the four basic dopants (phosphorus, arsenic, antimony, boron), mostly studied are phosphorus and boron, due to their wide application as $n-$ and $p-$ type dopants in silicon. Besides, phosphorus is used more often than other elements in constructing shallow $n^+ - p -$ junctions with a dopant of high surface concentration. These both conditions raise the density of defects in crystals that is particularly unwanted for shallow junctions.

The diffusion–induced dislocations can emerge in silicon because of thermomechanical stresses, a recombination of OSFs, a prismatic press-out of discharges of SiO_2; and finally there acts a fourth dislocation–producing mechanism – this is a diffusion of dopants, for example, phosphorus and boron, from the sources of highly–concentrated impurities.

The diffusion–induced dislocations usually give rise to some flat grid being parallel to the crystallization front, and thus to the $p - n -$ junction. The most portion of such dislocations is edge dislocations, with the Bŭrgers vector of the type $\vec{b} = \dfrac{a}{2} < 110 >$. A specificity of such dislocations lies in that that they are mainly distributed within a flat region inside a crystal (a position of this region relates to a diffusion profile), usually between a crystal surface and the $p - n -$ junction, at the distance from the surface equal to $1/3 - 2/3$ of the depth of the $p - n -$ junction. Besides, only few dislocations approach the $p - n -$ junction itself.

In general, it is possible to imagine the following typical procedure of oxidative and diffusive processes; they are able to change a nature and morphology of defects and finally create a complicated structure with high–density dislocations:

a)   generating OSFs either on $A-$ or $B-$ clusters or in the places mechanically damaged by oxidation;

b)   decorating stacking faults by dopants or fast diffusants (most often by copper or iron). Decorating by impurities may occur during oxidation or diffusion when a $p - n -$ junction is being constructed;

c) changing a structure of the decorated defects during thermal treatments related to further diffusive operations, for example, diffusing phosphorus to form an emitter in $n - p - n -$ transistors.

1.5
Ion-implanted defects

Last years have seen a wider and wider application of ionic doping or ion implantation to local doping. In this technique, the atoms of a dopant are introduced (implanted) through the silicon surface with the help of ions of the energy from some units to hundreds of keV. Upon getting into silicon, the ions are slowed down at some definite distance from the surface. This technique finds an extremely wide application in manufacturing superlarge ICs, since it provides a regulated doping resulting in a good uniformity of distributions of impurity concentrations. In contrast to the ordinary diffusive doping of silicon, ionic implantation has some advantages. As the implantation is usually performed at the indoor temperature, the areas not to be doped can be masked by a wide choice of materials including a photoresist impossible to be used in the case of diffusion. Ionic implantation makes it possible to regulate with high precision a doping level, i.e. an ionic nature of the implant allows a use of coloumbmeters for measuring concentrations. Besides, a doping process is possible to be very precisely regulated, via regulating the intensity of the ionic beam and its energy. Thin oxide films on the silicon surface reduce the depth of ion penetration, thus creating a possibility of supplementary control of the junction depth and impurity distribution in the area of junction. An impurity distribution profile created by implantation is characterized by two parameters: by the length of runs and by a spread of runs of the particles being incorporated.

However, high-energy irradiation of the crystal induces a large number of relatively stable structural distortions called irradiation defects (IDs). The simplest point irradiation defects include the defects like the Frenkel pairs (a vacancy and a self–interstitial atom). Such defects arise either as a result of the striking mechanism (i.e. during an elastic collision the energy, surpassing a threshold value necessary to replace the atom into an interstitial site, is transferred to a lattice atom) or as a result of the subthreshold mechanism during electronic excitations (ionic, impurity–ionized, plasmic and cumulative mechanisms).

Basic parameters describing the ionic runs (Fig. 1.8) are as follows:

- A length of the run (R) is a general trajectory of ionic motion;
- A projected length of the run (R_d) is the distance an ion is running prior to a full stop in the direction perpendicular to the surface of the target;
- A standard deviation of the projected run (ΔR_d) is a magnitude describing the fluctuations or a scattering of the projected run;
- A lateral scattering (ΔR_L) is the fluctuations of the final position of ions, perpendicular to the direction of the original motion;
- The magnitudes ΔR_d and ΔR_L depend on mass and energy of the ions being implanted (Table 1.3).

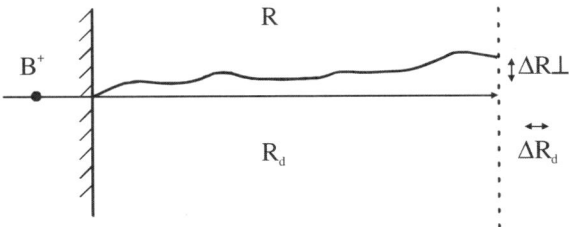

Fig. 1.8. Basic parameters of ionic runs.

Table 1.3

Values of R_r and ΔR_r for a silicon target

Ion	Parameters of runs in R_d ΔR_d	The accelerating stress (keV)									
		20	40	60	80	100	120	140	160	180	200
B	R_d	78	161	243	323	397	468	537	603	665	725
	ΔR_d	32	53	70	83	93	102	109	116	121	125
P	R_d	26	49	73	98	123	149	175	201	227	254
	ΔR_d	9	16	23	29	35	41	46	52	56	61
As	R_d	16	27	38	48	58	69	79	89	99	110
	ΔR_d	3.7	6.2	8.4	10	12	14	16	18	20	22
N	R_d	54	112	169	227	284	340	393	443	492	541
	ΔR_d	23	39	53	65	75	84	91	97	103	108
Al	R_d	29	56	85	114	144	175	205	236	266	297
	ΔR_d	11	19	27	35	42	48	54	60	65	703
Ga	R_d	16	28	39	50	61	72	83	94	105	117
	ΔR_d	4	6.7	9	11	14	16	18	20	22	24
In	R_d	14	23	31	39	46	53	61	68	75	83
	ΔR_d	2.5	4	5.3	6.6	7.7	8.9	10	11	12	13
Sb	R_d	14	23	31	38	46	53	60	67	74	81
	ΔR_d	2.4	3.8	5	6	7	8	9.5	10	11	12

High–energy irradiation of semiconductors is characterized by the following events and processes:

• in the presurface region there are available the fluxes of particles of assigned density and the electromagnetic field is constant;

• a presence of minor excitation processes on the electronic and atomic levels; an emergence of strongly–excited atoms and molecules and also unstable chemically–active ions and radicals;

• a multichannel spreading of excitations in a solid body; a mutual approach and overlapping of energy zones in impurity atoms and the wafer; and a characteristic time for relaxing of separate stages of physicochemical processes;

• a great number of fluctuations in the composition; an activation in the motions of boundaries; an existence of unbalanced states (in the Boltzmann sense).

In this connection, three dissipative stages for the irradiated solid body are possible to be distinguished: energy dissipation, kinetic dissipation and evolutionary dissipation (Fig. 1.9) [Kotov and Gromov (1988)]. The energy dissipation approach observes the diffusing processes of electrons, ions, neutrons and γ − quants, X-irradiation and minor δ − electrons induced by excitation and atomic ionization [Bethe (1964)]. The basic problems in this approach are estimating the penetration depth and the protective properties of this or that substance. The length of the energy dissipation stage is $10^{-15} - 10^{-9}$ s.

During the kinetic dissipation stage, strongly–excited atomic and molecular states are being aggregated and are relaxing fast; electron–hole pairs and the associated minor luminescence arise, migrate and recombine; this process takes $10^{-9} - 10^{-3}$ s [Smirnov (1977)]. Further, an atomic and molecular structures in a solid are reconstructed in a largely slower evolutionary way, by irradiation–stimulated chemical reactions or phase transitions [Kotov and Gromov (1988)].

Here, inside and on the surface of a solid there arise a great number of stable defects, like the Frenkel or Schottky pairs, which, in cooperation with electrons, create F, F', R_1, R_2, M and other light–absorbing centres and traps as well.

Thus, the nature and energy spectrum of irradiation–induced defects in semiconductors are determined by the energy transferred to the atoms of the irradiated substance, by the form of bombarding particles, by the extrinsic composition in crystals and by the implantation conditions (by the temperature of the target, its orientation with respect to the beam, the intensity of the beam, etc). The distribution of irradiation–induced defects is repeating the distribution of implanted atoms, however, the maximum of the defect distribution is shifted to the target surface, in contrast to the maximum of the ion implantation profile. The reason for this observed shift lies in the following. On its way an implanted ion creates the regions of highly–concentrated defects (their average depth is X_{dif}) which hold back a portion of ions, thus working as some kind of a specific screen with respect to the deep layers and reducing the probability of defect formation in these layers. The magnitude $\dfrac{X_{dif}}{R_d}$ for different ions varies within the range $0.8 - 0.93$.

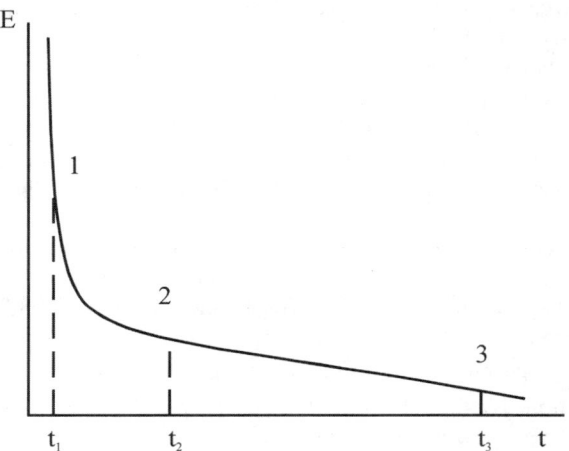

Fig. 1.9. A change of the falling irradiation flux in a solid.

Silicon doped by light ions (by boron) has basic distortions which are the isolated vacancies and atoms located interstitially; their migration give birth to clusters or aggregates with extrinsic atoms. Heavy ions incorporated into silicon induce local amorphous zones $3 - 5\,nm$ in diameter. The number of such zones grows linearly with the irradiation dose, until a complete amophization of the surface occurs. However, for boron atoms a slight increase in temperature of the wafer (by $323\ K$ and more) prevents a formation of amorphous material for any implantation dose, because a boron ion causes a shift in a large number of silicon atoms and the arising point defects will manage to recombine. On the basis of the temperature dependence of the threshold dose needed for creating an amorphous layer, one determines the activating energy for this process $(0.3 \pm 0.05)\ eV$; it is close to the energy of activating the vacancy migration in silicon.

The accumulation of irradiation defects is characterized by some dose corresponding, for example, to a phase junction or to some definite change of this or that property in the semiconductor: the AES signal, the IR–reflection, etc. The shape of the dependence of this distinctive dose upon the irradiation intensity and temperature can betray the dominating mechanisms of defect aggregations.

The following processes for stable defect formation are possible:
a)   generating the components of the Frenkel pairs;
b)   generating the original aggregates;

c)   generating the minor aggregates;

d)   directly recombining the aggregates of the Frenkel original pairs between themselves;

e)   recombining all the original and minor aggregates – "irradiation annealing";

f)   killing the original and minor aggregates by a thermal decomposition or by migration to the drains (a thermal annealing);

g)   the aggregates of Frenkel pairs vanishing on the unsaturated traps or an indirect recombination occurs.

Firstly, the ionically–implanted defect generation is determined by a primary and minor collisions of the implanted ions with the atoms of the original crystal. An ion falling onto the semiconductor wafer surface sequentially collides with the atoms in the target material, looses its energy and finally makes a halt. Mutual collisions of ions with atomic nuclei or electrons bring energy losses. During these collisions with the target material atoms, the dissipation angles are very considerable and the energy losses large. In these mutual collisions, the atoms having accepted the energy exceeding the bond energy are leaving their positions in the lattice and this results in primary (original) defects in the lattice. The simplest case generates the defects like a vacancy or an atom lying in the lattice interstitially; this case also produces small aggregates of the simplest defects and the disordering regions. Forming minor defects needs somewhat time for the point defects to migrate and to be united between themselves, with impurities or with other imperfections in the lattice. When colliding with an electron, the energy loss, calculated per a single collision, is small, and the scattering angle may be neglected. Since the implanted ions and the atoms having abandoned their places in the lattice (the recoil atoms) undergo multiple collisions prior to their halt, there always arises a region with a large number of aggregates of defects, most often of complicated nature. The number of the defects being generated in the crystal is considered to be proportional to a total loss of energy stipulated by the collisions with the atomic nuclei. Since point defects partially diffuse and vanish, the lattice defects arising immediately upon ionic implantation are determined by the temperature of the wafer. How the type and number of defects in the layer with implanted ions depend on a doping dose is seen from the following: small doses of ions of such average atomic mass elements (the doses being not more than 10^{14} $ions/cm^2$) as, for example, phosphorus create relatively simple PD aggregates which can be easily eliminated via annealing at $573-673$ K. An increase in the dose of implanted ions will result in a thin amorphous layer. With this dose being increased further, the regions of amorphous (disordered) layers are united to form a continuous amorphous layer which can be cancelled by a sequential anneal at ≈ 873 K. The disordered regions are supposed to involve a central zone enriched with vacancies and an encircling membrane enriched by the interstitial atoms. The distribution of such disordered regions throughout the depth of the damaged layer is Gaussian. The amorphous layer is formed according to temperature; besides, the threshold concentration of ions needed to form an amorphous phase is inversely proportional to the temperature of the specimen during implantation.

In order to eliminate the defects and correct the positions of the implanted ions, i.e. to transfer them into an electrically–active state in the interstitial sites, one makes use of annealing.

The reasons for annealing the implanted silicon may be different. First, if an amorphous zone is formed by high irradiation dose, then annealing is needed to reestablish a crystalline structure; and second, annealing reduces the density of defects, and hence the electrical properties in the implanted regions become better; and third, annealing activates the dopant atoms that thus take an electrically–active position in the lattice. A transfer from amorphous to a crystalline state terminates at $\approx 873\ K$ and higher and brings a change in the electrical properties of the implanted layer. The annealing at more than $873\ K$ will change the concentration of charge carriers and the specific resistance in the implanted regions. A short–term anneal induces a high density of small dislocation loops and PD clusters. A long–term anneal reduces a general density of dislocations, whereas the dislocation loops grow in size and the loops assume an irregular configuration. Though, even a long–term anneal fails to fully get rid of all the defects, and so dislocations and dislocation loops retain in the layer.

If the ionic implantation is followed with the annealing in the oxidizing ambient or with the diffusion of other dopants, then the defects in the implanted layers will change their structure accordingly. Oxidation usually gives rise to a large number of stacking faults, since the implanted regions have highly–concentrated point defects interacting with a flux of interstitial atoms born by oxidation. Here, the concentration of vacancy–type defects may be expected to increase.

1.6
Epitaxial layer constructing technology

The processes of forming the Si mono– and polycrystalline films on silicon and other wafers have found a wide application in the manufacture of various $Si-$ based devices. Depositing monocrystalline films onto monocrystalline wafers is called an epitaxial build-up or simply an epitaxy. In the technology of Si devices, the epitaxial technique should be exploited because of many reasons. In particular, this technique makes it possible to easily produce the films differing from a wafer both in type of the dopant and in concentration level (and consequently, by specific resistance). For example, when depositing an epitaxial boron–doped layer onto a phosphorus–doped wafer it becomes possible to construct a $p-n-$ junction without resorting to a hard–phase diffusion, i.e. the layers of any (different from the wafer) doping level can be constructed. Another important application of epitaxy is constructing in silicon some closed regions which are different from the environmental mass of silicon by specific resistance or type of conductivity. Such "latent" or "buried" layers are widely used in the IC technology, to create limited strongly–doped regions under the transistors; these transistors are constructed within the epitaxial layer being grown from above. The third application of epitaxy is a possibility to use isolating wafers. As wafers for epitaxial build up of silicon, one can use the monocrystalline wafers of sapphire (a SOS-structure) and definitely–oriented spinels (a SOI-structure). Here, their crystalline lattices are "compatible" with the silicon lattice in the sense that they allow to grow high–quality monocrystalline epitaxial films. When constructing high–density ICs in the upper layers of silicon, the wafer's isolating nature suggests a greater technological flexibility. Epitaxially growing the films onto a wafer of the same material, in particular, a building up of silicon, is called homoepitaxy. Epitaxially growing a material onto a foreign wafer is called heteroepitaxy (for example, growing silicon on sapphire).

In the manufacture of ultrahigh–speed ICs, epitaxial structures constitute a serious alternative to polished wafers, at least because of the growth microdefects being practically absent in epitaxial layers (due to the specificity of their deposition). But epitaxial technologies attracted the greatest attention thanks to a jump of microelectronics to a submicronic and nanomicronic level in forming device structures; and also there has appeared a real possibility to produce superhigh–speed ICs on the basis of epitaxial heterostructures in the system " $Si - SiGe$ hard solutions".

In the last years, the processes of epitaxial growth in combination with ionic implantation and impulsive irradiation of the material start to play greater role in constructing most complicated silicon device structures. A wider application in superhigh–speed ICs start to find single–layered and multilayered homo– and heteroepitaxial structures with separate very thin (on the nanometer level) layers, with sharp, planar $p - n -$ junctions and interphase boundaries, with a preset doping level (in some cases this level being very complicated). Severe requirements are imposed on the residual impurities and structural defects to be admissible in an epitaxial layer, and also on the uniformity of distribution of electrophysical properties within the epitaxial layer. As the geometry of device structures gets more and more complicated, the importance of the processes of local epitaxy rises [Mil'vidsky (1997)].

Growing silicon films epetaxially can be performed in various ways. More often, one employs a chemical sedimentation from a gaseous phase (for example, pyrolysis of monosilane), i.e. decomposing silicon gaseous compounds into elementary silicon and some other gaseous by-products. Here, monocrystalline silicon is being formed on the wafer immediately; and if the temperature and the sedimentation rate are regulated accordingly, then the relation between the growing layer and the wafer will be epitaxial. That is, if a wafer is of silicon, then a monocrystalline film will be grown with the same orientation as the wafer has, and if a wafer is a foreign one (for example, of sapphire), then with the orientation providing the best fit between a film lattice and a wafer lattice.

In performing a chemical sedimentation from the gaseous phase, temperature varies within the range of $773-1473\ K$, and here to construct a monocrystalline film we need high temperatures (not lower than $1273\ K$). In the reactor, the wafers are placed on some graphite pedestal and the epitaxial process is performed at the temperature assigned and in the gas–carrier chosen (H_2, He, Ar or N_2), together with any volatile compounds of silicon, for example, $SiH_4, SiCl_4, SiHCl_3$ or SiH_2Cl_2. Into the gaseous phase, there are also introduced the dopants, in the form of volatile compounds, for example, PH_3, B_2H_6, AsH_3 or SbH_3. Depending on the size of wafers, epitaxial chambers are usually made vertical or horizontal. The rate of the film epitaxial growth depends on temperature of the epitaxial process, and also on the concentration of the silicon volatile compound in the gaseous phase. The thickness of epitaxial films varies from 1 to $200\ mcm$, depending on the type of the device. In thicker films, a uniform distribution of the dopant is found to be regulated easier. In a general case, a doping level and a type of conductivity (if they are different) in the wafer–film interface are desirable to be as sharp as possible. However, a high–temperature epitaxy will cause a relatively fast diffusion of impurities in the wafer and in the growing epitaxial film, and this will "wash away" a sharp boundary between the epitaxial layer and the wafer, thanks to impurities mutually diffusing from one region into another. The impurity distribution within the layer–wafer interface is regulated by two overlapping profiles, one of which is determined by the diffusion of the dopant from the wafer into the growing film, and the other by the cross diffusion of the impurity from the film into the wafer. If the type of impurity in the wafer and in the film are different, then there emerges a $p-n-$junction in the region where these profiles are "intersected". If the impurities in the film and the wafer are similar and if the wafer has been doped heavier, then the diffusion from the wafer into the film (selfdoping) will substantially hamper a growth of very thin epitaxial layers with needed specific resistance.

A basic tendency in the development of gaseous epitaxy technology becomes a further substantial decrease of working temperatures. A success in applying low–temperature techniques of gas–phase epitaxy is mainly determined by the sterility gained in the technological process. In the first turn, it is achieved at the expense of vacuum–sealed equipment that provides a stable work in the low–pressure reactor and suggests a wide application of irradiative heating. Of great importance is a jump to exploiting new, more easily dissociating gaseous sources of silicon of special purity (SiH_4, Si_2H_6, SiH_2Cl_2, SiH_2F_2) and also to a use of nonthermal (optical, plasmic, electromagnetic, etc) ways of epitaxy stimulation. At present, substantial attention is paid to controlling a growth of epitaxial layers on the atomic level, to provide conditions for a stable deposition of layers [Voronkov and Zhoukov, et al. (1997)].

Difficulties of gaseous epitaxy are solved in the MBV technique, a growing of epitaxial films from molecular beams in vacuum. This technique is based on evaporating silicon from a liquid phase (a source) and its further crystallization on the surface of a heated wafer being placed near the source. Silicon is evaporated from a drop being formed by the electronic bombardment of the end of the silicon rod. This technique makes it possible to deposit silicon layers at the rate of $0.05 - 0.1 \; mcm/\min$.

As a variety of this technique there is a sublimation of silicon during its evaporation from the highly–heated (up to $\approx 1573 \; K$) surface of a hard specimen and then its crystallization on the wafer surface placed near the source; the surface having lower temperature at the pressure of residual gases being equal to $133 \cdot (10^{-7} - 10^{-10}) \; Pa$.

Constructing silicon layers by silicon sublimation technique is simpler from the point of view of the hardware needed; this technique is exploited in comparatively "pure" vacuum conditions, though it provides low depositing rates (the maximum rate is $0.3 \; mcm/\min$), a low efficiency and suggests limited capabilities in manufacturing semiconducting devices.

This technique is being developed not only along the way of constructing ultrathin multilayered homo– and heteroepitaxial structures on large–size wafers but also along the way of synthesizing – within a single technological cycle – the epitaxial MDS-structures, with a use of various variants of local epitaxy. The hardware designed for this purpose helps to combine – within a single technological cycle – the epitaxial process with the ionic implantation of implants needed in the layer being synthesized, and also with electronic irradiation or fast thermal annealing. All this drastically increase the possibilities of the molecular–beam epitaxy. High–vacuum chemical epitaxy is also being developed.

One more variety of epitaxy is the technique of molecular epitaxy based on interaction of silicon molecular beams with a heated monocrystalline wafer, with elementary components deposited onto it. Epitaxial growth is performed in ultrahigh vacuum through a simultaneous reaction of multiple molecular beams of different density and chemical composition with a heated wafer; this growth is characterized by a low – controlled rate of epitaxy $(1 \; mcm/h$ or $1 \; atomic \; layer/s)$ and this allows to modulate a molecular beam with the precision of up to a monolayer. Low temperature in wafer $(773 - 873 K)$ diminishes the number of thermodynamical defects (vacancies) in the epitaxial layer which has a high structural resolution exceeding the traditional techniques as much as two orders and has a negligibly low diffusion of the dopant inside an monoatomic layer.

At last, liquid–phase epitaxy implies a monocrystalline layer build-up from a metallic melt saturated with a semiconducting material being recrystallized on the wafer surface. With the wafer and the solution being thermally balanced, this build-up undergoes as a result of solution oversaturation. A lightly melting component of the growing compound is used as a dissolvent; this reduces the crystallization temperature, increases the purity of the growing layer and reduces the vacancy concentration.

It is already for more than 30 years that dielectric–based structures have been attracting careful attention of the designers of discrete devices and ICs. This attention rests first on the below advantages of these structures: they substantially reduce parasitic capacitances; provide a reliable dielectric isolation for the IC devices; make the current regulation comparatively simple; ensure a drop of working stresses and powers. Initially, the SOD-structures were oriented upon constructing high–temperature, irradiation–stable ICs for electronic equipment to be exploited by aerocosmic, automobile and atomic industries. However, the most interesting, especially in the last decade, are considered the perspectives of these structures in the manufacture of low–voltage and low–powerful, high–frequency superhigh–speed ICs widely used in portable electronic apparatus (cell telephones, portable computers, and the like). Such ICs need the structures where both a thin working layer of crystalline silicon and a dielectric layer could be practically defect–free and have no stresses, could be of the same thickness across the entire square of the structure, and the interface between them could have low density of surface states.

Nowadays, high–quality SOD-structures are constructed by the following three techniques [Mil'vidsky (1999)]:

• a dielectric isolation by implanting oxygen ions into monocrystalline silicon wafers (the SIMOX-process);

• a direct thermocompressive bonding of oxidized wafers, with a further thinning of one of them by polishing its reverse side (the BESOI-process);

• a direct thermocompressive bonding of the oxidized wafer with a thin monocrystalline silicon layer, this layer being separated from another wafer by ionically–implanted hydrogen (the Smart–Cut process) [Colinge (1998)].

In the SIMAX-process, oxygen ions are implanted into a monocrystalline wafer to form, together with silicon atoms; a latent oxidized layer, at some definite distance from the wafer's surface, above which a thin monocrystalline silicon layer is deposited.

The quality of a latent dielectric layer in the wafer is increased by supplementary oxidative annealing at $2623 \, K$. This annealing creates a layer of thermal oxide on the wafer surface and substantially improves the characteristics of the latent dielectric and microroughness of the surface layer in monocrystalline silicon. Such process was called an internal thermal oxidation (the JTOX-process) [Katsutoshi (1998)].

As a result, to-day there is the well–optimized low–dose SIMOX-process which for wafers $200 \, mm$ in diameter provides the SOD-structures of the below characteristics: the thickness of the silicon layer is $170 \pm 5 \, nm$ and that of the latent oxidative layer is $115 \pm 10 \, nm$; the density of dislocations in the silicon layer is $< 1 \cdot 10^3 \, cm^{-2}$; the surface microroughness is $0.4 \, nm$; the microroughness of the interface is $< 2\ nm$; and the pollution by metallic impurities is $< 5 \cdot 10^{10}\ cm^{-2}$. The structures of such quality are absolutely fit for creating submicronic ultrahigh–speed ICs. At present, intensive research is being made to further improve the quality of the structures created by the SIMOX-process and raise technicoeconomical features of this process.

Very attractive is manufacturing high–scale ICs where silicon epitaxially–grown films are used on the wafers of $\alpha - Al_2O_3$ habitually called sapphire. A crystallographic and chemical perfection of the silicon epitaxial layer grown on the insulating wafer is mainly determined by the below three aspects.

a)   Of great importance is a lack of coincidence between silicon crystalline lattices and the wafer. For the {100}-silicon – on – {1$\bar{1}$02}-sapphire structures this misfit is ≈ 12.5 percent if coincidence of Si and Al atoms is considered. For silicon on a spinel, a misfit of lattices is much less (≈ 1.9 percent).

b)   The extent of silicon film pollution by the wafer material is very important. With temperature increasing, a chemical interaction between silicon and the wafer increases. If sapphire is used as a wafer, such interaction may bring an incorporation of aluminium into a growing film of silicon. Of most importance here are the two reactions:

$$2Si + Al_2O_3 \rightarrow Al_2O + 2SiO$$
$$2H_2 + Al_2O_3 \rightarrow Al_2O + 2H_2O.$$

In the second reaction, a source of hydrogen is silane from which silicon is produced. The film pollution with aluminium is possible to be reduced by decreasing the growth temperature, but this is worsening the perfection of crystallographic films. Vice versa, high temperatures increase the crystallographic perfection of the film; however, the quality of the film is worsened by impurities. Therefore, the temperature $1273\ K$ is used as a low limit for temperatures employed for growing silicon from silane on sapphire.

c) The difference between the thermal extension coefficients for silicon and the wafer can give rise to residual elastic stresses in silicon films. The impact of this aspect usually drops when the films become thicker.

As a rule, the defects emerging in silicon films grown on isolating wafers arise mainly in the vicinity of the silicon–wafer interface. Here, basic defects are twins, dislocations and so–called tree pyramids. The films grown on the isolator are inconsistent and hold a great deal of impurities in the wafer–film interface.

Electrophysical parameters of these devices are determined to a large extent by the defects in epitaxial structures. The misfit dislocations are produced by the strains of the misfit periods of film and wafer lattices exceeding a critical strain in the beginning of plastic deformation. In homoepitaxial silicon structures the misfit of lattice periods $f = \dfrac{\Delta\alpha}{\alpha}$ arise in the n^+n- or $p^+ - (n^+p,\ p^+n) - $type structures and usually does not exceed $(2-4) \cdot 10^{-4}$. In the heteroepitaxial structures (the SOS ones) this misfit may reach some percents.

Maximal strains in the film appears at the initial stage of growth. They are proportional to the Young modulus E and to the deference between dopant concentrations in the film and the wafer.

Most often, misfit dislocations are located in the form of a grid consisting of straight dislocation lines in the plane of the interface. Their Bűrgers vector also lies within the interface plane.

With an epitaxial film being grown on a slight–dislocation or dislocation–free wafer, the misfit dislocations can arise as a result of dislocation semiloops being unstably generated on the surface of the growing film. If a radius of such semiloop exceeds some critical value, then this loop can expand by sliding under the action of the misfit strain and – approaching the interface – it can cause in it a segment serving as a misfit dislocation.

If dislocation–free wafers are used, sliding lines can appear in the film. If a wafer is adjacent to the pedestal closely, then a thermal gradient is possible to appear. Here, either a central portion of the wafer or its edge (depending on the bending sign) will have higher temperature. Activating the surface, peripheral and heterogeneous sources of dislocations by thermal strains gives rise to wide dislocation lines and sliding bands.

Hence, in epitaxy a special attention should be paid to a high flatness of the wafer.

Basic defects in the epitaxial films are dislocations and stacking faults (oxygen–induced stacking faults /OSFs/ and epitaxial stacking faults /ESFs/). The dislocations in these cases are generated by thermomechanical stresses determined by thermal gradients, mechanical treatment and by dislocations held in wafers. Thermal gradients for wafers may be eliminated by "soft" thermal treatment, specially–designed furnaces, and by heating the wafers by IR-irradiation, etc.

For homoepitaxial structures, especially for films with latent strongly–doped regions a misfit between atomic radii of silicon and a dopant may function as a source of stresses. With a dopant of high–concentration, the lattice constant α will change and this in the final end will expand or compress the lattice. These changes may be calculated in the following way:

$$\frac{\Delta \alpha}{\alpha} = \gamma \cdot N_{imp} ,$$

where N_{imp} is an impurity concentration (cm^{-3}); γ is a lattice expansion/compression coefficient determined by the difference of radii ($cm^3 \cdot at^{-1}$); for phosphorus the coefficient is $+1.12 \cdot 10^{-24}$, for antimony, $-3.8 \cdot 10^{-24}$, for boron, $+3.83 \cdot 10^{-24}$, and for arsenic, $-0.12 \cdot 10^{-24}$.

One of the ways to diminish the concentration of misfit dislocations is simultaneously incorporating an electrically – neutral or electrically–active impurity of the corresponding radius, to compensate the strains. In heteroepitaxial films, structural defects are generated more efficiently, since the sources for them are the differences both between the constants of silicon lattice and wafer and between thermal expansion coefficients.

Most part of ESFs arises in the places where a wafer gets into contact with an epitaxial layer. If a wafer surface has a lack of regularity, it may serve both as a crystallization centre and a reason of distortions in the ordering system of atoms. When the boundaries of rightly– or wrongly–packed regions converge, a break in the crystalline structure occurs. If further layers are rightly packed, then the atomic layers above the wrongly growing wafers will misfit the layers formed above the right crystallization centres. ESFs may be born by microdefects, mechanical impurities of metals, organic pollutions, residuals of the oxide film and pollutions in the atmosphere of epitaxial reactor. Thermal treatment in the oxidative environment gives rise to OSFs. The $A-$ and $B-$type microdefects are found to be a source of OSFs and ESFs.

ESFs are produced by the pollutions incorporated into the epitaxial layer from the reactor, the pollutions and mechanical distortions on the wafer surface and by various defects located in the wafer in the immediate vicinity to its working surface. Basic pollutants here are nitrogen, traces of oxides and of organic compounds retained after a wafer washing, particles of dust, scratches. The quality of chemicomechanical polishing drastically affects the formation of OSFs. Even the purest graphite graphite contains $10^{-4}-10^{-3}$ percent of such pollutants as iron, aluminium, copper, manganese that can pollute an epitaxial layer. Special attention should be paid to purity of graphite carbide–silicon pedestals.

In slightly–doped epitaxial layers, an accumulated layer of uncontrolled impurities will be, together with defects in wafers, a main source of stacking faults. In strongly–doped layers, a dopant aggregation may be a source of stacking faults.

PD clusters arise during a cooling of epitaxial structures grown on the $\{111\}-$planes on the wafers oriented in the $<112>-$direction.

When being cooled these structures are oversaturated with point defects migrating to the presurface drain. At the initial stage of cooling, at sufficiently high temperature the point defects are being incorporated into the fractures of the surface atomic steps. However, firstly attracted are the atoms of fast–diffusing impurities (metallic) being more flexible. At low temperature, these atoms will block the incorporation of point defects and this results in defect aggregation. The atoms of fast–diffusing impurities turn into centres where the PD clusters arise, their density being proportional to that of surface fractures related to the surface disorientation level.

Very often one observes such defects as shallow saucer–like pits being classified as $s-$pits. When heavily concentrated, these defects are usually called "haze" or "mist".

High–quality films epitaxially grown at $1223-1523\ K$ are produced with a use of silane, dichlorsilane, trichlorsilane and tetrachloride of silicon; for them superpure hydrogen is needed.

"Haze" is a result of overlapping and mutual influence of a number of processes: a reduced purity of gas in the reactor (due to pollution of the source of gases or the system itself), pollutions and defects in the wafers, wrongly chosen temperature or rate of the epitaxial build-up.

In thick (up to 100 mcm) epitaxial films, the defects arise on the surface of these films; these defects are seen in microscope at small magnification $(10-100^x)$. They are usually called "hillocks", "studs", "orange crust", "waviness", etc.

Alongside with structural perfection of epitaxial films, there exists an important problem of cancelling an autodoping, i.e. uncontrolled doping by diffusing from the wafer or from strongly–doped regions into the growing film. Autodoping can be efficiently reduced by reducing the epitaxy temperature.

In the last years there has been developed a technology producing superlarge ICs on the basis of the SOI planar structures, with implanting oxygen ions and nitrogen to synthesize dielectric SiO_2 and Si_3N_4 films under a Si layer. Ionic synthesis of these compounds in the bulk of the semiconductor being distant from the surface (hidden layers) must provide a monocrystalline surface $0.1-0.3$ mcm layer of silicon of high structural perfection above the dielectric, it must possess small leakage currents, must ensure small leakage currents, high breakthrough voltage and a sharp transient boundary. One of the requirements needed for this synthesis is a stechiometric relationship of components in the compound being synthesized. The concentration of oxygen and nitrogen atoms incorporated into silicon must be equal to $N_{Si} = 5 \cdot 10^{22}$ cm^{-3} that corresponds to the dose $(1-2) \cdot 10^{18} cm^{-2}$ at the ion energy 200 keV.

Planar structures (SOD) have a more perfect crystalline structure in a presurface layer and are more friendly to high–temperature thermal treatment as compared against heterostructures.

1.7
Lithographic processes

To selectively dope the necessary regions on the silicon wafer by diffusion or ionic implantation, the remaining parts of the wafer have to be protected by a corresponding "mask". Most often, oxide films are employed for this. A process of selective masking and etching the silicon dioxide or other dielectrics is called a lithography. At present, the theory and practice of lithography have advanced so far that the basic efforts are only oriented upon reducing the minimal sizes of structural components on wafers, for example, up to a thickness of the metallization lines to be sputtered.

Among the well–known lithographic techniques, the photolithographic technique is used widely. In its simplest form it involves the following. Onto an oxidized wafer there is deposited a film of light–sensitive polymer (of a photoresist) stable against the acids able to dissolve SiO_2 and Si ; upon the definite wavelength irradiation (usually the UV–irradiation is used) the photoresist polymerizes; the polymeric film is dissolved in several organic dissolvents, the photoresist is selectively exposed through the photopattern made by sequentially reducing large draughts or diagrams to the size of the wafer–held image. The exposed regions of the polymer are removed by a corresponding dissolvent, so that in these places an oxide's layer can be etched away (until a pure monocrystallic silicon). A nonpolimerized photoresist is removed by hot acids or by a plasmochemical burn-out. An oxidized layer retains on the silicon wafer; in this layer there has been etched a needed figure or diagram, and this allows to have the dopant diffused locally. The specificity of photolithography lies in a separate photopattern being calculated and produced for every device or diagram. The photolithographic process itself is performed sequentially: a photopattern is laid upon (is matched) a photoresist–covered wafer, and the resistor is selectively exposed to the UV-irradiation. Most devices and circuits for their manufacturing need a multiple use of lithography, and besides, in each operation the position of the pattern has to coincide (with very insignificant deviations) with its positions in previous and following operations. Habitually, photolithography makes use of a contact printing technique; and as a line width is reduced and a size of wafers increases the disadvantages of this technique are becoming evident more and more. In the case of multiple lithographic operations, the defects in the pattern and a misfit of the associated image lines on various patterns get accumulated to give rise to distortions. A sharpness of images is reduced by curvatures of wafers. A contact between a wafer and a pattern needed for high resolution leads to a damage of the pattern.

The disadvantages of contact printing may be reduced by numerous techniques of projection lithography; in these techniques a pattern is not brought into immediate contact with a wafer. In projection techniques, a pattern can serve for practically unlimited time.

The basic photolithographic defects may include the following:

• subetching a masking layer under the photoresist film, because of too aggressive etchant used;

• reducing the sizes of microimage elements, because of the overexposure;

• increasing the sizes of microimage elements, due to a small exposure or insufficient activity of the etchant;

• unetched points in the wafer surface brought about by failures in the technological regimes of etching, etc;

• a rise of surface pollutants when the residues of resistors are being removed in a plasmochemical way.

In particular, SiO_2 films are often polluted with sodium atoms when a photoresist is being removed in the oxygen plasma. Upon removing a photoresist mask, the surface concentration of sodium atoms on the wafer can become as high as $10^{14} cm^{-2}$.

The migration of Na atoms to the $SiO_2 - Si$ interface is determined by the ionic bombardment, which looks similar to the process of photoresist burning. Incorporating some portions of gas CF_4 into oxygen makes SiO_2 films less polluted. There occurs an etch-off of the surface SiO_2 layer mostly polluted with sodium.

The basic drawback of a photolithographic process lies in its limited resolution capability. The optical lithography provides a line width of up to 1 mcm; a higher (submicronic) resolution is limited by the optical diffraction and coherence effects.

To get a photolithographic resolution increased, a resistor is exposed to a lesser wavelength irradiation, for example, to electronic beams and X-rays (electronic, ionic and X-ray lithography).

The difficulties encountered in the electronic and X-ray lithography are determined by distortions in patterns and wafers, by difficulties of fitting and by pollutants in the utilized materials (especially the materials that emit electronic beams or are a source of X-ray irradiation and also are used as resistors).

There arises a necessity to increase a precision of lithographic operations because there is a great demand of high–density ICs and a further drop in the sizes of circuit components.

The precision of reproducing a topological structure of ultrahigh–density ICs is determined by the precision of forming an image and a resistor layer during exposition and also by the precision of other technological processes (diffusion, etching). In order to eliminate the errors of matching the images, one exploits a selfmatching technique; its idea is the following: the image obtained by a single exposure and etching will serve as a mask in the subsequent technological operations.

The precision of the vertical structure is determined by the precision in epitaxially–grown films, by thermal oxidation or chemical sedimentation from the gaseous phase, and also by the accuracy in calculating the etching depths, reproductiveness of the parameters of diffusion or ionic implantation.

1.8
Technological processes of etching

Chemical etching (CE) of semiconducting materials is not only one of the most important operations in chemically treating the wafers but an absence of this operation makes it impossible to study electrophysical, structural and other properties of semiconductors. Chemical etching is exploited:

- to prepare a wafer surface and clean it from pollutants and oxides;
- to remove a damaged layer retained on the wafer surface after various mechanical treatments (calibrating, shearing, slicing, etc) and to investigate a

nature of mechanical damages in the surface (caused by slicing, grinding, polishing, etc);

- produce the wafers of the assigned thickness and with a perfect surface;
- determine an orientation of crystals and detect structural defects in the crystal (dislocations, impurities, $p-n-$junctions, grain boundaries, etc);

- remove $p-$ and $n-$ layers; to construct mesastructures and runners on the wafers and multilayer structures, to rip up windows with an assigned tilt of walls in the semiconducting layers and dielectrics, etc.

Chemically etching and polishing the semiconductors suggest a number of substantial advantages, in contrast to other techniques of treating a wafer surface. First, these advantages are as follows:

- a relative simplicity and an easiness of chemical etching, needing no special complicated and expensive equipment;
- a quickness of its performance and its reliability make this technique universal for treating various materials;
- chemical etching gives rise to no structural distortions and deformations;
- a wafer surface of practically any complicated profile and configuration can be subjected to these technique, from a single to mass specimens;
- high resolution allows to implement various techniques and operations both on chemical treatment of specimens for research purposes and on chemically removing thin layers in mass production, to obtain a needed relief;
- high smoothing capability and efficient short–term polishing;
- an extraordinary visualization of the results of treatment;
- a comparatively low labour consumption of the process.

Chemical etching may be isotropic (crystallographically sensitive etching), anisotropic and selective.

An isotropic chemical etching is a process of dissolving a semiconductor, with all the edges of the crystal being etched at similar rates. A rate of chemical etching does not depend on crystallographic orientation of the etching surface.

An anisotropic chemical etching is a process of dissolving a semiconductor at nonsimilar rates of etching different edges of the monocrystal. A rate of chemical etching depends on crystallographic orientation of the surface to be etched.

A selective chemical etching is based on different rates of dissolving the semiconductor of one and the same crystallographic orientation of the wafer. This type of etching brings about etching pits (hillocks), heterogeneous impurity distribution and other defects, i.e. a microstructure of the surface.

A chemical polishing is a process of chemically etching a semiconductor in a polishing composition of the etchant where the roughness of the wafer relief is being smoothed and reduced, i.e. a surface purity grade is increased.

A chemicodynamical polishing is a process of chemically polishing in a polishing solution of the etchant; this operation is performed in the hydrodynamical conditions of a rotating disc and in a strictly laminarized movement of the etchant flux with respect to the surface of the specimens being polished [Perevostchikov (1995)].

If hydrodynamical and other conditions fail to provide smoothing (polishing) of a wafer surface, then a usual chemical etching is arranged; this process dissolves a specimen but somewhat worsens a quality of the earlier polished surface (for example, after diamond or chemicomechanical polishing) and the geometrical parameters of wafers.

Chemical etching, chemicodynamical polishing, anisotropic and selective etching are able to solve various problems in the technologies of microelectronics and research (Fig. 1.10).

Chemical etching and polishing of semiconductors

Isotropic
Chemical etching under uncontrolled hydrodynamical
    conditions:
- removing a damaged layer
- wafers thinning
- wafers cleaning

Chemicodynamical polishing under hydrodynamical conditions
    of the rotating disc:
- an increase of quality in wafers

Local chemical etching under turbulent hydrodynamical
    conditions:
- mesaetching
- contour etching (of windows, craters, pits, membranes, beams, etc)
- metallic sedimentation

Contents and compositions for chemicodynamical etching
    of semiconductors
- bromine–methanolic solutions
- solutions of hypochlorite
- ionic–exchange polishing
- peroxide mixtures, etc.

Anisotropic
Local chemical etching to create a desired relief on the surface
    and in the bulk of the wafer (thin wafers, membranes);
Oriented chemical etching, defect hunting, manufacturing
    of devices

Selective
Defect hunting, determining the boundaries of structures,
    layers, grains, $p - n -$ junctions and the like;

Chemical etching of the semiconductor porous structure;
Mesaetching, structure splitting, forming a desired profile.

Fig. 1.10. Classification of techniques for chemically etching and polishing the semiconductors.

Anisotropic chemical etching is employed in the metallographical and optical studies of structural surface defects and spatial defects; in microelectronics this technique is used in manufacturing a wide choice of devices and instruments, where there is a need in grooving, etching the windows, holes, membranes, masks and the like.

A selective etching makes it possible to reveal the defects in a crystalline structure of wafers (grain boundaries, small–angle and twin boundaries, dislocations, stacking faults, etc.); it helps to detect the planes of small indices for optical orientation of crystals and to prepare a specimen and split multilayer heteroepitaxial structures, etc.

Alongside with an integrated chemical etching where a semiconductor is being uniformly dissolved throughout the entire surface of the wafer; there is also used a local etching which is responsible for removing the material only from strictly limited and assigned regions of the wafer surface. For a local chemical etching, one uses the disguising or protective masks obtained via photolithography or other techniques. A local chemical etching can exploit both isotropic and anisotropic etchants.

There are other etching techniques as well: dimensional etching, by-layer etching, decorative etching, painting chemical etching of semiconductors, etc. A by-layer chemical etching is used to uniformly and sequentially remove thin surface layers of the semiconductor upon a diffusion of the dopant, an ionic implantation, in order to study surface and bulk defects of the wafer crystalline structure, epitaxial layers, etc. For this purpose, polishing etchants with low rate of chemical etching ($\leq 0.1\,mcm\,/\min$) and selective etchants are usually used.

In the planar–epitaxial technology of microelectronic units manufacturing, one of basic objectives of semiconductor chemical etching is forming technologically or physicochemically pure surfaces. High–quality surfaces of semiconductor wafers is a necessary condition for obtaining structurally–perfect epitaxial layers; here, the most important parameters of wafers (an absence of a structurally–damaged surface layer; a minimal roughness of the surface; a high surface flatness and chemical purity) are obtained by a chemicodynamical polishing and via finishing operations of wafer scrubbing.

Semiconducting devices are manufactured by various types of chemical etching. An etching operation logically follows a lithographic one. The liquid chemical etching makes it possible to provide a high selective action, and, besides, to etch anisotropically the silicon in accordance with the orientation of its surfaces. The dielectric most widely used in diffusion is silicon dioxide (SiO_2). Its etching is done by reactants holding fluorine atoms. The chemical etching rate of dielectric depends both on the composition of the etchant and the quality of the film (the density of the dielectric), the chemical composition and steichiometry of the dielectric, porousness and residual stresses in the film. The oxide and nitride films, being deposited at low temperature by such techniques as anodic oxidation, chemical gaseous–phase precipitation, plasmicochemical sedimentation and vacuum evaporation, are of lower density, and as a result of this have a higher etching rate. During thermal treatment, compressing these films (they were produced by thermal oxidation and high–temperature pyrolysis of silane and ammonia) will drop an etching rate.

In manufacturing semiconducting devices, etching should be extremely selective with respect to different materials and different regions of the wafer. This demand is determined by the silicon wafer being itself very complex in its composition; it holds various constant or temporary materials and components. For example, hydrofluoric acid selectively attacks only SiO_2, but does not react with silicon and slightly reacts with Si_3N_4. Likewise, when selectively etching the Si_3N_4 films by phosphorus acid, one may use SiO_2 or SiO as a mask : phosphorus acid dissolves Si_3N_4 without touching the SiO_2 or SiO films, etc.

The dielectric–etching rate can be increased through locally incorporating the irradiation damages by ionic or electronic bombardment; the irradiated regions have an increased rate of dissolution. So, through chemical etching in a corresponding etchant, it becomes possible to immediately establish a needed pattern in the oxide layer by following the traces of the ray, thus avoiding the operations of photoresist deposition and removal.

Mechanical and organic dissolvents are removed from the semiconductors' wafer surfaces (a degreasing process) by treating them upon each operation in the cleaning liquids, i.e. organic dissolvents or special washing solutions. An interoperation cleaning is intensified by boiling, reactant spraying under pressure, by treating in vapours of organic dissolvents, mechanical action by special brushes (a hydromechanical washing), ultrasonic oscillations and the like.

A use of organic dissolvents substantially affects the electrophysical properties of the semiconductor surface. The results of measuring the contact potential differences, the surface photo-emf and the differential effect of the field make it possible to consider how chemical treatments affect the magnitude of electronic yield (φ), the height of the potential barrier in the presurface exhausting layer (a bend of the bands φ_s), the drop of the potential in the surface oxide layer ($\Delta V = \Delta \cdot \varphi_O - \Delta \cdot \varphi_s$); these changes are induced by a change in the magnitudes and the nature of surface charge distribution and by a change in the densities of surface states (q_{ss}) brought about by chemical treatment.

Prior to chemical treatment, the original values of surface bending in the zones and those of surface state densities were $\psi_s^{\,o} \approx (-400 \pm 30)\ mV$, $q_{ss}^{\,o} \approx (2 - 5) \cdot 10^{13}\ cm^{-2}\ eV^{-1}$. It is seen in Table 1.4 that upon action of organic dissolvents these values change as much as some times. Irrespective of the specific nature of the dissolvent, organic dissolvents, as revealed by studies, always increased the absolute values of the surface bending of the zones (φ_s).

The value of the effect, however, substantially depends upon the dissolvent in the series:

$$(CH_3)_2 CO - CH_3CH_2OH - C_3H_7OH - CCl_4 - C_6H_5CH_3 .$$

Table 1.4

Effects of organic dissolvents on electrophysical parameters
of the $GaAs$ natural surface

Dissolvent	$\Delta\varphi_s \ (mV)$	$\Delta v_O \ (mV)$	$\dfrac{g_{ss}}{g^O_{ss}}$
Acetone	− 171	− 38	0.96
Ethyl alcohol (ethanol)	-118	-30	0.75
Isopropyl alcohol	-89	-92	2.2
Tetrachrorine	-67	-21	1.3
Toluene	-24	-105	1.5

The potential in the surface oxide was dropping faster. All types of treatment produced no significant effects on the density of surface states (SS). The reagents studied exerted substantial effect on the surface upon chemicodynamical polishing. A freshly–etched surface was found to be substantially different in its charge state from a natural surface; the former has higher values $|\psi_s|$, namely, $\approx (550 \pm 30) \, mV$. For all types of dissolvents, g_{ss} in its absolute magnitude did not change more than $(2-3) \, kT / e$, here, in contrast to a natural surface some organic dissolvents behaved as donors (the positive values of g_s) (Table 1.5).

Table 1.5

Effects of organic dissolvents on electrophysical parameters
of freshly–etched surface of $GaAs$

Dissolvent	$\Delta\varphi_s \ (mV)$	$\Delta v_O \ (mV)$	$\dfrac{g_{ss}}{g^O_{ss}}$
Tetrachrorine hydrogen	− 51	120	1.3
Acetone	-35	128	1.04
Toluene	22	92	4.0
Isopropyl alcohol	30	-48	5.2
Ethyl alcohol	68	-115	1.83

The results presented demonstrate that during cleaning and washing out the semiconductor wafer surfaces the organic pollutants are removed by organic dissolvents, and the surface potential is changed at the expense of dipole moments in the adsorbed molecules of the organic dissolvents. On a freshly–polished surface of $GaAs$ wafers one observes a substantially less quantity of the residual oxide and this explains a less quantity of physically adsorbed water in it; this, in its turn, evidently explains why, upon etching, the surface bending of zones becomes so sharply increased. In the chemically etched semiconductor, the physicochemical state of its surface naturally affects the nature of interaction of organic dissolvent molecules held in this surface. That some chemical treatments essentially rise the density of surface states betrays a participation of the associated impurity molecules in creating surface electronic states.

Chemical etching of semiconductors can also be accompanied with such negative impacts for the material under treatment as possible pollution of the reaction surface by impurity atoms adsorbed by the impurity atoms from the etchant; the reaction products can also establish so–called island–like films. The changes induced by chemical etching in the extrinsic defect composition and in the morphology of crystal surface affect the electrophysical and mechanical properties of the presurface layers, the defect–forming processes at low– and high–temperature treatment of the material, and in the final end they become one of the reasons for degradation and drop of the operating reliability of semiconducting devices. Therefore, at present there are severe demands to a purity of reagents for chemical etching, to the technology and the quality of a crystal wash-out prior to and after etching, to the technological regulations to be strictly observed in these operations.

In addition, there exists one more aspect of cleaning and chemically etching the crystals; this aspect is little studied so far. This is a defect formation in crystals, particularly in presurface layers; this process is initiated by the very chemical destruction of the material. In this aspect, the consequences of chemical etching are manifested not only as doping the surface by impurity atoms but also as a result of active effects of the external environment upon the crystal, these effects yielding thermal impacts and intensive gas emission.

Etching the semiconductors chemically may be considered as a set of basic processes in the micro– and nanoelectronical technology inducing changes in real structures of crystals at significant distances from the reaction surface [Perevostchikov and Skoupov (2004); Perevostchikov and Skoupov (1987a)]. The components of extrinsic defect composition (alongside with possible uncontrolled pollutants in the surface) can be transformed, for example, during chemicomechanical or chemicodynamical polishing, because of unbalanced IP defects, elastic waves and hydrogen accumulation in the places where lattice damages are localized. A local rise of temperature caused by etching can also contribute to defect recombination.

It is known that an excessive concentration of IP defects (vacancies, in particular) may be explained by an unbalanced state of crystal during its by-atom "evaporation" into the external environment (etchant) [Fistoul' and Sinder (1981)]. If crystal is balanced with an etchant, then above its surface there is established a balanced concentration of the off–etched substance; this concentration is equal to its solubility in the solution, and crystal itself will accordingly have a balanced concentration of vacancies.

Here, the flux of the off-etched atoms and the flux of the atoms back condensed onto the surface are equal. If an off-etched substance is transported from the reaction surface dynamically or its solubility is high, then the above fluxes are not coincident. In this case, vacancies emerge on the surface; they are diffusing into the bulk of the crystal and make the solution oversaturated. Under constant external conditions, the vacancy profile in the presurface layer becomes stable. If crystal has fast–diffusing impurities (FDI), they decorate the vacancy distribution profile; this was observed with sodium atoms in the etched silicon [Fistoul' and Tchernova, et al. (1980)].

During chemicodynamical polishing, unbalanced IP defects are able to transform the microdefect structure in crystals, as well as during abrasive treatment. This has been detected in the experiments on chemically polishing $n - Si\,(111)\,CZ$ ($\rho = 15\ Ohm \cdot cm$, phosphorus–doped) and $p - Si\,(111)\,CZ$ ($\rho = 15\ Ohm \cdot cm$, boron–doped) in the solutions of the composition (vol. portions): HNO_3 (70%) : HF (49%) : CH_3COOH (99.8 percent), with the components having various content and etching rates 7, 1 and $0.5\ mcm/\min$. Microdefectiveness was measured by selective chemical etching in the Sirtl solution and through atomic–force microscopic topograms at the "Accurex" unit in the non-contact regime.

It was experimentally shown that irrespective of the conductivity type the density of microdefects drops with a rise of the etching rate, approximately linearly, with tangent bias $a = -\,(1.4 - 1.5) \cdot 10^{-5}\ mcm^3/\min$ for $n - Si\,(111)\,CZ$ ($\rho = 15\ Ohm \cdot cm$, phosphorus–doped) and $a = -\,(2.4 - 2.5) \cdot 10^{-6}\ mcm^3/\min$ for $p - Si\,(111)\,CZ$, ($\rho = 15\ Ohm \cdot cm$, boron–doped). Surface microroughness depends on the polishing rate in a similar way; this roughness is produced by selective etching of small microdefects ($C-$ and $D-$types) during chemicodynamicalal polishing: the selectivity rises with a fall of the polishing rate. With the material removing rate being equal to $7\ mcm/\min$, the parameter R_z will be $16\ nm$, and at the rate $v = 0.5\ mcm/\min$ it will approach $19.4\ nm$. What is specific is that after chemicodynamical polishing the low–temperature 7-hour annealing of the specimens at $453\ K$ increases the concentration of small microdefects up to $10^{10} - 10^{12}\ cm^{-2}$.

In particular, structural changes in silicon chemically–etched wafers (upon cutting them from an ingot) were detected by the X-ray three–crystal spectrometer, through the magnitude of the residual deformation of the crystal lattice, the in-depth dispersion of deformation and within a cutting plane, and also through a distortion of the planes parallel to the surface under question, and also through measuring microhardness by the Russian–made hardness tester, with the indenter being loaded from 5 to $200\ cN$.

Table 1.6 exhibits the results of X-ray measurements for the structural characteristics of specimens in which a $25\ mcm$ layer damaged by the slicing process was removed at different etching rates. The error of deformation in the direction being normal with respect to the surface did not exceed $\pm 1.2 \cdot 10^{-6}\ rel.u.$ and that of curvature $0.5 \cdot 10^{-3}\ m^{-1}$. A negative sign of the curvature stands for the convexity of the surface under study, and a negative sign for deformation implies the actions of extending stresses within the specimen surface plane in the layer of the order of extinctive $8\ mcm$ depth of the X-ray K_{α} – irradiation.

Table 1.6

Residual deformation and curvature of silicon wafers
upon chemical etching as deep as $25\ mcm$

Etching rate $(mcm/\min)$	Residual deformation of lattice $(10^{6}$ conv. u.)	Variance of deformation $(10^{6}$ conv. u.)	Curvature of wafers $(10^{-3}\ m^{-1})$
1.0	- 0.5	7.8	- 0.7
5.0	- 1.9	6.2	- 1.1
8.3	- 3.1	5.1	- 2.7
14.0	- 4.4	4.6	- 3.8

As seen from Table 1.6, a rise of the etching rate brings a rise in the average amplitude of the lattice residual deformation in the presurface layer and simultaneously a drop in the deformation variance. Matching the bending signs and the relative increment of the lattice period allows to conclude that during chemical etching primarily the positive dilatation centers (responsible for extending elastic stresses) are aggregated in the presurface region (the region less than $8\ mcm$ thick). That each such centre overlaps the elastic fields causes a drop in the deformation variance; this drop is detected immediately upon a cease of etching. However, it was also noticed that the specimens etched at large rates (more than $5\ mcm/\min$) and when maintained at the room temperature for some days obtain a constantly growing variance, against a drop of average deformation and curvature. In studying the effects of chemical etching on the defectiveness of abrasive–polishing crystals, it was found that structural post-etching parameters are relaxing with an efficient activation energy, $0.6 \pm 0.2\ eV$.

All these facts can be explained with a use of considerations in [Perevostchikov and Skoupov (1992); Perevostchikov and Skoupov (1987a); Fistoul' and Sinder

(1981); Fistoul' and Tchernova, et al. (1980)] concerning the aggregating of IP defects near the surface in the shape of the simplest aggregates; during further annealing these aggregates will coagulate into small–size clusters (the oversaturated solution of IP defects is decomposing), and the fluxes of unbalanced vacancies or interstitial atoms are dissolving the ambients around the original microdefects. This is indirectly supported by the following results of the acoustic emission of etching crystals with structurally–damaged layers [Perevostchikov and Skoupov (1992); Gorshkov and Perevostchikov, et al. (1989)]: a rise of etching rate makes the emission signals more intensive, i.e. there will be a drop in the probability for creating metastable aggregates of IP defects and that of further dissolution of microdefects, since the elastic wave field will contribute to the surface–bound escape of point defects. In chemicodynamical polishing, elastic waves emerge each time when a structural defect (say, a dislocation or a microdefect) is approaching a surface where its motion is accelerated by the mirror image forces and this brings an impulsive relaxation of local mechanical stresses [Hirth and Lothe (1970); Alekhin (1983)].

In addition to the above–listed aspects, there is still one more that determines changes in extrinsic defect composition; this aspect is decorating the crystals by atomic hydrogen arising when crystals are being dissolved in the mixtures of nitric and hydrofluoric acids. Hydrogen being accumulated in the molecular shape in the vicinity of the structural inhomogeneities is able (upon chemicodynamical polishing) to change a lattice period of the crystal and its microhardness [Perevostchikov and Skoupov (1987a)]; this favours the microdamages of the material in the subsequent thermal treatments (especially, when the material is subjected to sharp heating and cooling). It should be emphasized here that in contrast to abrasive treatments the chemically–induced defects are still in the beginning of their study. The study of these processes is extremely urgent for manufacturing microcircuits with submicronic and nanomicronic sizes of their components.

Selective etching of dielectrics has been significantly improved through a plasmicochemical etching, i.e. by selectively etching away the material when it is reacting with chemically active radicals formed in a smouldering discharge. The plasma employed for etching is an ionized gas involving the ions of selective reaction capability, free electrons and free radicals. A plasmic etching (a dry treatment, in contrast to a liquid chemical etching) is usually performed at low temperature ($323 - 523 \ K$) and low pressure ($0.1 - 3 \ torr$).

Plasma is initiated by a high–frequency (HF) generator (its frequency is about $13 \ MHz$; an inductive or capacitance excitation of the discharge is used) in a vacuum chamber through which a plasma–forming gas is passed. For plasmicochemical etching one usually uses the halogen substituting hydrocarbons (CF_4 and the like). Ionized reagents interact with the etched material to form volatile compounds being evacuated from the reaction zone by the vacuum system.

The advantages of the plasmicochemical etching lie in that that it is a "dry" process where no fluid reagents are employed (for example, acids) and no further washing and drying of wafers is needed. In addition, this way of etching suggests a higher resolution and a lesser resistor subetching.

An ionic–plasmic treatment in vacuum is also widely used in manufacturing electronic equipment and items of other designation. First, these processes are used for material mass removal. In one case, this is a local or integral removal of the material from the entire surface of the treated wafer, in order to clean and polish it and construct a needed relief on its surface via etching or spraying. In another case, spraying a material is used for film depositing onto a wafer. In particular, etching by spraying inertial gas in plasma is employed to remove entirely or locally the surface layers of the material, in order to clear it from pollutants or construct a relief on its surface. An ionic–plasmic deposition as a thin film constructing technique is equivalent or superior to a traditional technique of vacuum spraying. The material needed for a film is transferred into a gaseous phase (is sprayed) by evaporating it from the resistive heater or by electron–beam evaporation. Then, the sprayed material is placed into the target–wafer space and becomes deposited on the wafer. The advantages of such film–constructing technique are as follows:

• the films constructed are more adhesive than those produced by other techniques, due to high energy of the sprayed particles arriving to the wafer;

• the material of the film is deposited without a change in its stechiometric composition;

• it becomes possible to obtain films of high–melting and nonmelting materials with a higher regulation of the gaseous composition and properties of the films to be obtained;

• there is a possibility to overlap the cleaning operations with the operations of preparing a wafer prior to, during the process and upon finishing the ionic deposition.

A reactive ionic–plasmic deposition of films is also used, with a use of reactive gases chemically interacting with the target material being sprayed, for example, oxygen or nitrogen, for constructing oxide, nitride or oxinitride films. Here, a material depositing rate is reduced and the types of the sprayed particles and the compositions of the films being deposited onto the wafers change. Table 1.7 shows the characteristics of the SiO_2 films obtained by various regimes of HF spraying a silicon target by Ar ions. Despite a considerable content of argon in the films, their electrophysical properties and resistance to chemical etchants are slightly different from the properties and resistance of the films produced by thermal oxidation.

Table 1.7

Properties of SiO_2 films constructed by thermal oxidation
and ionic–plasmic deposition

Film characteristics	Thermally grown SiO_2	SiO_2 produced by ionic-plasmic deposition in different regimes of HF spraying		
		I	II	III
Content of argon ($\dfrac{Ar}{SiO_2}$)	–	0.03	0.09	0.01
Density (gr/cm^2)	2.2	2.25	2.16	1.98
Refraction coefficient for 540 nm wavelength	1.461	1.47	–	1.465

Silicon nitride (Si_3N_4) films are produced through a reactive ion-beam depositing, with spraying the silicon target by argon ions and incorporating nitrogen into the region where the film is depositing onto the wafer.

One of the major reasons of why the ionic–plasmic deposition still finds a restricted application to the manufacturing of electronic items is an induced charge arising in the dielectric films; this charge is degrading dielectric parameters. Such degradation may cause an uncontrolled escape of parameters of the devices based on these films. When treating the SiO_2 films 0.1 mcm thick by argon ions of the dose $2 \cdot 10^{13}$ $ions/cm^2$, an electrical field of $9.3 \cdot 10^6$ V/cm emerges on the surface; this may cause a break-through of dielectric, within a small region or across the entire surface. During ion neutralization, there are formed on the wafer surface the electron–hole pairs, and the pollutants in dielectric are ionized; these processes bring a later drift of holes and ions in the electrical field to dielectric–wafer ($SiO_2 - Si$) interface. Here, a fixed charge and a density of surface states on Si are increasing (Fig. 1.11) [Ivanovsky and Petrov (1986)].

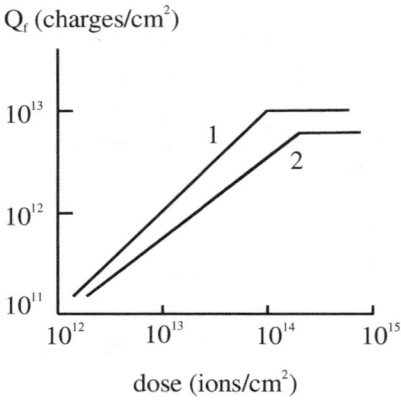

Fig. 1.11. A dependence of a fixed charge in a SiO_2 film on the ionic dose: **a** the ionic energy $9.6 \cdot 10^{-17}$ j; **b** the ionic energy $8.0 \cdot 10^{-18}$ j.

The basic pollutants responsible for dielectric degradation are univalent atoms of alkaline metals Na and K. Degradation is also caused by ionic–chemical and plasmicochemical etching, since the potentials of ionic ionization CF_3^+, CF_2^+ and F^+, treating the wafer surfaces, are sufficiently high. Upon ionic irradiation in the plasma of C_2F_6, the density of surface states in the SiO_2 films becomes equal to $10^{12} - 10^{13}$ $cm^{-2} \cdot eV^{-1}$ (with the admissible value $5 \cdot 10^{10}$ $cm^{-2} \cdot eV^{-1}$) if the energy of the bombarding ions is $800\,eV$ and the ionic current density is 0.4 mA/cm^2.

Plasmic–chemical and ionic–chemical etching of materials also damage a wafer surface. Upon etching in the plasma of CF_4 or CF_4/O_2, roughness of SiO_2 surface (with the original surface being smooth) increases less than in the case of the Si surface. Plasmic–chemical etching, with a corresponding ionic bombardment, gives birth to defects on the surface in the form of local regions of increased internal mechanical stresses arising in the vicinity of pollutants. The surface layers of the material under treatment are also subjected to amorphousness and polycrystallization. One of the types of pollutants induced by plasmic–chemical etching of the surface are the components of the electrode material (a wafer holder) sprayed by the ionic bombardment and back–scattered onto the surface (Fig. 1.12) [Ivanovsky and Petrov (1986)].

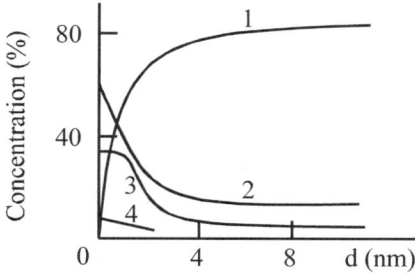

Fig. 1.12.   1 In-depth distribution of pollutants on the silicon surface: **2** nitrogen, **3** oxygen, **4** fluorine upon treatment in the plasma of CF_4 / O_2.

1.9
Technological processes of metallization

Metallization is the last, among the basic, technological processes in manufacturing semiconducting devices and ICs. This is a process of constructing metallic compounds in the form of thin metallic films connecting different regions of an IC or serving as contacting pads for electrical leads to be subsequently connected to them. A needed pattern of metallization is usually put onto the wafer surface by lithographic techniques.

Metallic films of silicon wafers can be deposited by physical techniques: by thermovacuum evoporation, electronic evaporation, cathode spraying, impulsive laser evaporation, ionic-plasmic sedimentation, plasmic and magnetronic techniques, etc. Small–width lines, i.e. high resolution, in particular, for ICs and the most of discrete devices, are, almost exclusively, produced by vacuum evaporation or ionic–plasmic and laser spraying.

The requirements to be met by metallization are various but the main of them are determined by the designation of contact pads (an ohmic pad or of the Schottky type forming a barrier with a semiconductor), the power to be consumed, the integration density in ICs (i.e. by the width of lines), the depth of the $p - n -$ junction, the manner of connecting the external electrical outputs, and by a high reliability.

The pads to silicon are most often made of aluminium. It is highly conductive and can be produced by a vacuum sedimentation, with a use of resistive or electronic heating. Silicon dissolves in aluminium to give very good ohmic pads. However, in the structures with rather shallow $p - n -$ junctions, aluminium may penetrate into the junction area to deteriorate the junction. Depositing the films of silicon alloy saturated with aluminium helps to escape from this difficulty. To efficiently fight the aluminium diffusion into silicon, one also uses diffusive barriers, i.e. a so–called "barrier" metal, say, platinum, is deposited, which interacts with silicon during thermal operation to form silicide. It is the latter that hampers the pad material to penetrate into silicon.

Metallizing the contact joints gives rise to distortions in the dielectrical coating of wafers. Degradation of dielectrics during metallization (the process of ionic–plasmic irradiation) implies the fact that a wafer is being somehow subjected to ionic and electronic irradiation. The irradiated dielectric is partially restored by the anneal at $873\ K$ in the atmosphere of nitrogen. The pollutants in the dielectric film are fully killed by the annealing at about $1173\ K$.

The reliability of metallic joints may be worsened by changes in the structure of the metallic film during its exploitation; it may be caused by a break down of internal connections accounted for electromigration and by the corrosion of metallic films, etc.

Electromigration represents itself a mass transfer of the conducting material. It is performed by electronic impulses running (under the action of the electrical field applied to the conductor) to the positive ions of the metal. This results in pores or gaps and nearly lying hillocks or other manifestations of metallic accumulation.

Numerous types of metallization corrode in the environment with a high content of moisture. The problem of corrosion becomes more complicated for the metallic paths being located closely to each other, especially when electrical field is applied to them.

The metallization techniques are being continuously perfected both in the aspect of constructing cheap (economically efficient) metallized contact pads and providing their high reliability and no-failure operation.

2 Effects of defects on electrophysical and functional parameters in semiconducting structures and devices

This chapter discusses the effects of point defects and their aggregates, dislocations and stacking faults on electrophysical and functional parameters of semiconducting structures and devices including the characteristics of device layers in the SOS–, SOD–, SOI– structures, porous silicon layers, etc.

2.1
General

This chapter systematizes sufficiently well studied peculiarities of semiconducting materials and device compositions; these peculiarities are determined by crystallographical distortions, concentrations whose sizes and electrical activity can be regulated by gettering. The last implies that not all the types of even single–type defects can be removed by gettering or their electrical activity can be suppressed. For example, the dislocations induced by mechanical treatment (by scribing) are possible to be removed from a crystal by a thermoirradiating gettering, whereas the misfit dislocations in homo– and heteroepitaxial structures or in heavily–doped diffusive layers are practically insensitive to gettering actions, as the defects minimizing free energy in the system. The aggregates of intrinsic point (IP) defects or the clusters near the heterophase boundary of the film–wafer interface (here maximal elastic stresses are located) are also impossible to be removed by gettering, though they can immediately vanish in the course of gettering but can arise some time again upon a cease of gettering, because their presence reduces free energy in the system. On the other hand, practically always liable to gettering are analogous types of defects, say, held in the wafer, induced by thermodynamically–unbalanced wafer formation process during a growth of monocrystal and its later treatment – these process defects are firstly induced because of final length of the technological processes when the extrinsic defect composition does not manage to relax to a balanced state.

The extrinsic defect composition in the active regions of the device compositions starts to be formed during growing a semiconducting ingot and then this process proceeds during all stages of the technological line and also during the exploitation of the manufactured item – thus determining a degradation of the performance of the device or its accidental malfunctions. An original (or growth) extrinsic defect composition is initiated by external and internal factors. The external factors are the technological effects resulting in increased temperatures and thermal gradients, final heating and cooling rates for structures, varying external pressures, chemical activity of technological ambients, corpuscular and photonic fields, polluting by uncontrolled impurities from technological liquids, gaseous ambients or attachments. The internal factors chiefly embrace the specificities of extrinsic defect composition retained after technological operations performed before a given operation; these specificities include a spectrum and concentration of the components of the extrinsic defect composition, the profiles of their spatial distribution within the bulk of the structure, a charge state, a temporary stability and a chemical activity of defects. Here should be also included the internal mechanical stresses induced by a manufacture of the device elements; in each case these stresses have individual topology and local amplitudes of the field interacting with an individual intrinsic deformation field of each crystallographic defect in the structure.

Of interest is the phenomenon of interaction between impurities and dislocations occupied by various impurities. The impurity–free dislocations and the dislocations without specific defects are not so active as the emission–free recombination centres; and therefore they do not hamper the functioning of solar batteries. However, if the impurities (for example, Ni or Cu) occupy a dislocation, the dislocation gets activated thus reducing the lifetime of minor charge carriers and, consequently, the efficiency of a solar battery. The behaviour of the dislocation with the impurity depends on the position and the state of the occupying impurity. High–rate recombination on a "dirty" dislocation is determined by the ability of the impurity able to occupy not only free electrons and holes but also electrons and holes from one–dimension zones determined by the dislocation itself – and this increases many times the probability of recombination. Fig. 2.1 shows how the recombinating activity of the dislocation (an EBIC contrast of a single dislocation) is impurity–dependent prior to and upon gettering this dislocation. In the figure it is seen that gettering at room temperature has dropped the recombining activity almost down to zero. The behaviour of impurities on dislocations is an interesting fundamental problem whose applications are evident.

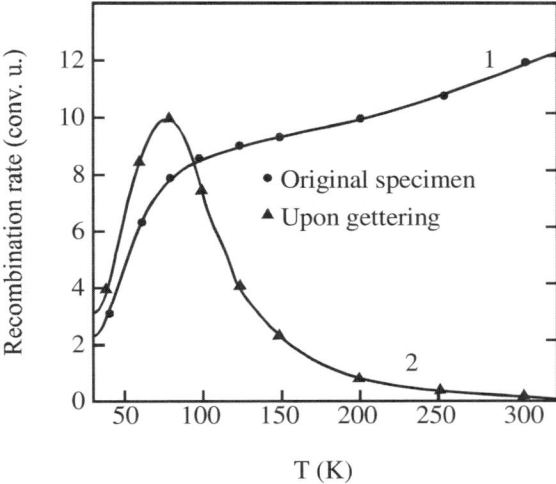

Fig. 2.1. Recombination rate (conv. u.).

Effects of temperatures on the recombination rates measured for elastically–deformed $FZ - Si^{-8}$, prior to (curve 1) and upon (curve 2) gettering. Here, the points stand for the experimental data and continuous lines for calculations according to the theoretical model [Kveder and Kittler, et al. (2001)].

In general, an extrinsic defect composition of device structures (more exactly, the composition of their monocrystalline base, leaving out of the account the dielectric layers, metallization and their interface with the semiconductor) represents itself a set of different combinations of three types of structural defects – point defects, clusters of point defects (microdefects) and dislocations; these types having a metastable balance between themselves and with a field of elastic stresses. Each specific type of the semiconducting devices is characterized by its own ratios of concentration of its defects and by its own distribution profiles for specific regions of the structure; this is determined by the design and topological specificities of the work piece and by its manufacturing technology. As of to–day, most of the studies has been performed on silicon devices, this is natural because silicon is most popular in contemporary microelectronics. These studies demonstrate in a vivid fashion how the defects affect the performance of the devices; that is why our main attention in this chapter will be given to these defects. Alongside with this, we shall specify these effects detected in the devices exploiting some other semiconducting materials.

2.2
Point defects, aggregates and microdefects

A change in device's performance caused by structural distortions (point defects included), the simplest aggregates and clusters of point defects is determined by their electrical activity; under this we imply a capability of defects as generating–recombinating centres to affect the transportation properties of the charge carriers, i.e. their concentration, mobility and lifetime [Matare (1971); Ravi (1981)]. Depending on the charge state, the intrinsic point (IP) defects are capable of creating in the semiconductor prohibited region both donar and acceptor levels located at different distances from the conductivity and valency regions.

However, the thermodynamically–balanced IP defects practically exert no impact on the characteristics of the devices produced on the basis of elementary semiconductors, because the concentration of IP defects at normal temperature is low (less than $10^{10} \, cm^{-3}$) and they are more mobile, as compared against the impurity atoms. In the semiconducting compounds, the IP defects determined by the deviations in the stechiometric composition, are of higher concentration (up to $10^{17} \pm 10^{19} \, cm^{-3}$) than germanium and silicon; and, due to their low mobility, they may retain in an isolated fashion in the active regions of the normally running devices and may function as efficient centres recombining with the deep levels in a prohibited region [Matare (1971); Mil'vidsky and Osvensky (1984)].

The concentration and, accordingly, an extent of the effects exerted by IP defects on electrophysical performance of the semiconducting materials and devices, produced on their basis, will grow if in them there emerge the unbalanced vacancies and intrinsic interstitial atoms induced by thermal, barric or radioactive actions. However, the unbalanced Frenkel pairs and an associated change in the structurally–sensitive properties of semiconductors occur only during actions of external factors. When these factors cease to act, some time upon this (the relaxation time) the semiconductor assumes a new metastable state caused by transformations in the initial extrinsic defect composition and by the emerging new PD aggregates or larger aggregates like clusters of IP defects, impurities, extrinsic precipitates and the like. In the majority of cases, simple and complicated aggregates of point defects are born by the heterogeneous mechanism when the crystal is sufficiently highly oversaturated with vacancies and/or interstitial atoms on its original defects–centres, i.e. the nucleating centres; both dopants and background impurities can work as such centres. Such well–known aggregates in silicon as $V - P$ (E is a centre), $V - O$ (A is a centre), $V - B$, $V - Al$ and the like are examples of such centres. If the crystal is highly oversaturated with IP defects, the aggregates of vacancies and of intrinsic interstitial atoms are possible to emerge homogeneously. Practically all PD aggregates create deep acceptor and/or donor levels in the prohibited region of semiconductors and hence affect the electrophysical parameters of devices, in particular, determine the degradation phenomena in them during the actions of the external destabilizing factors. In more details these issues are discussed below.

Besides the IP defects and their simplest aggregates, the deep levels in the semiconductor can be induced by metallic pollutants from the ingot growing process and from later technological machining. The most specific impurities in silicon and their levels, being accordingly calculated from the edge in the conductivity (c) or valency (v) band, are given in Table 2.1 taken from [Milis (1973); Yemtzev and Mashovetz (1981)]. This table also holds the results [Yenisherlova and Marounina (1984)] of measuring the concentrations of some impurities in commercial CZ–grown silicon ingots by the neutron–activation analysis.

Table 2.1

Characteristics of metallic impurities in silicon
[Milns (1973); Yemtzev and Mashovetz (1981); Yenisherlova and Marounina (1984)]
(A – acceptor; D – donor)

Impurity	Level position (eV)	Level type	Concentration (cm^{-3})
Ag	0.29 (c)	A	–
	0.32 (v)	D	
Au	0.54 (c)	A	$7 \cdot 10^9$
	0.36 (v)	D	
Cu	0.52 (v)	A	$2 \cdot 10^{13}$
	0.37 (v)	A	
	0.24 (v)	A	
Cd	0.45 (c)	A	–
	0.55 (v)	A	
Zn	0.55 (c)	A	–
	0.31 (v)	A	
Co	0.53 (c)	A	–
	0.35 (v)	A	
Fe	0.40 (v)	D	$1 \cdot 10^{15}$
Ni	0.35 (c)	A	$1 \cdot 10^{16}$
	0.23 (v)	A	
Cr	0,22 (c)	D	$2 \cdot 10^{13}$
	0.41 (c)	D	
Mo	0.33 (c)	D	10^{14}
W	0.22	A	10^{11}
	0.30	A	
	0.37	A	

The most specificity of the majority of metallic impurities lies in their high diffusive mobility in semiconductors, in particular, the diffusion parameters of such impurities as copper (Cu), cobalt (Co) and nickel (Ni) are comparable against the similar parameters for light elements (hydrogen and lithium) [Ravi (1981); Vavilov and Kiselev, et al. (1990)]. The values of the preexponential factor D_0 and those of diffusion activation energy E_g for some impurities [Ravi (1981)] are presented, as examples, in Table 2.2.

Table 2.2

The diffusion parameters of metallic impurities in silicon

Impurity	$D_0 \ (cm^2 \cdot c^{-1})$	E_g (eV)
Cu	$4.7 \cdot 10^{-3}$	0.43
Ag	$2.0 \cdot 10^{-3}$	1.54
Au	$(0.244 - 2.75) \cdot 10^{-3}$	$0.38 - 2.0$
Fe	$6.2 \cdot 10^{-3}$	0.86

The behaviour of fast diffusing impurities in silicon is mainly determined by their below properties [Ravi (1981); Mil'vidsky and Osvensky (1984); Vavilov and Kiselev, et al. (1990)]:

a) by a fall of impurity solubility in the crystalline lattice at technological temperatures. For example, a solubility limit for copper at $873 \ K$ makes up $2 \cdot 10^{15} \ cm^{-3}$, at $1473 - 1573 \ K$ it reaches $10^{18} \ cm^{-3}$ and then at $1573 - 1673 \ K$. sharply declines to $10^{16} \ cm^{-3}$. The solubilities of gold, zinc and iron are similarly temperature–dependent;

b) by a rapid fall in solubility in a solid state when temperature falls; and also high values of diffusion coefficients determine a heterogeneous build-up of impurities on crystalline defects;

c) by copper and iron that form discharges at very small concentrations; and, with their compositions in silicon exceeding $2 \cdot 10^{-4} \ at \, m \cdot \%$, these impurities can induce crucial structural distortions in thermally–treated crystals. For example, Cu discharges represent themselves the colonies of precipitates encircled by dislocations.

The discharges of fast diffusing impurities are often distributed in the way similar to a swirl–type distribution of growth microdefects, epitaxial or oxygen–induced stacking faults whose local fields of mechanical stresses are initiating a precipitation process. However, not only dislocations or stacking faults may serve as nuclei for creating metallic discharges or silicides of metals. In the capacity of such precipitating centres there can work the inclusions of second–phase particles, for example, SiO_2 or SiC. These are the latters that are responsible for star–like defects with a central nucleus from which the dislocations of complicated form are spreading [Ravi (1981)].

It should be noted here that the silicides $(MeSi_2)$ and the precipitates of transient metals, in contrast to isolated atoms, do not bring about deep levels and do not deform a band structure in silicon. As these inclusions are of the highest electroconductivity, they be evacuated from active regions of devices [Vavilov and Kiselev, et al. (1990)]. According to [Vavilov and Kiselev, et al. (1990); Nemtzev and Peckarev, et al. (1981)], the concentrations of copper or nickel close to 10^{16} cm^{-3} induce significant changes in the recombination parameters of silicon. Whereas the impurities with a lower diffusion coefficient, for example, titanium and vanadium (at $t = 1373\,K$), the ratios of diffusion coefficients make up $\dfrac{D(Ti)}{D(Cu)} \approx 10^{-4} - 10^{-3}$ and $\dfrac{D(V)}{D(Cu)} \approx 10^{-3} - 10^{-2}$; similar changes are observed at the concentration 10^{13} cm^{-3}. In other words, the defect levels related to titanium and vanadium are the most efficient recombination centres in silicon.

The metallic impurities in silicon also affect the properties of crystals; these properties are conditioned by background low–diffusion impurities, for example, by oxygen. As experimentally shown in [Bakchadyrkchanov and Zainobodinov, et al. (1976)], the concentration of optically active oxygen is dropped by nickel diffusing in silicon, this concentration being calculated by the coefficient of IR-absorption in the 9.1–mcm band. In the dislocation specimens, this effect was detected to be less than that in the dislocation–free specimens, i.e. one portion of nickel atoms deposits on dislocations and does not form any aggregates with oxygen responsible for diminishing the absorption peaks in the 9.1–mcm waves. The absorption, i.e. the concentration of the optically active interstitial oxygen, does practically change if crystals are slowly cooled starting from the diffusion temperature $1423 - 1623\,K$. When being cooled fast, crystals loose the concentration of oxygen markedly. In [Bakchadyrkchanov and Zainobodinov, et al. (1976)] this is explained by the fact that in slowly cooled specimens the large diffusion coefficient of nickel in silicon helps nickel atoms to leave their original interstitial positions and deposit onto the decomposition centres. With silicon crystals being cooled fast, the nickel atoms stay in the interstitial sites and react with the interstitial atoms of oxygen to form aggregates either with atoms of oxygen or vacancies. A structure of these aggregates remains still unknown but they may be active electrically. The latter was proved [Liblich and Nassibian (1973); Volkov and Zaitzev, et al. (1983)] by the measurements of volt–ampere characteristics in MDS-structures being previously doped by nickel of concentration $5 \cdot 10^{15}$ cm^{-3}. The presence of nickel was found to decrease an efficient positive charge in the MDS-structure. This is explained by the energy nickel–oxygen aggregate representing itself an acceptor level in silicon which is charged negatively if filled or neutral.

During a silicon oxidation, precipitation of metallic impurities was also observed. Through IR–spectroscopy, selective chemical etching and by measuring the volt–ampere characteristics, it was shown that a slow cooling of the $Si - SiO_2$ structures from the oxidation temperature $1423\ K$ will increase the number of background precipitates in the silicon presurface layer on the stacking faults; the impurities are captured by the growing SiO_2 film and the bonds $Si - O - Si$ become broken [Zaitzev and Pavlov, et al. (1981)]. This results in a change of the charge in the $Si - SiO_2$ interface, i.e. in the characteristics of the MDS–structures. There have been also studied the effects of the diffusing palladium, zirconium and nickel on the electrophysical properties in the $Si - SiO_2$ structures [Roumack and Soukchanova (1982)]. It was demonstrated that fast–diffusing cationic impurities affect the charge states of the $Si - SiO_2$ interface and the specificity of these effects lies in their ability to build small discrete acceptor–type levels in the Si presurface layer and positively–charged centres in the SiO_2 film. The position of the energy levels established by the impurities in the prohibited band depends on the cooling rate in the structures from the diffusion temperature $1323\ K$. It is supposed that a necessary condition for the charged centres to appear in a dielectric film is a SiO_2 low–temperature crystallization stimulated by metallic impurities.

How various aspects of gold doping affect the parameters of silicon transistor structures was analysed in [Mineyev (1982)], on the basis of the publications and his experimental results. It was shown that supplementary recombination centres when incorporated drop the transistor's amplification coefficients and increase the resistance in the collector regions. This is explained both by the arising $Si - Au - P$ aggregates which help the phosphorus atoms to become electrically inactive and by the gold acceptor levels compensating the donor properties of phosphorus. A presence of metallic impurities explains high leakage currents in $Si\ \ p - n -$junctions and a degradation of transistor and diode structures [Nemtzev and Peckarev, et al. (1981); Volkov and Zaitzev, et al. (1983)]. Within the presurface layers of such structures, there occur the precipitates which can slow down the diffusion of acceptor impurities (boron) during the formation of basic regions and can bend a diffusion front. Here, the precipitates are dissolving, and after the emitting diffusion there arises a diffusing tube between the collector and emitter. Formation of such tubes in the presence of gold is explained by the gold phosphide (Au_2P_3) precipitation on dislocations. Incorporating nitrogen (up to the concentrations $10^{19}\ cm^{-3}$) into the Si epitaxial layers diminishes the reversive currents in the basic and insulating $p - n -$junctions in ICs; this is thought to relate to gettering the fast–diffusing metallic impurities by nitrogen precipitates like fine dispersive particles of silicon carbide [Biryukov and Zotov, et al. (1976)].

In addition to the doping and metallic background impurities, silicon monocrystals always contain oxygen and nitrogen as well. The concentration of these elements is dependent on the ingot growing technique (Table 2.3) [Ravi (1981); Yemtzev and Mashovetz (1981); Nemtzev and Peckarev, et al. (1981)].

How oxygen behaves in silicon has been studied sufficiently well and its main behavioural specificities include the following [Ravi (1981); Volkov and Zaitzev, et al. (1983); Batavin (1970)]. In thermally treated silicon crystals with a high content of oxygen, at $623 - 773 \ K$ there emerge small–size donor centres, so–called thermodonors, creating a wide choice of energy levels within the silicon's prohibited band from $0.06 - 0.07$ to $0.13 - 0.16 \ eV$ [Yemtzev and Mashovetz (1981); Vavilov and Kiselev, et al. (1990)]. Higher temperatures kill these centres and oxygen becomes electrically inactive. The activation energy of thermodonor formation and decomposition processes makes up $2.5 - 2.8 \ eV$ [Vavilov and Kiselev, et al. (1990)].

Table 2.3

Concentration of the background impurities
in monocrystalline silicon

The growing technique	Oxygen	Hydrogen	Metals
CZ growing	$(0.2 - 2.0) \cdot 10^{18}$	$(0.4 - 5.0) \cdot 10^{17}$	$\leq 10^{15}$
FZ melting	$(0.01 - 2.0) \cdot 10^{16}$	$(0.2 - 1.0) \cdot 10^{17}$	$\leq 10^{13}$

Thermal treatment at more than $1273 \ K$ causes irreversible changes in silicon and this results in no thermodonars appearing at a later heating up to $723 \ K$. A higher–temperature annealing is supposed to generate precipitates from silicon and oxygen which dissolve at $T \geq 1573 \ K$ [Ravi (1981)]. After such thermal treatment, annealing at $573 - 773 \ K$ will again cause oxygen to become electrically active. This electrical activity of oxygen during a low–temperature annealing is thought [Ravi (1981); Yackovencko and Gvelesiani (1975)] to be determined by its transition from an interstitial to a substituting position in the silicon lattice. As was proved experimentally, an growth of concentration of thermodonors during annealing reduces concentration of oxygen in the interstitial sites. The latter is measured through the IR-spectra of absorption by the wave length $9.1 \ mcm$; this absorption is determined by asymmetrical fluctuations in the molecular group $Si - O - Si$. One more band of absorption at the wave length $8.4 \ mcm$ relates to SiO_2 precipitates. The analysis of the IR-spectra in these two basic absorption bands makes it possible to study a state of oxygen atoms in silicon during various actions upon it. A transition of oxygen from interstitial sites to precipitates may occur through a number of intermediate forms of $SiO_x (x \geq 2)$, i.e. according to the scheme

$$Si - O - Si \rightarrow SiO_x \rightarrow SiO_2 \ .$$

Intermediates states of SiO_x can be electrically active and introduce donor levels into the prohibited zone.

Donors are possible to emerge as a result of direct interaction of oxygen atoms with the vacancies in such reaction as

$$O_i + V_{Si} \rightarrow O_S$$

$$O_S + V \rightarrow O^+{}_S V^- \ ,$$

where O_i, O_S is an oxygen atom in the interstitial site and in the Si lattice site.

The first reaction corresponds to a transition of oxygen to a site when the vacancy vanishes. In the second reaction, site–held oxygen is combined with a neibouring vacancy releasing to it one of its valent electrons, and it may function as a termodonor, in accordance with the below reaction

$$O_S{}^+ V^- \rightarrow O_S{}^{2+} V^- + e \ .$$

According to this model, in general the donors may be supposed to be generated as

$$O_S{}^{2+} EA + e \quad \text{or} \quad O_S{}^{2+} EA_1{}^- EA_2{}^- + e \ ,$$

where EA stands for any acceptor centre, say, a Si vacancy or acceptor metallic impurities of Cu, Al, etc. This model shows that the acceptors bound to fast–diffusing impurities may become electrically inactive, provided the conditions for combining them with oxygen donor centres are available. Actually, as shown [Ravi (1981)] by experimentally measuring the lifetime of the silicon minor charge carriers during thermal treatments, annealing at thermal ranges needed to build donor centres increases the lifetime and sharply drops it if the temperature grows up to $873 \ K$ at which such centres start to collapse.

It is worth noticing here that oxygen precipitation in silicon also diminishes the lifetime of minor charge carriers; this is thought to happen due to the capturing levels $E_C - 0.17(0.16) \ eV$ in the prohibited zone. Besides, the oxygen precipitates in silicon make the internal electrical field in crystal locally inhomogeneous and promote a microplasmic breakthrough of $p - n -$junctions to appear [Batavin (1970a); Batavin (1970)]. During a formation of precipitates, large distortions in the lattice near the SiO_2 particles caused by a change in the volume $\left(\dfrac{V_{SiO_2}}{V_{Si}} \approx 2 \right)$ [Ravi (1981)] give rise to excessive concentration of the intrinsic interstitial atoms in silicon. These atoms can diffuse into the bulk and condensate to form the Frank dislocation loops or to promote a growth of the defects encircling a second–phase particle. It is supposed [Vavilov and Kiselev, et al. (1990)] that during a growth of the precipitates the silicon interstitial atoms being emitted from the clusters–thermodonors make the oxygen holding defects electrically less active.

Carbon usually held in silicon in quantities exceeding a solubility limit is an electrically neutral substitutional impurity. Its effects on crystals are observed through various electrically active aggregates (being formed, in particular, with oxygen, metallic impurities or IP defects in silicon) and through the inclusions of silicon carbide as well [Ravi (1981); Yackovencko and Gvelesiani (1975)]; Smirnov (1980)]. A growth of carbon was detected to drop the concentration of oxygen thermodonors.

Carbon atoms may serve as centres for heterogeneously generating the oxygen microprecipitates, dislocation loops and extrinsic clusters locally deforming the silicon lattice.

As has been investigated, oxygen and carbon in silicon epitaxial layers may cause microdefects [Bazhinov and Nickoulov, et al. (1986)]. These elements penetrate into a growing layer from the wafer and from the gaseous phase in which an epitaxial process is performed. Most intensively the defects emerge at low–temperature epitaxy ($1223\ K$) and at low pressure in the reactor. With temperature increasing up to $1473\ K$, microdefects were observed only in a narrow region of the epitaxial layer lying close to the wafer – this is explained by an efficient removal of oxygen and carbon into an external ambient and by the defect generating centres retained only on the impurities having diffused from the wafer. On the wafers boron–doped up to the concentration $1 \cdot 10^{19}\ cm^{-3}$ and higher, the epitaxial films were observed to be of low concentration of microdefects. Boron atoms are supposed to form aggregates with background impurities hampering the penetration of impurities into a growing layer. The same results may be obtained if wafer's presurface layers, in which a background impurity segregates during preapitaxial technological treatment, undergo a preliminary etching off.

The effects of carbon on a structural perfection and electrophysical characteristics of silicon epitaxial layers is determined by the silicon carbide (SiC) particles which hamper an edge spreading of the growth steps, and this gives rise to stacking faults, twins and polycrystalline inclusions [Ravi (1981)]. Carbon was detected to promote a generation of epitaxial stacking faults on the growth microdefects in the wafers of dislocation–free silicon and also carbon enriches stacking faults with fast–diffusing impurities that drop the lifetime of the minor charge carriers in the silicon epitaxial layers.

Microdefects. As was noted in Chapter 1, dislocations in the bulk of monocrystal serve as efficient drains for unbalanced point defects, and primarily, for intrinsic interstitial atoms and vacancies which can emerge at any technological stage of the manufacture, starting from the ingot growing process. If there are no dislocations ($N_d < 10^2 cm^{-2}$) and no other structural distortions, then a surface of the ingot becomes a single drain for point defects. Though, if the diameters of the crystals are great and if the crystals are cooled from the melting temperature at the final rate, then the most majority of point defects do not manage to diffuse into the ingot external surface and, as a result of this, a oversaturated solution of interstitial atoms and vacancies arises in the bulk of the ingot. In dislocation–free crystals, such oversaturated solution decomposes to form specific structural distortions (microdefects) which were seemingly first described in detail in [Kock (1973)].

Nowadays, microdefects are commonly called the local distortions in the periodicity of the crystalline lattice being of the size from a hundred of fractions to dozens of micrometers; these distortions can be the aggregates of IP defects, extrinsic defect clusters, dislocation microloops and the like [Ravi (1981); Mil'vidsky and Osvensky (1984); Vavilov and Kiselev, et al. (1990); Mil'vidsky (1986)].

In dislocation–free crystals of silicon and germanium, as was noted in Chapter 1, microdefects are often distributed on a layer-by-layer basis within the bulk, in the form of concentric or spiral–like bands of variable thickness. Therefore, initially the microdefects were called swirl or swirl–defects. Though, it should be born in mind that such name denotes rather the specificity of their spatial distribution in an ingot than describes the nature and structure of such defects. A swirl–like picture of microdefect distribution is determined by periodic fluctuations in the semiconductor crystallization rates during the growth process. The contemporary techniques of monocrystal growth make it possible to suppress the fluctuations of the growth rate and this results in a practically homogeneous distribution of microdefects across the entire bulk of the ingot [Mil'vidsky (1986)]. Mostly studied have been the microdefects in the silicon crystals whose dislocation–free ingots of large diameters are produced commercially.

It was also noted in Chapter 1 that as of to–day there have been detected in silicon at least four types of microdefects, indicated as $A-$, $B-$, $C-$ and $D-$defects. The $A-$defects are the largest and their spatial concentration usually makes up $10^6 - 10^7\ cm^{-3}$ but may change depending on the ingot growing conditions. Via the electronic microscopy [Ravi (1981)], the $A-$defects were found to be some separate dislocation loops or colonies of loops of the interstitial type $1-5\ mcm$ in size, primarily with the Bürgers vector $b = \dfrac{a}{2} < 110 >$, i.e. they are full. The Frank loops with $b = \dfrac{a}{3} < 111 >$ are also observed sometimes, constraining the stacking faults. These defects are detected by a selective chemical etching and look like sufficiently big flat–

bottomed wells or etching hillocks [Ravi (1981)]. Next in size go the $B-$ defects $(0.05-0.1\,mcm)$ whose concentration may approach $10^7-10^9\,cm^{-3}$. According to the electron microscopy, the $B-$ microdefects (especially PD clusters) induce local fields of elastic stresses; these fields may also hold extrinsic (most often carbon) micropricipitates. The $C-$ and $D-$ defects are less than $0.01\,mcm$ in size and their concentration may be $10^9-10^{10}\,cm^{-3}$. These defects are mainly distinguished between themselves by the fact that the $C-$ defects appear in the crystals grown in quartz crucibles, and the $D-$ defects are induced by an increase in growth rates. No principal morphological difference between them has been found. Their nature is not yet known, though they are thought to be small–size vacancy clusters. The spatial concentration of $C-$ defects has been observed to drop with a drop of carbon in crystals [Mil'vidsky (1986)].

There are usually no microdefects in the narrow (about $2-3$ mcm) band close to the ingot's edge surface and also close to the growth dislocations. Depending on the growth regimes, the zero–defect band can change its thickness. Here, the $B-$ defects lie closer to the ingot surface than the $A-$ defects, and the $C-$ and $D-$ defects are distributed across the ingot cross–section uniformly. The annealing of Si zero–dislocation crystals at $1473-1523\,K$ was shown to dissolve microdefects; though this dissolution does not pass traceless, since their birth during a growth process was heterogeneous most often [Milevscky and Vysotzkaya, et al. (1980)]. The latter implies the particles of chemical compounds like SiO_2, SiC, etc left in the crystal by the dissolvable dislocation loops during annealing.

The growth microdefects as centres for charge carrier recombination are thought to determine a spatial inhomogeneity in electrical properties of Si monocrystals, even if they have not undergone special high–temperature treatments [Ravi (1981)]. Microdefects are usually considered to work as drains for minor charge carriers, because they have a developed surface, i.e. the interface with the monocrystalline matrix [Kontzevoy (1970); Kontzevoy and Filatov (1987)].

A power of such drains is determined by their sizes and by the parameter similar to a rate of a surface recombination; this parameter describes a rate of charge carrier recombination in the defect – crystal interface.

With the content of defects (precipitates, etc) being high, a total square of the internal surface in the interface was shown to be comparable or exceed the square of the external surface in the crystal under measurements. Here, during recombination on the defects of spherical radius r_{sph}, the lifetime of charge carriers is defined by the ratio

$$\tau_{sph} = \frac{1}{4\pi N_{sph}} \left(\frac{1}{Dr_{sph}} + \frac{1}{Sr^2_{sph}} \right),$$

and for cylindrical defects of the radius τ_{cyl} (the dislocations decorated by impurities) by the ratio

$$\tau_{cyl} = \frac{1}{2\pi DN_{cyl}} \left[\ln(\frac{1}{r_{cyl}\sqrt{\pi N_{cyl}}}) + \frac{D}{Sr_{cyl}} \right],$$

where

N_{sph} is a density of spherical defects (cm^{-3});

N_{cyl} is a density of cylindrical defects (cm^{-2});

S is a rate of recombination in the defect – matrix interface $(cm \cdot c^{-1})$;

D is a coefficient of diffusion for charge carriers $(cm^2 \cdot c^{-1})$.

A rate of recombination S on the interface depends on the processes of decorating the defect with impurities and is determined by the electrical activity of the defect [Kontzevoy (1970)].

It is worth mentioning here that the electrical activity in the growth microdefects in silicon is far from a simple process and up to now it has no single–valued solution. From the one hand, it is shown that a swirl distribution of microdefects correlates with a inhomogeneous change in lifetime of minor charge carriers and with a conductivity of silicon dislocation–free crystals, and also with the dislocation pictures for local regions with increased leakage currents in diode structures and generation currents in the CCD-matrices on the surfaces of silicon structures [Ravi (1981); Kontzevoy and Filatov (1987); Smoul'scky (1979); Lefevre (1982); Kubena and Hlavka (1985); Graff and Fisher (1982)]. On the other hand, the microdefects are stated not to substantially affect the electrophysical characteristics of silicon [Grishin and Goulyaeva, et al. (1982); Vysotskaya and Gorin, et al. (1984); Litvinenko and Troshin, et al. (1986)]. It may be supposed that such contradictory decisions on the effects of microdefects on silicon electronic properties stem, first, from a dual nature of their effects on charge carriers functioning as recombination centres (a decrease of lifetime) and a dual nature of the gettering defects (metallic impurities as getters) increasing the lifetime of charge carriers. Second, the difference in ideas may be explained by inconsistency of the spectrum and concentration of microdefects in the specimens under study. A general extrinsic defect background in crystal should certainly affect the results of such experiments, since in each of these experiments there were used the thermal treatments this or that way activating the transformations in structures and in original growth defects converting some of them into the stacking faults or gettering centres. Moreover, a concentration, a type and profiles of growth microdefect distributions are possible to be affected by slicing operations, and the nature of such effects will be regulated by specific conditions and regimes assigned for abrasive–chemical treatment of each individual slice [Alekhin (1983); Perevostchickov and Skoupov (1992)]. These issues will be discussed in detail in the below section dedicated to low–temperature gettering.

Microdefects have been also detected (though investigated yet slightly) in the dislocation–free regions in the crystals of superpure germanium with the concentration of electrically active impurities $10^9 - 10^{10}\,cm^{-3}$ [Zeeger and Fell, et al. (1979)]. In such crystals grown in the hydrogen atmosphere there have been revealed – by selective chemical etching – the semispherical etching wells that do not appear in the dislocation regions or in the material whose ingots have been produced in the atmosphere of nitrogen, argon or helium. In contrast to silicon, these etching wells are slightly apt to a swirl distribution. The electronic microscopy facilities failed to reveal the defects in germanium whose contrast could be comparable with the contrast of the $A-$defects in silicon, i.e. the contrast created by the dislocation loops. On this basis, the semispherical etching wells in the zero–dislocation germanium are bound to the local accumulations of vacancies and hydrogen, the latter being possible to be in a molecular fashion [Zeeger and Fell, et al. (1979)].

The crystals of semiconducting compounds of the type in $A^{III}B^V$ also can include microdefects which , during a selective chemical etching, give rise to specific shallow wells of the nondislocation nature [Mil'vidsky and Osvensky (1984)]. The number of such etching wells is dropping sharply in the vicinity of dislocations working as drains for the unbalanced point defects. Zero–dislocation heavily–doped crystals often turn out to have a swirl distribution of microdefects, similar to that observed in silicon. More specifically the microdefects were studied in the $GaAs$ monocrystals doped by the VI-group impurities (tellurium, selenium, sulphur). Via a selective etching, such defects have been also revealed in gallium phosphide, indium arsenide, gallium antimonide doped also by the VI-group impurities, in the postgrowth state or upon thermal treatment. Microdefects have been also found in the layers of tellurium–doped gallium arsenide grown by liquid epitaxy [Mil'vidsky and Osvensky (1984)]. With a use of electronic microscopy, it was shown that the heavily–doped crystals involve the below basic types of defects [Mil'vidsky and Osvensky (1984)]:

- single– and/or multilayer stacking faults $0.1 - 1.0\,mcm$ in size;
- small (up to dozens of nanometers) prismatic dislocation loops, both full and partial;
- the defects located within the square of a stacking fault or on a dislocation loop of the particle of the small–size extrinsic discharges.

Thermally treating the crystals of semiconducting compounds (in the regimes of annealing, hardening or their combination) induces substantial changes in the structure of their microdefects, their spectrum and concentration. These changes are caused by a rich choice of point defects in these materials which are recombined during thermal treatment. The study of the crystals of semiisolating $GaAs$ (CZ–grown from under the layer of the B_2O_3 flux), performed by the "projective" etching technique at $t \leq 1173\ K$, made it possible to conclude that annealing gives birth to aggregates of point defects; this aggregation is constrained in the extensive regions around the "high–temperature" growth dislocations emerging at the temperatures higher than $1373\ K$ [Markov and Grishina, et al. (1984)]. It was also shown that the thermally–induced aggregates make electrons less mobile and more concentrated, i.e. these aggregates lead to a lack of thermostability in the semiisolating material. It comes from here that when growing the semiisolating gallium arsenide and when there is no control of the ensemble of point defects, the instable electrophysical parameters in thermally–treated crystal (this instability was determined by aggregation) may be eliminated by growing the crystals with a relatively high density of "high–temperature" dislocations ($N_d > 5\cdot 10^4\ cm^{-2}$) [Markov and Grishina, et al. (1984)]. This means that microdefects in this case will be less active, since their concentrations and sizes will be reduced.

As was noted above, a presence of microdefects in original wafers may bring about structural disturbances in the layers being epitaxially grown on them. Through selective chemical etching and electronic microscopy of the gallium arsenide silicon–doped monocrystals and homoepitaxial aggregates grown by a gaseous epitaxy on the wafers (produced from these materials), it was found that initially the microdefects in ingots are represented by rare (the concentration less than $10^9\ cm^{-3}$) interstitial dislocation loops not more than 50 nm in size [Kalinin and Kolobova, et al. (1990)]. All loops were decorated with monocrystalline second–phase inclusions. In the ingot faces heavily doped by silicon, the dislocation loops increase up to 0.5 mcm in size and their concentration grows up to $3\cdot 10^{12}\ cm^{-3}$. In the epitaxial layers grown on the wafers and where no microdefects were detected via selective etching, growth dislocations were observed to be inherited according to the known mechanisms [Ravi (1981); Mil'vidsky and Osvensky (1984)]. In the regions with the wafers having dislocation loops, the epitaxial layer had the dislocation density 3 – 5 times exceeding that in the wafer. In the epitaxial films deposited on heavily–doped $GaAs$ crystals (up to $N_{Si} = (3-6)\cdot 10^{18}\ cm^{-3}$), the dislocation density has grown, against the wafer, nearly as much as two orders of magnitude. Hence, not all the microdefects in GaAs may be thought to work as additional sources of dislocations in the growing epilayer, but only those microdefects which look like large dislocation loops or aggregates of point defects decorated with second–phase inclusions [Kalinin and Kolobova, et al. (1990)]. High density of microdefects and dislocations in the wafer and in the epilayer give rise to negative changes in the performance of the devices manufactured on the basis of the $A^{III}B^{V}$ semiconductors. In particular, these defects are thought responsible for the inhomogenity of luminescence in injection GaAs–based lasers, high leakage currents in the Schottky diodes, a fall of concentration in free charge carriers in the Gunn diodes and various degradating phenomena in the devices [Mil'vidsky and Osvensky (1985)].

It should be emphasized here that the effects of microdefects (and other extended distortions, by the way) on the electrophysical performance of the devices most often do not relate to their being the delation centres disturbing a crystal band structure but rather their ability to getter and create the local regions of high conductivity or charge carrier recombination. Besides, it may be supposed that in contrast to such stable defects as dislocations, the microdefects are more mobile and easily transformable in a varying external ambient, and, therefore, they may be electrically more dynamical.

2.3
Dislocations and stacking faults

Dislocations and stacking faults decorating the dislocations may exert both direct and indirect impact on electrophysical performance of semiconducting materials and structures. In the first case, electrophysical properties of crystals undergo changes because of new levels and a change in the width of the prohibited band determined by dislocations; and in the second, determined by the extrinsic atmospheres encircling them. As far as various aspects of electrical activity of dislocations in the semiconductors were studies rather thoroughly (see, for example, [Matare (1971); Ravi (1981); Mil'vidsky and Osvensky (1984); Mil'vidsky and Osvensky (1985); Leikin and Zelenov, et al. (1978)], only basic of them will be discussed in this subsection.

The aperiodicity of atoms near the dislocations in the lattice gives rise to broken (unsaturated) interatomic bonds, i.e. unpaired electrons. These electrons can capture free electrons from the conductivity band to have dislocations negatively charged. Such situation is characteristic for monatomic donor–doped semiconductors. In these materials, the electrons captured by the broken bonds introduce the energy levels, lying in the vicinity of the valency band, into the prohibited band, i.e. the dislocation demonstrate their acceptor properties. Vice versa, in $p-$ crystals the dislocations are charged positively, i.e. they are donors.

Since the condition of electroneutrality should be fulfilled in the crystal, the dislocation lines should attract the ionized extrinsic atoms and mobile charge carriers of the opposite sign. In the $n-$ crystals, donor impurities are attracted by dislocations, and in $p-$ crystals the acceptor impurities, respectively. This results in cylindrical regions of a spatial charge encircling the dislocations. In such regions, a sign of the charge is antithetical to that of the charge in the dislocation line. A radius of the spatial charge cylinder can be calculated [Matare (1971)] as:

$$R = \sqrt{\frac{f_0}{\pi c (N_d - N_a)}} \ ,$$

where f_0 is the Fermi distribution function; c is a distance between the broken bonds on the dislocation line; N_d and N_a are the concentrations of donors and acceptors in the crystal. The function f_0 is approximated by

$$f_0 = \left[1 + \exp\left(\frac{E_g - F}{kT}\right)\right]^{-1},$$

where

E_g is an energy level in the dislocation; F is the Fermi level.

For silicon at $293\ K$ $f_0 \approx 0.1$ and at the concentration of carriers $\sim 10^{15}\ cm^{-3}$ the radius of the spatial charge region will be about 1 mcm.

A presence of the spatial charge around the dislocation causes a potential barrier for the carriers; a value of this barrier for n − silicon may be calculated from the ratio [Matare (1971); Ravi (1981)]:

$$e\varphi = E_0 f \left[3\ln(\frac{f}{f_c}) - 1.232\right],$$

where

$E_0 = \dfrac{e^2}{4\pi\varepsilon\alpha}$ is an electrostatical energy of interacting neighbouring acceptor states;

e is an electron charge;

a is a distance between broken bonds ($a = 0.334\ nm$ for a purely edge dislocation in silicon);

$\varepsilon = 1.02 \cdot 10^{12} \left[\dfrac{\Phi}{m}\right]$ is a dielectrical constant for silicon;

f is a degree of filling the acceptor states;

$f_c = a\left[\pi(N_d - N_a)\right]^{1/3}$ is a constant being determined by a crystal doping level.

In order to determine a form of the charge carrier potential well established by a tube of spatial dislocation charge and also to determine a structure in the dislocation nucleus, there have been studied the effects of the external electrical field on the height of the barrier ($\Delta\varphi$) of the potential well and nucleus conductivity were studied in [Rzaev (1991)]. Here, the form of the curve (a shape of the potential well) for potential energy of charge carriers interacting with the dislocation was estimated through thermally stimulated currents; this technique is based on the Frenkel − Pool effect which implies a growing probability of thermal ionization in deep centers in the electrical field, as a result of the drop in the potential barrier. In [Rzaev (1991)], subjected to studying were the silicon epitaxially–planar $n^+ - p -$junctions of small square; this square allowed to hold a single electrically–active dislocation in the $p -$ region. Through electrical and electronomicroscopic measurements there have been determined a type, a form, an orientation and a region of a cylindrical spatial charge and a nucleus of electrically–active dislocations. No Cottrell ambients (so–called minor dislocations) were found in the dislocations under study. In the experiments, the thermally–stimulated currents through $p - n -$ junction were measured at various values of reversive stresses when the dislocation passes across the spatial charge region perpendicular to the junction plane and is located between the basic and collecting contacts of the junction. In this way, the external field applied to the junction will decrease the emission barrier for carriers captured on the levels generated by unsaturated (broken) bonds of the dislocation nucleus.

It was found that a growth of the external field is shifting the thermal dependences of junction currents into the field of low temperatures. This speaks about a change in the height of the barrier for carrier emission. A drop in the potential barrier ($\Delta\varphi$) was calculated as a difference of the activation energy values estimated through a position of the peak in the temperature scale for various values of the external field. As measurements showed, the deep centres related to dislocations are the donors with the level $\varphi = 0.38\,eV$ (with the field of the $p - n -$ junction being equal to $6 \cdot 10^4 \, V / cm$), their concentration is $5 \cdot 10^{16} \, cm^{-3}$ and the emission coefficient $5.33 \cdot 10^3 \, c^{-1} \cdot K^{-2}$. The result obtained was proved by the capacitance relaxation spectroscopy of deep levels. Using the potential emission barrier as a function of the of the field there was obtained a coordinate dependence of the potential energy of interaction of carriers with a dislocation, i.e. a shape of the potential well was found; it turned out to be close to parabolic. Studying a type of emission as a function of the field and temperature, it was revealed that small fields and low temperatures create the regularity – $\ln j = \sqrt{E}$. This manifests conductivity through separate coulomb centres with nonoverlapping potential wells. With the field ($E \approx 8.4 \cdot 10^4 \, V / cm$) increasing, this dependence becomes close to linear – $\ln j = E$; this corresponds to the current running through coulomb centres with overlapping potential wells. As the temperature increasing, this overlapping vanishes and the dependence $\ln j = \sqrt{E}$ is reduced. The temperature at which the overlapping ceases will drop linearly if the level of filling the unsaturated bonds in the nucleus increases, and the temperature at which the overlapping falls linearly; whereas, with an increase of the external field, this temperature will grow parabolically. How the nucleus of the dislocation changes its conductivity in accordance with temperature and the field was explained by the extension and compression of the dislocation and also by the interaction of the charges captured on the unsaturated bonds [Rzaev (1991)]. Under the action of these aspects, a nucleus of dislocations may undergo the metal–isolator transfer, i.e. a transfer from overlapping potential wells for charge carriers (low temperature and high external electrical fields) to their discrete distribution along the dislocation line.

All the above said pertains to the dislocations having a shift vector edge component, since according to the model the purely swirl dislocations, having no broken bonds, must be electrically inactive. The swirl dislocations can affect the electrical properties of semiconductors, thanks to the activity of the extrinsic ambients encircling these dislocations. Though, in the latters the concentration is less than in the edge dislocations, since these ambients less active with the point defects, because of the zero hydrostatical component in the associated field of elastic stresses. Therefore, below are discussed only the dislocations with the Bűrgers vector edge component. The model of broken bonds is qualitatively proved by experiments on silicon and germanium, but it fails to fully explain a nature of electrical activity in dislocations. The point is that the model deals with full dislocations only, whereas such dislocations are actually available in semiconductors in a splitted form, and so the broken bonds may vanish as a result of the atomic rearrangement. In this case, the energy levels in the prohibited band must be determined by heavily deformed bonds in the nucleus of partial dislocations and stacking faults being encircled by them [Ravi (1981); Mil'vidsky and Osvensky (1984)]. In addition, the spectrum of energy levels induced by crystal deformation and the electrical activity of dislocations are substantially determined by the extrinsic composition and temperature–time parameters of the deformation process. In other words, the effects exerted by the dislocations on the electrophysical properties of crystals turn out to be indirect, they act through the implemented electrical activity of the point defects, i.e. the components of the extrinsic ambients encircling the dislocations. In some cases, the contribution of the indirect effects may exceed the importance of changes in the crystals' properties caused by the dislocation nucleus. The role of the last factor becomes evident at high densities of dislocations ($10^7 - 10^8 \ cm^{-2}$). Alongside with this, in spite of difficulties in interpreting the experimental data, there exists to–day a number of specific regularities describing the electrical properties of dislocations. For example, the measurements of the specific resistance in elastically deformed silicon have detected a sharp anisotropy of conductivity in the perpendicular ($\rho_\perp$) and parallel (ρ_{II}) dislocation lines of directions, and also a difference of these values from the resistance of zero–dislocation material (ρ_0) [Leikin and Zelenov, et al. (1978)]. At less than room temperatures, the inequality $\rho_\perp > \rho_{II} > \rho_0$ is valid. With the temperature rising to $473 \ K$, these differences vanish. The effects of anisotropy depend on the deformation temperature: at the temperatures close to $973 \ K$ the above inequality is valid, and if the deformation occurs at $1273 \ K$, then $\rho_\perp = \rho_{II} = \rho_0$ [Glaenzer and Yordn (1968)]. The impurities diffusing at high temperatures to the dislocations and a saturation of the broken bonds are supposed to cancel the spatial charge region giving rise to the conductivity anisotropy.

The effects of dislocations on the lifetime of minor charge carriers and the anisotropy of the carriers' mobility when orienting the electrical field parallel or perpendicular to dislocation lines [Matare (1971); Ravi (1984)] – these processes are thought to be explained by the presence of spatial charge regions. Though, the experimental results here retain very contradictory, since they depend on what mechanism, direct or indirect, governs the interaction of dislocations with charge carriers or dominates in the material having a given doping level and background impurities. For example, a change in the dislocation charge states, caused by extrinsic ambients, can create different dependencies of the lifetime of charge carriers on the dislocation density ($\tau \approx N_d^{-1}$ or $\tau \approx N_d$) or may lead to an absence of such dependence [Ravi (1981); Leikin and Zelenov, et al. (1978)].

Very often, the dislocations and stacking faults exert a nonunique effect on the performance of the devices, for this effect is determined by the manufacturing technology, the type and the design, and also by the functioning conditions of the apparatus.

For example, the effects of dislocations on electrophysical characteristics of the MDS-structures are usually explained by a crystalline imperfection in the $Si - SiO_2$ interface. It was found that silicon thermal oxidation, usually giving rise to dislocations of the density $10^3 - 10^5\ cm^{-2}$, increases in average by $2.6 - 3.6\ V$ the threshold voltage and decreases the hole mobility in the inversion channel of the MDS-structures [Velchev and Toncheva, et al. (1980)]. On the other hand, in studying the effects of dislocations within the density range from 0 to $10^9\ cm^{-2}$ on the charges and the density of surface states in the MDS-structures these characteristics were found to change sharply at $N_d \approx 10^9\ cm^{-2}$ (the state density rises up to $10^{13}\ cm^{-2}$). The further thermal treatment at $723\ K$ in the hydrogen atmosphere drops the state density in the $Si - SiO_2$ interface. A lack of coincidence of the results obtained in [Velchev and Toncheva, et al. (1980)] and in [Mc Canghan and Wonsiewicz (1974)] is sooner accounted for a difference of extrinsic ambients around the dislocations in the structures under study.

The dislocations become highly dense and, as a result of this, the density of surface states increases when highly–doped source and drain regions in MOS-transistors are formed via diffusion or ion implantation. In the first case, the dislocations are primarily the misfit dislocations and so–called out-of-contour dislocations. The misfit dislocations being almost up to the density of $10^8\ cm^{-2}$ slightly affect the transistor performance, in contrast to out-of-contour dislocations already capable of disturbing the running of the devices at the dislocation density $10^2\ cm^{-2}$ [Ravi (1981); Volkov and Zaitzev, et al. (1983)]. When using an ionic doping, the spatial lifetime of charge carriers and the rate of surface generation in MOS-structures will depend on the dose of implanted ions: the lifetime will drop and the generation rate will grow. In this case, an observable change in the performance of the MOS-structures is explained by the electrical activity of stacking faults induced by a transformation process in irradiation–induced defects during silicon oxidation [Ravi (1981)]. A negative role of stacking faults is mostly visible in charge coupled devices (CCD), chains of MOS–condensers, where the charge storing time practically drops exponentially with a growth of density of oxygen–induced stacking faults (OSFs).

Though, it should be noted here that the defect concentrations, which generally worsen the performance of MOS-devices, turn out to be higher than those for the bipolar devices being more sensitive to a structural perfection in the crystal regions of different conductivity and interface, i.e. sensitive to $p - n -$ junctions. A presence of defects brings into bipolar transistors such undesired phenomena as a rise of leakage currents and a fall of breakthrough voltages in $p - n -$ junctions, a drop of the current amplification coefficient, a formation of conducting regions (tubes) short–circuiting the emitting and collector regions. The dislocations and stacking faults in this case will serve as the ways for accelerated diffusion of impurities and centres of precipitate generation, i.e. they will work as local regions of higher–conductivity. The defects within the zone of the $p - n -$ junctions causing a drop in breakthrough voltage and an inhomogeneous nature of the breakthrough will give rise to a "soft" section on a descending slope of the volt–ampere function (i.e. they will increase leakage currents), high fluctuations in the current amplitude on a prebreakthrough section and also to a luminescence in some separate sections of the $p - n -$ junction [Ravi (1981); Mil'vidsky and Osvensky (1985)]. The above light emission, called a microplasma emission, is usually related to a local amplification of the electrical field on the dislocation–like defects, dislocation aggregates or stacking faults. All the listed facts cover the dislocations transversing the $p - n -$ junctions and lying within the spatial charge region, i.e. the dislocations are passing through the active elements in the device. Besides, these dislocations and stacking faults are most often decorated with impurities, mainly, those of transient metals. The stacking faults, as noted in [Mil'vidsky and Osvensky (1985)], may affect the yield of good devices more than the individual dislocations. This is provided by stacking faults being more efficient drains, and therefore are easier to be decorated with extrinsic discharges on the partial dislocations (restraining these stacking faults) or across the entire square of the stacking faults.

The defects located beyond the active elements in the device can, on the contrary, play a positive role – they are gettering the undesired impurities and in this way bettering the performance of the devices and increasing the yield of products [Vorobyev and Ignatyeva, et al. (1979); Kontzevoy amd Litvinov, et al. (1982)]. This is proved by directly matching the dislocation distributions on X-ray topograms (the topograms were shot prior to cutting into crystals with ready devices) against the planar distributions of fit and unfit devices [Mil'vidsky and Osvensky (1985); Kontzevoy and Litvinov, et al. (1982)].

All the said equally refers both to discrete devices and ICs. Though, it should be kept in mind that more often the ICs turn out to be more sensitive to defects, due to their design specificity and manufacturing technology. For example, bipolar circuits usually have more shallow junctions and less sizes of the active elements as compared against the discrete devices; this increases a probability of defect–induced faults, especially when defects are distributed across the bulk of the crystal inhomogeneously. Besides, an IC–manufacturing technology involves a great number of thermal operations required, say, to electrically isolate the active elements in the bipolar circuits, and this leads to supplementary structural distortions and, consequently, to a fall in the yield. The MOS (SMOS) ICs functioning on the basic charge carriers are less sensitive to defects, and as their technological line includes less thermal operations the yield of the products is usually more than for bipolar circuits.

In complicated semiconducting $A^{III}B^{V}$–type compounds, a lattice polarity and a wider choice of point defects complicate a picture of structural distortions in the electrophysical characteristics of these materials and in the devices produced on their basis. As for the edge dislocations, a charge state of their nucleus depends on whether the shift vector intersects the $\alpha-$ or $\beta-$surface [Matare (1971)]. The same dislocations in $p-$crystals attract positive charges and also create a barrier, a nucleus having a donor property. Alongside with this, the $\beta-$dislocations always carry donor properties, irrespective of electronic or hole conductivity of the material. A donor action by $\beta-$dislocations in the $p-$type indium antimonide may create an inversion of the conductivity type [Mil'vidsky and Osvensky, et al. (1984)]. As well as in the elementary semiconductors, the defects in compounds may exert direct and indirect effects on the performance of crystals. The importance of indirect mechanisms in crystals rises if the dislocation density falls and the concentrations of impurities and IP defects rise. As in the semiconducting compounds the concentration of point defects at $N_d \approx 10^4 - 10^6 \, cm^{-2}$ is high, the mechanism of indirect effects of dislocations on the electrophysical performance of crystals seems superior.

In CZ $GaAs$ crystals, such electrical parameters as surface resistance, threshold voltage between a source and drain in transistors, volume specific resistance and the Hall mobility of charge carriers demonstrate a direct or reciprocal correlation with a dislocation density distribution profile [Ainskrook and Wissman (1988)]. A correlation was also found between a crystal dislocation structure and an inhomogeneous point luminescence in the active regions in the injector $GaAs$–based lasers; this correlation is explained by indirect effects of dislocations through the extrinsic ambients encircling them; these ambients serve as local centres of emitting recombination [Mil'vidsky and Osvensky, et al. (1985)]. Through comparing the watt–ampere characteristics in $GaAs$ (Zn) laser diodes, the effects of the $\alpha-$ or $\beta-$dislocations on the emitting power were found to be different; this difference depends on the type of dislocations introduced and on their orientation with respect to the $p-n-$junctions [Osvensky and Proshko, et al. (1967)]. It turned out that with $\alpha-$dislocations made excessive, the emission power is larger when the $p-n-$junction is perpendicular to dislocation lines, and for $\beta-$dislocations parallel to them. With the dislocation density growing up to $2\cdot 10^6\ cm^{-2}$, this orientation dependence vanishes. Incorporating the $\beta-$ or $\alpha-$dislocations parallel to the junction plane will worsen the diode characteristics. Though, the emission power will increase if the $\alpha-$dislocations are perpendicular to the $p-n-$junction and their density is $(0.5-1.0)\cdot 10^6\ cm^{-2}$. The regularities unveiled are explained by different types of dislocations interacting with zinc atoms and by the extrinsic discharges primarily established on the $\beta-$dislocations; in consequence of this, a width and planarity of the $p-n-$junctions are used [Osvensky and Proshko, et al. (1967)]. However, here it is not excluded that the dislocations themselves will directly change the characteristics of recombinational emission if the defect density approaches the level of $10^6\ cm^{-2}$. A complete review of various effects of dislocations in $GaAs$ and GaP, and also in their hard solutions, on the performance of lasers and light–emitting diodes may be found in [Mil'vidsky and Osvensky (1985)], where the parameters of field transistors and those of the Gunn diodes produced on crystals of various dislocation density are described.

Upon studying experimentally the $GaAs$–based transistor structures, it becomes possible to derive the following general conclusion: a yield of products is determined rather by inhomogeneity of dislocation distribution in the material under study but not by dislocation density as such [Ainskrook and Wissman (1988)]. Electrophysical characteristics are greatly affected by extrinsic defect discharges induced by the very thermal stresses that are responsible for a generation and growth of dislocations. These discharges have the same spatial distribution as the dislocations have, though they are not bounded with them immediately. The problem concerning a nature and elements of such discharges should be solved individually, depending on the type of the material used, the way it is being grown, the type of the dopant and background impurities, their concentrations and the nature of their distribution across a crystal cross–section. Here, the role of dislocations in the precipitation process should be not seemingly neglected, for they can activate a disintegration of hard solutions by their intrinsic field of elastic stresses [Bullough and Newman (1970)].

2.4
Defects in the device layers of the silicon-on-dielectric (SOD) structures

The analysis of main tendencies in the contemporary microelectronics definitely shows a perspective in investigating and industrially implementing the technologies of the SOD-based ICs. Replacing a conducting wafer in an IC by a dielectric it becomes possible to fully eliminate parasitic bonds between the circuits elements or diminish negative external thermal, barometrical and irradiative actions, significantly increase the integration and high–speed levels in ICs [Vavilov and Kiselev, et al. (1990); Kravtchenko and Boud'ko, et al. (1989); Baranov (1989); Weber (1987)]. Since the microcircuit elements are isolated from each other and produced in thin device Si layers with a vertical geometry of small–square $p - n -$ junctions, they become resistant to the impulsive ionic irradiation (because of small ionic currents), in contrast to the circuits on monolithic silicon. The SOD-structures are mostly used in LSICs with the help of the LOCMOS-technology; a transfer of basic charge carriers exploited in the ICs increases their resistance to neutron irradiation, as compared to bipolar devices.

The variants of the SOD-technology exploited nowadays may be conventionally divided into three basic groups, depending on the structure of the dielectric and its spatial position in the structure. For example, the bipolar ICs, with the lifetime for minor charge carriers being the most crucial parameter in them, exploit a so–called epiprocess. In this case, the SOD-structure is built by etching on the surface of the monocrystalline Si wafer some grooves between the regions intended for the IC components and then oxidizing the surface and growing on it a thick polycrystalline Si layer. Then, this structure is grinded off, from the side of the wafer, until monocrystalline islands become exposed; these islands are isolated from each other and in them the circuit elements are established. There are many variants of this technology, in particular, the dielectric layers and the ways of their deposition may be numerous. Such techniques for producing the SOD-structures are described in details in [Yefimov and Kozyr', et al. (1986); Tchernyaev (1987)]. Here, we leave them beyond our discussion, since in the last years their application was constantly narrowed – because of their technological complexity and low yield. This technology is giving way to the silicon-on-sapphire (SOS) and silicon-on-isolator (SOI) technologies.

2.4.1
Silicon-on-sapphire (SOS) structures

As was noted in subsection 1.6, Si films are epitaxially grown on monocrystalline wafers of sapphire (Al_2O_3) or on magnesium–aluminium spinel ($MgO \cdot Al_2O_3$) by decomposing or thermally restoring the Si hydrides and chlorides in hydrogen at the atmospheric or lowered pressure [Papckov and Tzyboul'nickov (1979)]. A continuous Si layer is usually formed by such operations as nucleation, growth and coalescence of nuclei, this process being largely dependent on the growth technique. As compared against the chloride process, depositing from monosilane brings a higher density of nuclei, gives an earlier continuity; such nuclei emerge already primarily oriented and possess lower defectiveness. When grown epitaxially, Si atoms are supposed to substitute Al atoms on the surface and get bound to oxygen atoms in the wafer lattice to form a first epitaxial layer. A misfit in the lattice periods in the growing layer and the wafer depends on their mutual orientation. For example, for the most popular commercial SOS-structures like the silicon (001) − oriented film on the sapphire $(01\bar{1}2)$ − oriented wafer, this misfit makes up 12.5 and 4.2 per cent for the $[11\bar{2}0]$− and $[1\bar{1}01]$−orientations, respectively. A spinel has a cubic lattice allowing the Si layers to be grown with the orientation parallel to the wafer orientation, i.e. with similar Miller indices − this provides a misfit of film lattice period and wafer period to be equal not more than 1 per cent [Mil'vidsky and Osvensky (1985); Papckov and Tzyboul'nickov (1979)]. From this viewpoint, a use of spinel is more preferable than sapphire. Nevertheless, employing a spinel as a wafer is thought not perspective, since, first, to obtain structurally−perfect large− size crystals of $MgO \cdot Al_2O_3$ is technologically difficult, and, second, in their mechanical characteristics they are junior to sapphire [Kravtchenko and Boud'ko (1989); Papckov and Tzyboul'nickov (1979)].

A misfit of lattice periods in the conjugative layers and a difference in thermal expansion coefficients in the films of silicon $(4.2 \cdot 10^{-6} \ K^{-1})$, sapphire $(8.3 \cdot 10^{-6} \ K^{-1})$ or spinel $(8.6 \cdot 10^{-6} \ K^{-1})$ give rise to crucial amplitude− contracting stresses in the device layers of SOS-structures. These stresses partially relax due to misfit dislocations whose density may approach $10^7 - 10^9 \ cm^{-2}$ [Kravtchenko and Boud'ko (1989); Papckov and Tzyboul'nickov (1979)]. Except the misfit dislocations, the Si films on sapphire were found by the electronic microscopy to contain also spatial chaotic aggregates of dislocations (of the density up to $10^{12} \ cm^{-2}$), stacking faults and microtwin lamellae whose density in the vicinity of the epilayer−sapphire wafer interface is equal to $10^9 - 10^{10} \ cm^{-2}$ [Papckov and Tzyboul'nickov (1979)]. The concentration of structural defects in the device (001) − oriented layers is usually higher than in (111) − oriented layers, and it drops when the Si film becomes thicker.

Highly−concentrated structural defects and residual elastic stresses in the device Si layers naturally affect their electrophysical properties. A rise in specific resistance in the films was found to increase the differences between the charge carrier concentrations (at similar ρ) in them and in the spatial Si specimens; this is explained by a misfit in the mobility of carriers in the SOS-structures and monocrystals [Papckov and Tzyboul'nickov (1979)]. The impurity distribution profiles change inhomogeneously across the thickness of Si films and this also determines inhomogeneous profiles for spatial distributions of charge carrier concentrations. It was noticed that as soon as we are approaching a wafer closer and closer a by-layer etching off will diminish the concentration of electrons and holes in the film. This is explained by the carriers being captured on traps and recombination centres whose density rises near the $Si - Al_2O_3$ interface. In the

SOS-structures with device n − type layers, the presence of aluminium, being an acceptor impurity in silicon, becomes very crucial. Its presence in a Si layer is thought to make the SOS-structures stable thermally [Papckov and Tzyboul'nickov (1979)]. Aluminium gets into an epitaxial layer during its growth and later thermal treatments. The Si layers grown epitaxially from pure silane have a high specific resistance and electrical conductivity, when deposited onto monocrystalline Si wafers. However, their conductivity may be converted to a hole conductivity when the layers are grown on sapphire wafers; this occurs due to aluminium autodoping. The aluminium concentration in the Si film in the SOS-structures depends on temperature and epitaxy rate. Hence, there may come the situation when at the given temperature and varying growth rate (or vice versa) the electronic conductivity in films is compensated by the hole conductivity created by aluminium atoms. Generally, aluminium spreads across the entire bulk of the epitaxial layer inhomogeneously, primarily settling near structural defects, and this results in an additional scattering of charge carriers and a fall of mobility.

The heteroepitaxial layers on sapphire, as was shown by electrical and optical measurements, involve deep levels in a prohibited region of silicon; this leads to additional leakage currents, diminishes a mobility of carriers in the channel and shifts the threshold voltage in the MDS-transistors produced on the SOS-structures [Dumin and Robinson (1969); Heiman (1967); Tikhomirov and Kititchenko, et al. (1984)]. The spectroscopy of deep levels has made it possible to find in Si films such impurities as cobalt, copper, gold, chromium, oxygen and oxygen–holding aggregates, with energies of activation, concentration and capturing cross–sections being dependent on the type of epitaxial layer conductivity, thermal history of specimens, and, in particular, depending on chemical activity of the gaseous ambient during a thermal treatment. A complicated extrinsic defect composition of films determines a considerable variance of their electrophysical parameters across the surface of the SOS-structures. It was shown that specific resistance scatters across the square of the commercial n − type structures as much as 200 per cent (a 5-mm edge zone not considered) and may become mobile up to 340 per cent [Tikhomirov and Kititchenko, et al. (1984)]. Here, the mobility values are distributed along the diameter of structures, they reach their maximum ($\leq 500 \ cm^2 / V \cdot s$) in the central region and drop down to $80 - 160 \ cm^2 / V \cdot s$ in the peripheral regions of the wafers. Such distribution is also correlated with the inhomogeneous positions of fit SDS-devices on the SOS-surface; the most yield of such devices is provided by the central regions of these structures.

The parameters of the SOS-based devices are crucially dependent on the structure and properties of the silicon–sapphire interface [Mc Greivy (1977)]. A transient layer, $30 \ nm$ thick or more, was found to vary in its composition, involving silicon and aluminiumselicate compounds [Kuhl and Schlotterer, et al. (1976); Gastev and Soukhoroukov, et al. (1985)]. According to [Kuhl and Schlotterer, et al. (1976)], a transient region is similar to a mosaic monocrystal. In contrast to this conclusion, in [Gastev and Soukhoroukov, et al. (1985)], where the thermal effects on the charge of the silicon–sapphire interface were explained, this interface was supposed to be amorphous. A similar model for the transient region is used to analyse spectral characteristics of reflection from the SOS-structures [Lagowski and Jastrzebski, et al. (1983)]. Studying the silicon–sapphire interface by ellipsometry and through fast electron diffraction has showed the transient region to be not less than $20-30$ nm thick and to hold mostly alumosilicates with a considerable content of amorphous phase [Gastev and Soukhoroukov, et al. (1988)]. In addition to this, the transient region also contains silicon in the form of mosaic monocrystals and polycrystals. It has been revealed that a quantitative ratio between the amorphous and polycrystalline phases in the Si film near the interface varies not only from one specimen to another but also in different regions of one and the same structure. Such structure of the transient region determines a high density of the localized energy states for charge carriers in the prohibited zone of the material forming the silicon–sapphire interface. This zone is wider than the prohibited zone in the monocrystalline silicon. Redistributing the charge carriers between localized states and the zones in silicon will create, if a thermodynamic balance is available, in the epitaxial layer a spatial charge region. Such supposition on the nature of the charge in the $Si - Al_2O_3$ interface agrees well with the effects of technological aspects upon the SOS-properties [Gastev and Soukhoroukov, et al. (1985); Gastev and Soukhoroukov, et al. (1988); Gastev and Mikhailov, et al. (1984)]. In the field transistors, a disordered interface works as an additional evacuation channel between a source and a drain. In such a system, a superior mechanism of current passage involves an injection of charge carriers from the contact regions and their transportation across the bulk, with the traps participating in this process.

Localized states will work as traps [Lampert Murray and Mark Peter (1970)]. A channel conductivity depends both on the density of localized states in the transient region and also on the injecting characteristics of the contact pad, and so may be considerably different for the $n-$ and $p-$ type contact pads. It has to be expected that at small stresses the source–drain conductivity of the channel will be determined by the contact characteristics, and at high stresses will be constrained by the spatial charge captured on the traps. How the leakage currents related to a transient region arise was experimentally observed in [Vasudev (1983)].

The structural specificity of the SOS- heterostructures is seen in behaviour during irradiation. With a use of low–temperature luminescence, there have been studied the specificities of irradiation–induced defects, their identification in the Si epitaxial layers in the SOS-structures being irradiated by $\gamma-$ quantums from the source ^{60}Co, electrons ($E = 3\ MeV$), $\alpha-$ particles ($E = 5\ MeV$) and also by ions of helium, neon, argon and krypton of $E = 20-300\ keV$ [Zouev and Larionova, et al. (1988)]. In original structures, through optical spectrums, the films $0.6\ mcm$ thick were found to be more perfect than $3\ mcm$ films which had a large number of bands of extrinsic impurity irradiation. Besides, the bands in the film luminescence spectrums are supposed to emerge due to an emitting recombination of unbalanced charge carriers on the extended defects like microtwins, stacking faults or epitaxy–induced dislocations. It is not excluded that some of the luminescence bands relate to hydrogen as impurities contaminating the Si films during their sedimentation by monosilane pyrolysis. That the photoluminescence spectrums in the films were effected by the $\gamma-$ quantum or electron irradiation was detected only at large doses (up to $10^{19}\, cm^{-2}$) and was explained by an emitting recombination of unbalanced charge carriers on the defects containing aluminium; it is brought into the films from the sapphire wafer by the autodoping process [Papckov and Tzyboul'nickov (1979)]. The implanted atoms of inertial gases were found to actively participate in the defect formation process. For example, it was found that the spectrum lines for the extrinsic luminescence centres containing atoms of inertial gases are shifted, as compared to the spectrums from the monocrystalline silicon. This shift is thought to be governed by a high amplitude of the intrinsic mechanical stresses in SOS-films and by different sensitivity of extrinsic luminescence centres to the lattice deformations.

A high density and a wide repertoire of structural defects in the Si epitaxial layers on sapphire are responsible for a "natural" aging of the films and a degradation of their electrophysical properties when the SOS-structures are being kept under normal conditions prior to technological operations. Until now, this issue has not yet properly been discussed in the studies dedicated to the SOS-structure degradation under technological actions [Papckov and Tzyboul'nickov (1979)]. Though, it may be rightfully supposed that even at room temperatures the elastic stresses, considerable in values, are able to activate such processes as partial polygonization of the structure, formation of extrinsic ambients around the extended defects, redistribution of doping and background impurities across the bulk of the epitaxial layer, etc. Consequently, these processes will change a topology of spatial charge regions in the film, i.e. its electrophysical characteristics will be changed.

In order to drop a negative effect of film defectiveness on the parameters in the devices produced by the SMOS – SOS technologies, one traditionally employs the Si epitaxial layers not less than $0.5\, mcm$ thick; and in the LSICs constructed on these layers the elements are made $3-4\, mcm$ in size. A jump to manufacturing the ICs of less topological norms, i.e. with a larger integration and high–speed, demands a substantial bettering of films' quality – the films should be not more than $0.3\, mcm$ in thickness, with sapphire wafers being $100\, mm$ and more in diameter. Recrystallization of the epitaxial films is considered to be the most perspective way for diminishing the defectiveness of device layers in the SOS-structures [Kravtchenko and Boud'ko (1989); Baranov (1989)]. For this purpose, the film deposited from gaseous phase is subjected to a two–stage irradiation by silicon ions, first of a high (up to $400\, keV$) and then low ($80-200\, keV$) energy. Upon each irradiation, a recrystallization annealing is arranged, during which a Si monocrystalline structure starts to recover from the surface and proceeds towards a buried amorphous layer in the silicon–sapphire interface. Such SOS-technology makes it possible to diminish by one–two orders the concentration of defects, in particular, microtwins, in the device layers – this results in a fall of leakage currents and a rise of mobility of charge carriers in the MOS-transistors. However, the empirical data obtained is not yet sufficient to find the optimal regimes for ionic implantation and later annealing, to build high–quality heterostructures [Vavilov and Kiselev, et al. (1990)]. Deeper investigations are, therefore, needed to unveil the effects of these processes on electrophysical and structural properties of epitaxial films and also on the silicon–sapphire interface where the autodoping effect is most noticeable. The serious difficulties, up to now available in the SOS-technology, restrain the applications of these structures; they are chiefly exploited in manufacturing irradiation–resistant ICs of low integration.

2.4.2
Silicon-on-isolator (SOI) structures

Rising an integration and high–speed in ICs with retaining them resistant to external destabilizing factors is thought real if these ICs exploit the SOI-structures. The SOI-technology makes the SHSIC elements efficiently isolated, this results in a fall of parasitic capacitance and a rise of circuit high–speed. A comparatively low cost of the SOI-structures is worth mentioning here, as compared against the SOS-structures. As for the SOI-based SMOS-circuits being radiation–resistant, a rise of their resistance is guided by four aspects:

a)   a hidden dielectric layer hampers a travell of unbalanced charge carriers and structural defects (induced in the wafer by corpuscular or photonic irradiation) to a thin presurface device region;

b)   since the IC active elements become less in size, this drops the concentration of radiation–induced unbalanced charge carriers thus determining a rise of leakage currents;

c)   A dielectric isolation between the circuit elements allows to get rid of the "click" effect, i.e. an emerge of parasitic 4–layer structures induced by impulsive irradiation.

The SOI–structures are produced by several basic technologies [Kravtchenko and Boud'ko (1989); Weber (1987); Wittkower (1999)]: 1) the technology of zone–melting recrystallization (ZMR); 2) the technology of full isolation by porous oxidized silicon (FIPOS); 3) the bond etching silicon–on–insulator (BESOI) technology; 4) the separation by implantation of oxygen (SIMOX) technology.

1) The technology of zone–melting recrystallization (ZMR)
This is a high–temperature recrystallization of polycrystalline or amorphous silicon layers deposited on an oxidized monocrystalline wafer. A zone

recrystallization is performed either by a set of graphite heating elements or by electronic or laser irradiation [Kravtchenko and Boud'ko (1989)]. In implementing this technology, first, the SiO_2 bands of some hundred nanometers in thickness, separated by gaps up to $10\ mcm$ wide, are deposited on a silicon wafer. These gaps between SiO_2 provide an immediate access to the wafer surface; this surface serves as a monocrystalline nucleator during high–temperature recrystallization of the poly– or amorphous silicon which deposits over the entire surface of the $Si - SiO_2$ structure. To make the process more efficient, the layers to be recrystallized are coated with an antireflective Si_3N_4 film. The technology, say, for recrystallizing by a directed high–energy irradiation includes the following: a laser beam (Fig. 2.2) or an electronic bundle is first focused on one of the gaps and then is being shifted (with the rate needed for recrystallization) from the gap towards the SiO_2 film. A Si_3N_4 coating provides a uniform and homogeneous distribution in the thermal field across the entire activated surface of the structure.

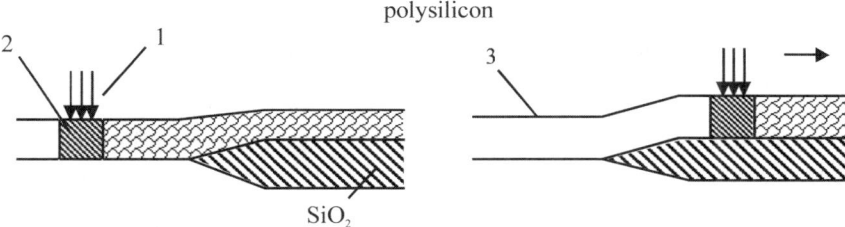

Fig. 2.2. Recrystallizing by a scanning laser beam with an edge nucleation from the monocrystalline silicon of the wafer: **1** laser beam; **2** recrystallization zone (a melting zone); **3** silicon monocrystal.

In detail, various variants of this technology may be seen in [Kravtchenko and Boud'ko (1989)]. The electrophysical prameters obtained by this technique for device layers may be found in [Kravtchenko and Boud'ko (1989); Roudenko T. and Roudenko A., et al. (1994); Lysenko and Nazarov, et al. (1994)]. The last research is interesting due to the Si films built by high–temperature laser recrystallization of polysilicon deposited onto multiplayer dielectrics with the gaps of nitride and silicon oxynitride. The SOI-transistors produced on such structures are shown to have good parameters and slightly accumulate a charge when being exposed to the gamma–irradiation of the dose up to $10^6\ rad$.

The disadvantages of the ZMR-technique involve a low productivity of the crystallization process and technological complexities arising in constructing qualitative device layers on large–diameter wafers. Besides, the thermal gradient present in this technology gives birth to thermomechanical stresses causing in Si films the sliding lines, small–edge boundaries and dislocations and also an uncontrolled redistribution of impurities worsening the electrophysical performance of device layers. Therefore, despite its relative simplicity this technique has not yet found a wide commercial application in manufacturing the SOI-structures intended for superhigh–speed ICs. This technology, however, may become a basic one in manufacturing the discrete devices, i.e. the active elements for sensors of physical parameters.

2) The technology of full isolation by porous oxidized silicon (FIPOS)

This technology implies constructing around the device the regions or the islands of porous silicon and their later annealing. It rests on the following basic physical processes [Nikolaev and Nemirovsky, et al. (1989)]: the local regions of monocrystalline silicon, being electrochemically treated in the solution of hydrofluoric acid, turn to porous silicon; here, in the $p-$type silicon the rate of this process is higher than for the $n-$type silicon. At such regimes and at low rates of monocrystalline silicon oxidation, the porous silicon is oxidized easily.

The priority of processes employed in the FIPOS-technology is given in Figs. 2.3 and 2.4.

Except diffusion or epitaxy, the device monocrystalline islands usually constructed by the FIPOS-technique may be also built by a protonic irradiation [Imai (1981)]. In this case, the process starts with depositing the Si_3N_4 film, about 0.1 mcm thick, onto a surface of the $p-$type wafer and forming on it by photolithography a pattern consisting of the sections masking the device islands. Upon this, the surface is irradiated by boron ions ($E = 40-50\ keV$, a dose of up to $10^{13}\ cm^{-2}$) and annealed in the inertial ambient at $1273-1373\ K$; this results in a heavily–doped $p^+ -$ silicon in the regions not protected by Si_3N_4.

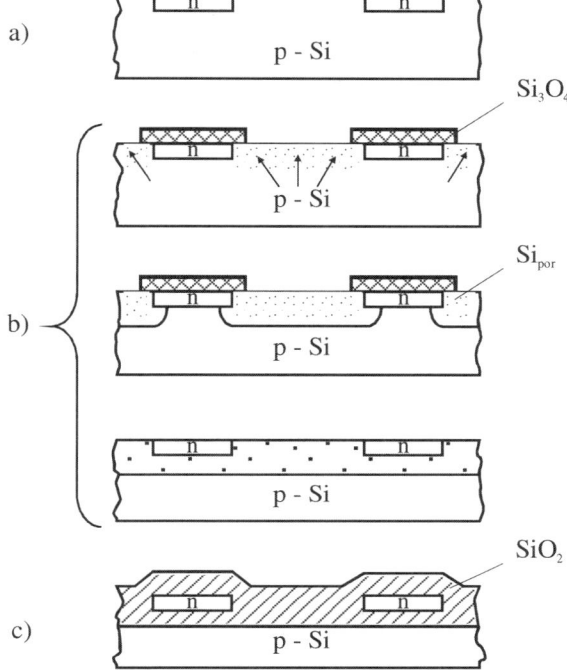

Fig. 2.3. The priority of operations in the FIPOS-technology: **a** forming the $n-$ conductivity regions by implantation or diffusion; **b** forming a porous silicon layer through the Si_3N_4 mask by an electrochemical treatment and removing the Si_3N_4 mask; **c** thermal oxidation of porous silicon.

Then, the structure is irradiated by protons whose energy must be sufficient for them to penetrate through the Si_3N_4 film into the silicon wafer as deep as the thickness of the device layer. A later low–temperature annealing at $623-723\ K$ results in the regions under Si_3N_4 becoming electrically conductive and remaining monocrystalline during the electrochemical process. In the latter process, the porous silicon is formed only in the p^- and p^+ silicon. In the course of treatment, when passing the electrical current from the back (without Si_3N_4) surface to the front surface of the structure first the p^+ regions between the $n-$ silicon islands become porous and then the $p-$ type regions, located under these islands. The next stage of high–temperature oxidation of porous silicon removes hydrogen and reduces a hole $p-$ type conductivity in the $n-$type regions. The Si_3N_4 film having been removed, the structure is now ready to have the $n-$ and $p-$channel MOS-transistors formed in the monocrystalline $p-$silicon islands isolated from each other and from the wafer by a layer of oxidized porous silicon.

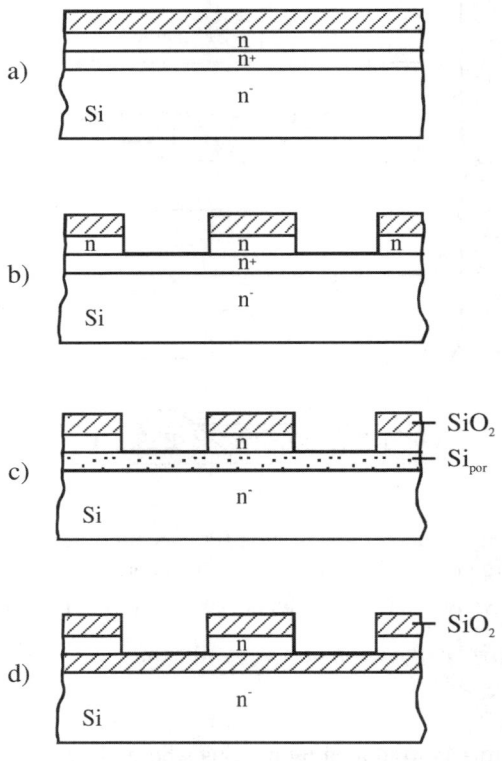

Fig. 2.4. The priority of operations in the FIPOS-technology of the IBM company (the photopicture taken from [Kormishina and Perevostchikov, et al. (1997)]: **a** an epitaxial sequential growing of the $n^+ -$, $n-$layers and constructing a thermal SiO_2 layer; **b** a reactive ionic etching of the Si layer through the SiO_2 mask; **c** constructing a porous Si layer electrochemically; **d** oxidizing thermally.

The specificity of the FIPOS-technology is in its ability to obtain the active elements completely isolated, provided the depth of porous silicon regions increases with an increase of the width in the monocrystalline island. Therefore, the islands are usually $15-20$ mcm in width, though their length is unlimited. However, there is a technology able to build the device islands about 40 mcm

wide; this technology makes use of a highly selective self–ceasing anodization of a "buried" $n^+ - Si$ layer in the $n - n^+ - n$ structure with a thermal oxidation of porous silicon [Bondarenko and Yakovtzeva, et al. (1994)]. In such structures, porous silicon is first vertically formed in the $n^+ -$ layer in the open (not protected by Si_3N_4) regions where the layer under the epitaxial $n -$ islands underwent a slight edge anodization. With the front of pore forming process approaching the $n -$ wafer, there starts an edge spreading of the process under the islands, until the porous $n^+ -$ layer under them closes completely. Since anodization under epitaxial islands is done horizontally, then the pore channels under them are oriented similarly, and this makes a supply of oxygen easier in the later oxidation of porous silicon. This process is well controlled by varying a composition of electrolythic and anodization regimes, and is compatible with the standard SMOS-technology.

The electrophysical parameters of the device layers in the SOI-structures are negatively affected by some factors basic of which are the monocrystalline junctions and a uncontrollable pollution.

a)   A retaining of the monocrystalline junctions between the wafer and the device layer; this is specific for the Si crystals of high specific resistance (starting from $\rho = 12 \ Om \cdot cm$ and higher) and relates to an inhomogeneous distribution of current density during an in-depth anodization [Moushnitchenko and Zhourankov (1990)]. Such inhomogeneity depends on the ratio of specific resistances in the original wafer and the electrolith available in the porous space of the region under anodization. The current density in the surface of the junction approaches its maximal value near its low base connected with the wafer. An opposite terminal of the junction may have current density practically equal to zero. It was detected that in monocrystalline silicon of the charge carrier concentration less than $6 \cdot 10^{15} \ cm^{-3}$ any local inhomogeneity in electroconductivity on the way of the anodization front creates junctions even in the bulk porous silicon [Moushnitchenko and Zhourankov (1990)]. In particular, here lies the reason of why upon oxidizing the porous silicon its interface with the monocrystal looks dull. A presence of junctions lessens a polishing ability of the oxidized porous silicon, determines a formation of leakage currents through dielectric and lessens a break-through voltage.

b) The pollution of the porous Si layer by dopants and background impurities from the bulk of the crystal and external environment also leads to worsening its dielectric properties. The pollution was supposed to be chiefly induced by porous silicon oxidation, due to an accelerated thermodiffusion; and through this, there have been explained an increase of leakage currents along the boundary "isolator – device islands" [Nikolaev and Nemirovsky (1989)]. Studying layerwise the porous silicon elements by the secondary–ion mass spectroscopy (SIMS) technique has revealed that despite the type of the original wafer the impurities, upon anodization, are redistributed near and within the limits of the porous region [Perevostchikov and Skoupov, et al. (1994)]. It is remarkable that monocrystal except the products of electrochemical reactions (H, F, C, OH, etc.) was found to have both dopants (B, Al, P) and background (Na, Mg, Ca, Cu and other) impurities in the porous Si layer and near it; their in-depth concentration profile may be of the oscillating nature. This result, if combined with the results gained by other techniques, made it possible to suppose that a four–layer model may most efficiently simulate the real structures "porous silicon – monocrystalline silicon" (Fig. 2.5 [Perevostchikov and Skoupov, et al. (1998)]).

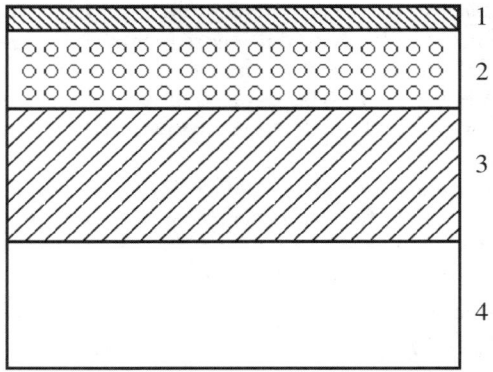

Fig. 2.5. The structure "porous silicon – monocrystalline silicon".
Layer 1 : a thin surface layer of amorphous silicon and silicon dioxide;
Layer 2 : porous silicon itself;
Layer 3 : a transient region of impurities, intrinsic point defects and their aggregates, of high concentration and inhomogeneous in-depth distribution;
Layer 4 : original elastically deformed monocrystal.

The impurity diffusion and aggregation in the transient layer 3, as well as gettering by porous Si at the anodizing temperature are closely associated with unbalanced IP defects, mainly, vacancies arising in the zone of monocrystal electrochemical dissolution [Perevostchickov and Skoupov (1992); Fistoul' and Sinder (1981)]. That the impurity diffusion is conditioned by a counterflux of unbalanced vacancies is supported by the results in [Koulikov and Perevostchickov, et al. (1997)] where the microdefects (aggregates and clusters of silicon interstitial atoms) were reported to dissolve during the dislocation–free crystal anodization. Most probably, it is a presence of the transient monocrystalline region 3 (Fig. 2.5) that determines a change in the surface electrical resistance of silicon (after a porous layer has been removed from it [Koulikov and Perevostchickov, et al. (1997)] and the drain channels additionally emerging upon oxidation [Nikolaev and Nemirovsky (1989)].

Therefore, to better the isolating properties in the oxidized porous Si layers, the original wafers need a preliminary treatment by a special technology, which, alongside with traditional abrasive–chemical operations, must also involve a killing of undesired impurities and larger defects in the crystal presurface layer.

3) The bond etching silicon-on-insulator (BESOI) technology

This technology exploits a thermocompressive bonding of silicon wafers (BW). The idea of the BESOI-technology lies in connecting a base Si wafer (mechanically loaded) to a working (device) wafer through a dielectric layer and thinning the working wafer to the thickness needed ($1-10\ mcm$), by a precisive mechanical, chemical or electrochemical polishing. Different variants of wafer bonding and its application to forming structures are described in various papers (for example, see Weber (1987); Stengl and Tan, et al. (1989); Volle and Voronkov, et al. (1991); Krymko and Yenisherlova, et al. (1998); Yenisherlova and Batchourin, et al. (1991); Petrova and Koshelev, et al. (1998); Koshelev and Yermolaev (1994); Kormishina and Perevostchikov, et al. (1997); Yenisherlova and Rousak, et al. (1994); Krymko and Yenisherlova, et al. (1998)]). Therefore, in this subsection only basic specificities of this process will be considered in brief.

At the first stage of direct bonding, silicon wafers with thoroughly cleaned hydrophilic surfaces are got into optical contact to obtain a sufficiently tight semiadhesive bonding, through forming hydrogen bonds between molecular layers of the water physically adsorbed on crystal surfaces. A later gradual heating changes the structures of bonds and makes the compounds harder. At first, at $T \approx 473\,K$ water molecules are partially separated, the wafers come close to each other and there are formed hydrogen bonds between silane groups of contacting surfaces:

$$Si - OH - (H_2O)_2 : (H_2O)_2 - OH - Si \rightarrow$$
$$Si - OH : OH - Si + 4H_2O$$

Further, at heating temperatures $973-1273\ K$ the hydrogen bonds between wafers are replaced with stronger silane chemical bonds:

$$Si - OH : OH - Si \quad \rightarrow \quad Si - O - Si \ .$$

As the temperature (or duration of holding) grows, the silane bonds between wafers can be transformed to direct covalent bonds between silicon atoms:

$$Si - O - Si \quad \rightarrow \quad Si - Si + O \ .$$

This bonding technique may be performed both through a SiO_2 layer and without it [Stengl and Tan, et al. (1989)]. The structures of the last type are able to successfully compete with the epitaxial structures used in the manufacture of powerful electronic devices [Volle and Voronkov, et al. (1991); Krymko and Yenisherlova, et al. (1998); Yenisherlova and Batchourin, et al. (1991)].

There is a way of connecting the wafers in filtered deionized water. Upon washing and tight bonding, the wafers are extracted from water and subjected to drying first on the centrifuge and then in the thermostat at $373\ K$ [Stengl and Tan, et al. (1989)]. Such technique, upon water evaporation, provides an intimate mating of wafers and practically cancells microcavities (induced by dust particles) on the interface. To get the hydrophilic surfaces cleaned, the Si wafers, prior to bonding, have to be cleaned in ammonia peroxide solution of the composition: $H_2O_2 : NH_4OH : H_2O = 3:1:1$ (vol. parts) [Yenisherlova and Batchourin, et al. (1991)]. Such a wash-off results in a hydrated natural SiO_2 emerging on the surface with a large number of silane groups. The thermocompressive bonding is performed at the working temperatures $1373 - 1573\ K$ providing a wafer bonding at the level of $10 - 20\ MPa$.

A wafer thinning procedure, down to a device layer thickness, may be illustrated with one of the variants described in [Petrova and Koshelev, et al. (1998)]. Here, the thermocompressive bonding (by the pressure of $\approx 2 \cdot 10^3\ Pa$) of Si wafers was performed through the formed layers of specially synthesized superpure multicomponent glass–like dielectrics. Thinning was exercised with a use of selective electrochemical and chemical etching of silicon with a stop–layer. For a selective electrochemical etching, the heavily doped $n^+ - Si$ monocrystals with a high–ohmic epitaxial $n -$ type layer 3 – 10 mcm thick were used as device wafers. Bonding the wafers was performed through dielectric layers $1 - 4\ mcm$ thick deposited by HF-magnetronic spraying. The working epitaxial wafer was preliminarily thinned by grinding and chemicomechanical polishing. The remaining $n^+ -$ layer in the silicon wafer $10 - 30\ mcm$ thick was removed via electrochemical etching, with a use of different rates of anodic dissolution of $n^+ -$ and $n -$ silicon in water–ethylene glycol solutions of hydrofluoric acid. There were used the etching regimes at which the anodic current of the $n -$ silicon dissolution was two orders less than for the $n^+ -$ silicon [Koshelev and Yermolaev (1994)].

Completely etching off the $n^+ -$ layer from the entire surface of the working wafer was provided by electrochemical etching, the structure was being gradually sunken into the electrolyte at a given rate. This way of thinning brought a layer in-depth spread throughout the entire surface not more than 10 per cent. In manufacturing the SOI-structures by chemicomechanical polishing, the basic difficulty is obtaining a thickness homogeneity for a device layer. For this purpose, there may be formed subsidiary grooves with a stop–layer of SiO_2 or Si_3N_4 being filled with dielectric or polysilicon and also the grooves simultaneously serving as stoppers for polishing and edge isolation of the devices being later formed [Petrova and Koshelev, et al. (1998); Koshelev and Yermolaev (1994)].

The BESOI-technology seemingly holds a leading place among other SOI producing technologies, for it allows to produce the structures with a wide choice of geometrical, electrophysical and functional parameters of device layers, to solve the problems of integrated microelectronics, optoelectronics and micromechanics. This is achieved by varying regimes and conditions of specific operations in the structure manufacturing line route, by varying the types of silicon wafers and dielectric layers, and also by a use of technological treatments additional to basic. Though, in any case, a basic scheme of the technology itself involves a number of operations potentially bringing an uncontrollable change in the extrinsic defect composition in each component of the SOI-structure. These operations embrace a number of abrasive–chemical treatments at the stage of preparing original wafers and thinning of the working wafer, thermal oxidation and thermocompressive wafer bonding, and also the processes of cleaning (liquid and "dry" cleaning) the wafer surfaces before bonding.

It is known that in mechanical grinding and polishing and also in chemicomechanical and chemical polishing of semiconducting monocrystals, silicon included, a relief of the surface is not only smoothed and the material removed, but its original (growth) extrinsic defect composition is transformed, and this occurs at the depths significantly exceeding the thickness of the residual damaged layer with dislocations or microcracks [Alekhin (1983); Perevostchickov and Skoupov (1992); Mil'vidsky and Osvensky (1985)]. Responsible for a change in the extrinsic defect composition are the unbalanced IP defects, elastic waves and the fields of static mechanical stresses arising during abrasive and chemical damage of presurface layers in monocrystals. An interaction of the unbalanced IP defects between themselves and with original defects usually increase a concentration of structural distortions near the surface under treatment. Dynamical and static elastic stresses stimulate these processes and may produce new defects, for example, dislocation loops in the vicinity of the second–phase particles [Alekhin (1983)]. And conversely, at large distances from the surface under treatment (exceeding the diffusion length of IP defects) the concentration of point defects like clusters or small dislocation loops may drop, due to their dissolution by elastic waves [Perevostchickov and Skoupov (1992)]. Since the device layers constructed by the BESOI-technique usually are of the thickness less than or comparable with the diffusion length of IP defects in silicon (for example, $L_v = 10 - 20 \ mcm$), their defectiveness should be expected to rise against the defects in the original wafer if a chemicomechanical polishing is used as a finishing operation. This supposition was experimentally proved on the SOI-structures with the device layer $10 - 15 \ mcm$ thick and dielectric layer of phosphorus–silicate glass $4 \ mcm$ thick [Kormishina and Perevostchickov, et al. (1997)]. When etching them selectively, it has been found that upon a subsequent abrasive grinding (by diamond synthetic paste, $40 / 28 - mcm$ grains), polishing (diamond synthetic paste, 3-mcm grains) and chemicomechanical polishing (by Cab-O-Sil suspension) of the original working dislocation–free wafer of $p - Si \ (001) \ CZ$ ($\rho = 12 \ ohm \cdot cm$) 460 mcm thick will crucially increase the microdefect density. The density of original microdefects made up $(6.72 \pm 2.2) \cdot 10^5 \ cm^{-2}$. Upon a thermocompressive 2-hour annealing at $1000°C$ in vacuum and thinning, the density of small microdefects in the working wafer of the device layer (these defects produce the etching wells about 1 mcm in diameter) has grown up to $10^7 - 10^8 \ cm^{-2}$ and the density of large microdefects (the etching wells of $1.5 - 10 \ mcm$) has dropped to $10^2 \ cm^{-2}$. There has been noticed a tendency of microdefect ordering, mostly along the <110> orientation (Figs. 2.6a and 2.6b [Kormishina and Perevostchickov, et al. (1997)]).

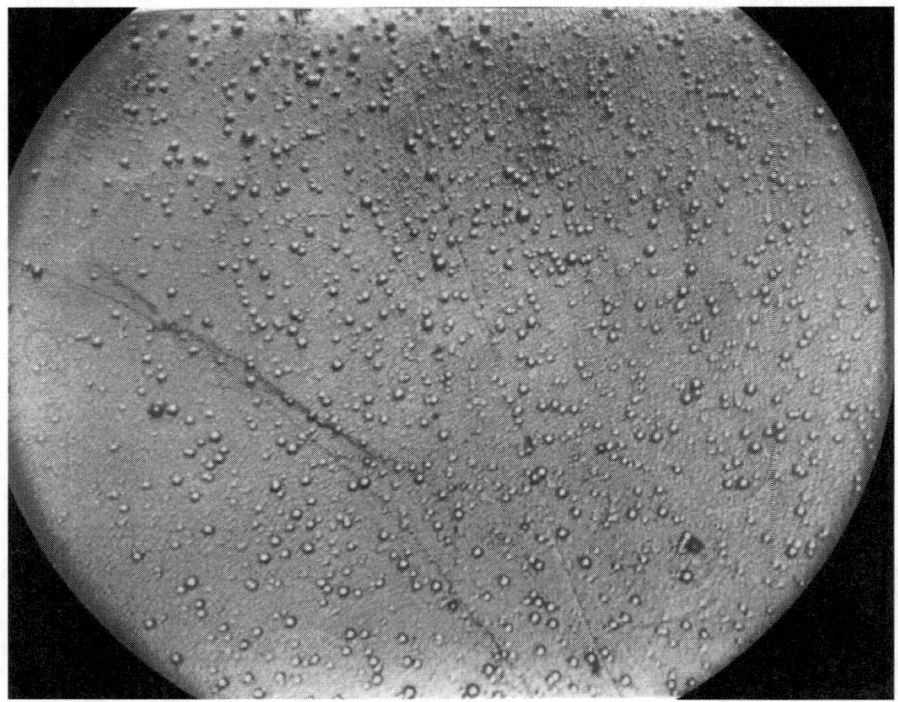

Fig. 2.6a. The surface of original silicon upon a selective etching (x250).

The X-ray and UR-spectrometrical studies of the structures have revealed that in the surfaces, though being formed by similar technological regimes, the surfaces in the working and base wafers may be disoriented from $20 - 30$ to $240 \ ang. \cdot s$ [Kormishina and Perevostchickov, et al. (1997)]. This speaks about a considerable spread in geometric parameters of the original wafers (bending, nonplanarity, microroughness); upon bonding of wafers, this spread partially retains determining their disorientation. The residual deformation in the device layers was $10^{-5} - 2 \cdot 10^{-4}$ conv. u. and corresponded to an increase of the layer lattice period with respect to that in the base wafer lattice – this betrays an aggregation in the layer, mostly, of the cluster–like defects from the interstitial atoms and interstitial dislocation loops.

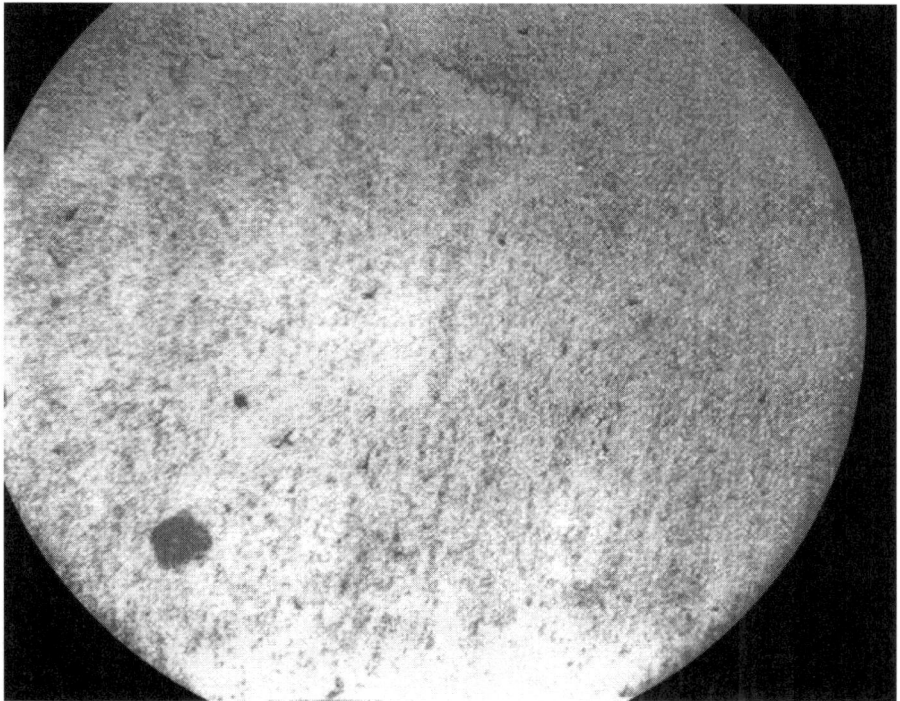

Fig. 2.6b. The device layer of the BW–SOI-structure upon selective etching (x250).

Upon thermocompression and thinning, the structures, as was shown by the UR absorption spectrums, increased SiO_2 concentration; it exceeded the concentration of such inclusions both in original wafers and specimens. They were annealed in the free state, without loads (during a formation process, the load upon the wafers was about $100\ MPa$), and their surface was additionally treated mechanically. The second–phase particles may be supposed to be formed at various rates, but they are formed at all stages of the structure manufacture because of the decomposition of hydrogen hard solution (CZ–grown crystals) stimulated by thermal, dynamical and static mechanical stresses and unbalanced IP defects as well.

Interesting data about the extrinsic–defect composition of the SOI-structures obtained by bonding is given in [Yenisherlova and Rousak, et al. (1994)]. Here, there were studied the effects of chemical treatment of the original Si wafers (performed prior to thermocompressive bonding) on the quality of adhesive low–temperature bonding ($443-473\ K$), high–temperature annealing ($1423-1473\ K$) and on thinning a working wafer down to $3-10\ mcm$ by a mechanical and chemical polishing. Prior to bonding, the wafers were treated in various compositions with oxidizing and alkaline ambients, with pH being from 1 to 10. As oxidizing ambients there were used the peroxide–acidic $HCl-$based solution $HCl:H_2O_2:H_2O=1:1:5$ (vol. parts) ($pH=1$) and the heavily–diluted solution of hydrofluoric acid ($pH=3$). As alkaline ambients there have been used the ammonia – peroxide (AP) solutions of various compositions starting from $NH_4OH:H_2O_2:H_2O=1:8:1$ (vol. parts) ($pH=9$) and finishing by the same composition with the ratio 1:2:1 ($pH=10$). Besides, an ammonia–water solution ($pH=10$) was used. A SiO_2 layer 0.2 and $0.6\ mcm$ thick was thermally grown on a portion of original wafers; such $Si-SiO_2$ structures also underwent a chemical treatment. Upon chemical treatment, an extent of water–repellent treatment of the surfaces was estimated by a wetting angle, and a quality of wafer bonding was estimated by the mechanical wedge-out technique and by counting the unbonded local regions ("bubbles"), by a laser and X-ray (the Lang technique) techniques. Also, mass–spectrums of minor ions were studied, the charge carrier concentration was measured and analysed metallographically, via selectively etching the surface, the transversal spalls and spherical edges.

The minimal value of the wetting angle (less than 7 per cent), i.e. the greatest manifestation of the hydrophilic properties of silicon and SiO_2, was experimentally found to be available upon using high–acidity solutions ($pH=9-10$) [Yenisherlova and Rousak, et al. (1994)]. After low–temperature annealing, the bonding strength analysed by the mechanical wedge-out technique showed that no bonding occurs between hydrophobic surfaces (of the wetting angle more than $28°$). The less the wetting angle is, the more intensive the bonding becomes. If hydrophilic surfaces are incorporated into a contact region, then even a low–temperature annealing brings an intimate mating of wafers.

A study of the extrinsic–defect composition in SOI-structures has detected a presence of pollutants and a drift of dopants into SiO_2 near the interface. In the wafers of $n-Si(001)CZ$ (antimony–doped) and $p-Si\ (001)\ CZ$, boron–doped, the SiO_2 films had an aggregation of boron and antimony, and also high concentrations of aluminium (up to 10^{19} cm^{-3}) and sodium (up to $4.5 \cdot 10^{18}$ cm^{-3}); especially this occurred in the structures subjected to a prebonding alkalization (in the ammonia – peroxide solution, 1:1:8 (vol. parts)). Measuring the spreading resistance has revealed a high–ohmic layer to be present near the $Si - SiO_2$ interface in the working Si layer of some specimens; this layer may be induced by the aluminium pollutant (in $n - Si$ (001) CZ ($\rho = 20$ $Ohm \cdot cm$, phosphorus–doped) and by a drive of boron into dielectric (in $p - Si$ (001) CZ ($\rho = 12$ $Ohm \cdot cm$, boron–doped) [Yenisherlova and Rousak, et al. (1994)]. The metallographical studies have demonstrated no significant structural changes in the device layers; this is sooner accounted for soft mechanical treatment and a great thickness of the layer chemically removed from the working wafer. This kind of treatment results in lesser changes in the extrinsic–defect composition and these changes may be recorded only through precise X-ray and diffractometrical measurements [Perevostchickov and Skoupov (1992)]. Upon thermocompressive bonding and thinning by the technology from [Yenisherlova and Rousak, et al. (1994)], the residual defects in the device layers of the SOI-structures were found through the halfwidths of the diffractional peaks on a two–crystal X-ray spectrometer [Krymcko and Yenisherlova, et al. (1998)]. Through the SIMS-technique, wafer bonding was shown to result in a dielectric pollution, by atoms of natrium and aluminium; this bonding also causes an oscillating redistribution of background and doping impurities near the interface "SiO_2 – working wafer" and "SiO_2 – base wafer"; such redistribution may negatively affect the electrophysical parameters of the device layer, in particular, it may increase leakage currents through SiO_2 [Krymcko and Yenisherlova, et al. (1998)].

In this connection, a preliminary cleaning of original wafers from impurities and suppressing the active channels of their pollution in the course of SOI manufacture becomes an urgent problem to be attacked. One of such cleaning procedures is a set of abrasive–chemical and electrochemical operations for thinning a working wafer; during these operations the wafer is contaminated through diffusion and impurities drifted from technological materials and ambients (abrasives, suspensions, electrolytes, an ambient, etc.). This disadvantage is not present in the Smart Cut technique (Fig. 2.7), a new fast–developing technique for producing the SOI-structures. This technology involves a wafer bonding through a thermal oxide, with protons preliminary implanted into one of these wafers in doses sufficient to form hydrogen–filled pores (a so–called blustering) by a low–temperature annealing [Bruel and Aspar, et al. (1997)].

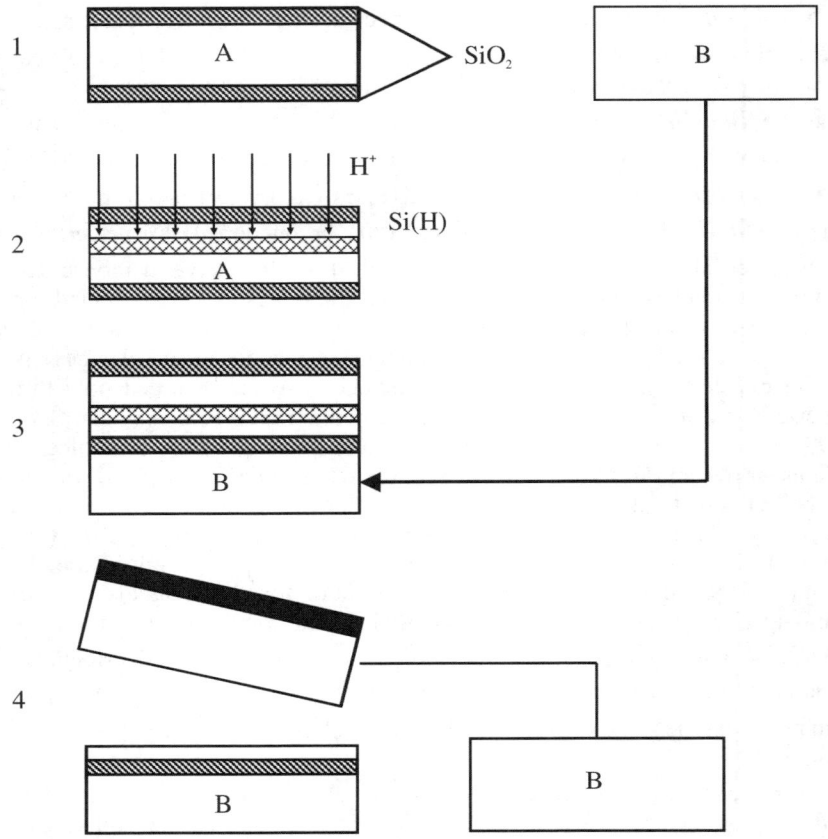

Fig. 2.7. Scheme of the Smart Cut technology.

The basic technological stages (Fig. 2.7.) in this technique are the following:

1) a thermal oxidation of the working wafer (A);

2) implanting protons into the wafer A through the SiO_2 film, by doses of $3 \cdot 10^{16} - 1 \cdot 10^{17}$, the energy of protons being assigned in accordance with the device layer thickness needed;

3) cleaning the surface and a low–temperature bonding of wafers ($T \approx 573 \ K$);

4) annealing the structure at $623 - 723 \ K$; during this process the hydrogen – filled micropores are generated and the working wafer A is detached from the base wafer B by the pressure of molecular hydrogen in the pores being more than

a silicon strength limit. Hydrogen contributes to a growth and coalescence of pores along the planes parallel to the surface of the wafer. At the final stage, the structures are annealed at $T = 1273 - 1473\ K$ and the surface of the device layer undergoes a precise chemicomechanical polishing. This latter treatment is caused by wafer blistering (A) into micropores; it may occur along several parallel crystallographical planes, because of the spread of runs of the incorporated protons.

The portion of the wafer A, separated because of blistering, may be used again for producing new structures. A thickness of dielectric and device layers may be widely varied in the Smart Cut technique through choosing needed regimes and oxidation conditions and also through setting a proton energy determining a depth of their penetration into a working wafer and a maximum of the hydrogen distribution profile within which there occur a blistering and a brittle spalling.

According to the data obtained by electronic microscopy [Popov (1998)], the device layer in the SOI-structures produced by this technique practically has no extended defects and in it there are observed only local regions seemingly induced by the inhomogeneity of residual mechanical stresses born by etching (polishing) and high–temperature treatment. In this study, the electrophysical measurements have shown a good quality of device layers: the charge state density within the $Si - SiO_2$ interface was found not to exceed $10^{12}\ eV^{-1} \cdot cm^{-2}$, the electron mobility to be not lower than $450\ cm^2 \cdot V^{-1} \cdot s^{-1}$, the breakthrough voltage for the dielectric $0.2 - 0.4\ mcm$ thick to exceed $100\ V$ and the leakage currents were found to be at the level $0.1\ nA$.

Alongside with this, the Smart Cut technology, in contrast to other SOI producing techniques, has not so far gained sufficient experimental and theoretical knowledge concerning the specificity of extrinsic defect compositions formed by this technique, the effects of the basic technological regimes (bonding, annealing and polishing) on the device layers. In this technology, there still remain unsolved such problems as constructing the structures of different crystalline orientations in the device layers; reducing the proton irradiation doses; increasing a structural perfection in the device layers, in particular, by excluding, upon spalling, the additional polishing.

To have the Smart Cut process efficiently implemented, the protonic dose to be implanted into silicon has to exceed some critical level, as was shown in [Kozlovscky and Kozlov, et al. (2000)]. With the energy of the implanted protons typically equal to $50\ keV$, the dose is usually $(4 - 8) \cdot 10^{16}\ cm^{-2}$, depending on the postimplantation annealing regime. A decrease of the required dose is possible if the helium ions are implanted additionally. Hence, the dose of implanted hydrogen may be reduced to $7.5 \cdot 10^{15}\ cm^{-2}$ if the protonic irradiation is followed with the gallium irradiation, of the dose $10^{16}\ cm^{-2}$; this in total will make up $1.75 \cdot 10^{16} \ cm^{-2}$, i.e. 3.4 times less than by usual Smart Cut technology [Kozlovscky and Kozlov, et al. (2000)].

In conclusion, it must be noted here that this process may be also exploited for producing structures with buried dielectric layers on other semiconducting materials. Such possibility was demonstrated on silicon carbide and helium arsenide crystals [Kozlovscky and Kozlov, et al. (2000)].

4) The separation by implantation of oxygen (SIMOX) technology

The SIMOX technology is synthesizing a SiO_2 layer during high–temperature annealing, with a locally highly concentrated atoms of oxygen implanted through ionic irradiation. As a variant to this technique, there exists the separation by implantation of nitrogen (SIMNI) technology and the separation by implantation of oxygen and nitrogen (SIMONI) technology. As of to–day, commercially there is exploited only the SIMOX technology [Mal'tzev and Tchaplygin, et al. (1998)].

For synthesizing a silicon dioxide, the crystal should be implanted with the oxygen atoms of the concentration $(4-5) \cdot 10^{22} \ cm^{-3}$ corresponding to the stechiometric ratio of the components in a new phase. This is achieved by irradiating doses $(1-2) \cdot 10^{18} \ cm^{-2}$, with the energy of the oxygen ions being equal to $200 \ keV$. Besides, the inequality $\dfrac{R_s}{\Delta R_s} \geq 3$ has to be obeyed, where R_s and ΔR_s are an average design scattering of runs [Vavilov and Kiselev, et al. (1990)]. This condition is true for the ionic energies higher than $150 \ keV$. In order to obtain a high quality of the device layer in the structures, the implantation is performed into the wafers heated up to $723-823 \ K$ that preserves the silicon crystalline structure. That the thermal range is narrow is determined by the fact that with a rise of temperature oxygen will be diffusing surface–bound, i.e. this will disrupt the stechiometric composition of the hard solution. A later postimplantation high–temperature anneal ($1423-1473 \ K$) will kill irradiation defects in the device layer and will activate a chemical interaction of silicon with oxygen to form a buried SiO_2 layer. Structural properties in the presurface monocrystalline silicon may be bettered through its amorphizing by silicon ions with a later recrystallization [Vavilov and Kiselev, et al. (1990)].

The basic drawback of the SIMOX technology is its low productivity determined by a need in high doses of oxygen ions. Therefore, the low–dose variants of this technique are being intensively studied to–day, in particular, with a use of additional high–temperature oxidation of the SOI-structures [Popov (1998)].

The Si epitaxial layers, $4.0-4.6 \ mcm$ thick, obtained on the structures with buried SiO_2 and Si_3N_4 layers formed by implanted oxygen and nitrogen ions ($E = 150$ and $200\ keV$, $F = 6 \cdot 10^{16} - 3 \cdot 10^{17}\ cm^{-2}$) and by a later annealing at $823 - 1373\ K$, were studied in [Litovtchencko and Romanyuk, et al. (1986)]. It was found that the density of dislocations and stacking faults in the epitaxial Si in the regions lying under the buried dielectric layer is largely lower than that above the nonirradiated zones in the wafer. Besides, the Si_3N_4 layers turn out to have the best gettering properties than the SiO_2 layers. According to the electroreflection, the Si surface layer implanted by nitrogen has no mechanical stresses, which were observed in the case of oxidation. During epitaxy, not only the Si_3N_4 film serves as a getter but also the interface between the implanted and nonirradiated regions (a planar gettering). The effect of spatial gettering by SiO_2 is expressed slightly, and a planar gettering retains in this case as well. It has been revealed that during the oxygen irradiation, in contrast to nitrogen irradiation, the results of annealing are affected by the gaseous atmosphere in which this annealing is done – the annealing in nitrogen produces a better quality of epitaxial layers. This is explained by oxynitride inclusions formed in the SiO_2 layer and exerting additional gettering action and thus reducing mechanical stresses in the structure. The planar gettering relates to the dislocations emerging in the interface between the regions and a buried dielectric layer and without it, the density in them may achieve $10^8\ cm^{-2}$. The point defects chiefly diffuse into the gettering region across a film surface for a substantial distance (up to $2\ mm$), whereas in the spatial gettering this distance is only $0.3\ mm$. Thus, the buried dielectrical layers manifest their gettering properties [Alekhin (1983)] and so there is a potential danger for their electrical strength to be reduced by pollutants, including metallic atoms. Hence, the results of this study unveil a need in the original crystals being primarily cleaned, prior to a formation of SOI-structures. It is most probable that a positive role may be played by the internal gettering layer formed within the bulk of the wafer at the depths exceeding a full run of the implanted ions; this layer was constructed by the thermocyclic treatment prior to oxygen or nitrogen implantation. Such getter is able to drop the concentration of not only the background impurities but also the irradiation–induced defects; this will make device layers better in quality. For this purpose, there may be also used some other gettering techniques considered in detail in the below chapters.

3 Techniques for high–temperature gettering

This chapter describes the basic ideas of the high–temperature gettering of impurities and defects from the semiconducting structures with a use of structurally damaged layers, striking and acoustic treatment, diffusion and ionic doping, laser irradiation; porous silicon layers, heterophase layers and metallic films, silicate glasses, and silicon nitride and silicide layers. Some other gettering mechanisms and models are also considered.

One of the most needed conditions to activate the gettering processes is to make point defects highly mobile, so that during a treatment they could leave in a diffusive–drifting way the region of the material for internal or external drains (discharges). Therefore, in most cases gettering is performed at sufficiently high temperatures, and most often is combined with basic operations of the technological line, i.e. with epitaxy, oxidation or diffusion. As drains for mobile impurities and IP defects there serve specially formed structurally damaged layers or heterophase films lying beyond the region to be gettered, usually on a back side of semiconducting structures. High–temperature techniques also embrace the techniques for processing the original wafers or semifinished items (the structures without special gettering layers previously created on them) in the gaseous ambient when gettering is done either at the expense of the chemical activity of the ambient or due to drains being constructed during annealing for impurities and IP defects inside the semiconductor itself. This chapter discusses these techniques.

3.1
Gettering by structurally–damaged layers

The techniques for removing the unbalanced IP defects and fast–diffusing backward impurities from the device layers in semiconducting structures by specially introduced crystallographical defects are most popular nowadays among technologists. This is understandable, for these techniques are universal to some extent, both from the point of view of the number of basic technological operations and a wide spectrum of the defects being gettered. Structurally–damaged layers on a front (working) – and mostly on a back (nonworking) – sides of slices can be formed by abrasive or impact–acoustical treatment, ionic implantation, laser irradiation and diffusive doping, up to such concentrations in the introduced liquid at which misfit dislocations emerge. Such layers, with a high density of defects, and, first of all, the dislocations with an edge component of the Bűrgers vector, are used as sufficiently powerful drains (discharges) of point defects in the later thermal treatment.

3.1.1
Gettering by mechanically–damaged layers

After abrasive treatment, cleaning the silicon crystals from fast–diffusing impurities and oxidation– or epitaxy–induced stacking faults by the layers of high–density structural defects is one of the first gettering techniques [Bookker and Stickler (1965); Pomerantz (1967); Laurence (1969); Nemtzev and Peckarev, et al. (1981); Labounov and Baranov, et al. (1983)]. Distortions may be incorporated practically by any kind of mechanical treatment, starting from an ingot slicing up to scribing them into chips. However, during the further high–temperature operations (due to locally inhomogeneous field of internal mechanical stresses) these defects (in particular, the dislocations) induced by the above mechanical operation, are able to penetrate– conservatively and unconservatively – very deep, even down to the device regions of the crystal. Therefore, mechanically–established getters are formed by special technological procedures that drop the probability of uncontrolled redistribution of defects. These procedures include [Nemtzev and Peckarev, et al. (1981); Labounov and Baranov, et al. (1983)]:

•   thoroughly choosing the abrasive materials and compositions of suspensions;

•   optimizing the processing regimes in accordance with the magnitude of contact pressure, the removal rate, the sequence of performing grinding–polishing operations, etc.;

•   using protective coats, for example, of silicon nitride (Si_3O_4), which simultaneously can partially compensate the amplitude of mechanical stresses;

•   alternating an abrasive treatment with a thermal annealing, etc.

One of the negative consequences of a mechanical getter when the sides of the semiconducting wafer had a nonequivalent abrasive treatment is its macroscopic bending. This bending increases with an increase of the density of the incorporated defects and a depth of the damaged layer. Due to the anisotropy of defects distribution, this bending is often inhomogeneous making the wafer look like a saddle, a cylinder or like any more complicated surface [Tarui (1985)]. This affects not only the quality of lithographical processes when forming a topology of devices but also causes an uncontrollable redistribution of the components in the extrinsic defect composition of the material. During technological operations, this is caused by even insufficient thermoelastic stresses. That is why, at the terminating stage of getter formation, this macroscopic bending in the structures is diminished by chemicodynamical or chemicomechanical polishing of its most defective layer, $0.5 - 2.0 \; mcm$, depending on the total thickness of the damaged region. Back sides of the wafers are also recommended to be also polished after the gettering anneal [US pat. No. 4144099]. In this case, a relief and a cracked zones in the structurally–damaged layer are removed. They favour an uncontrollable pollution of the structures during technological operations; but near the surface the gettering layer with the dislocation density of up to $10^8 - 10^9 \, cm^{-2}$ retains. In the course of a high–temperature treatment, a wafer's buckling caused by differences in the defectiveness of the front and back (with a getter) sides may be reduced by the damaged layer being formed with a circled concentric distribution of defects across the surface [US pat. Nos. 3905162 and 3923567]. Though, it should be said here that such symmetry of defects can be fast killed by thermal actions, due to the anisotropy of elastic–plastic properties of monocrystals; hence, this way can be efficient only when a getter is being built on the highly–symmetrical crystalline surfaces of wafers.

When building up a getter it is important to have in mind what kind of abrasive (i.e. a cohered or a loose one) is better to be used for incorporating structural distortions. There are diametrically opposite opinions on this point. The experts from the companies Wacker and Silicon Materials [US pat. Nos. 4144099, 3905162, 3923567, 2573464] prefer loose abrasives, namely, the grains of quartz, corundum, zirconium dioxide, silicates, silicon carbide of the size $5 - 200 \, mcm$ in the form of suspensions in water, glycerin, alcohol, petrol or their mixtures. On the other hand, the Japanese technologists [Japan pat. No. 53 – 1826] think a polishing by a cohered abrasive to be most efficient, because in this case no uncontrollable changes occur in the structure of the getter being built; these changes are determined by damaged abrasive grains and by their mixing up with the products of the semiconductor's treatment.

From the point of view of the damaged layer constructing mechanism, the difference in treating by a loose and cohered abrasive lies in the following [Karban' and Koi, et al. (1982); Karban' and Borzakov (1988)].

In the first case (*a loose abrasive*), the loose grains being plastically microforced through by a stress, normal with respect to the surface, will yield only compressing stresses within the contact zone. At some critical pressure, the compressing stresses will be followed with the extending stresses which are namely responsible for circled cracks (flaws). The moment of formation and the spreading depth of such cracks are dependent upon the size of the abrasive grains and the magnitude of the normal stress. Therefore, the depth of the damaged layer of the material ground by a loose abrasive cannot be less than some limit related to a start of brittle damage caused by the normal force only. A crack generating process is preceded by a microplastic deformation that results in positive and negative edge dislocations leaving the contact zone. In particular, their ability to escape to a free surface of the crystal is thought to be a reason of arising so–called "piles", i.e. a microrelief of the outdriven material around single grains or indenters.

In the second case (*a cohered abrasive*), a material surface is subjected to a scratching action by a set of cohered grains. Within a contact zone there work tangent and normal forces and this simultaneously gives rise to compressing and extending stresses. The latters are acting as shifting stresses and under some specific conditions initiate a formation and a travell of (primarily, one–sign) dislocations. The interactions of such dislocations under large loads usually accounts for microcracks and a damage of the material.

As shown in [Karban' and Koi, et al. (1982); Perevostchikov and Skoupov (1992); Karban' and Borzakov (1988)], a specificity of these two techniques for semiconductor surface processing lies in the fact that in similar regimes and conditions of actions the cohered grains drive away the material more (in average, by 30–50 per cent) than when grinding is done by a loose abrasive. However, a use of the latter gives a greater depth of the damaged layer. This depth may exceed by 45 – 100 per cent the thickness of the layer created by a cohered abrasive; besides, this cohered abrasive processing usually intensifies a cracking process and a material dispersion, i.e. it brings about a great number of gettering centres.

Alongside with the specificities in the crystallographical structure of the damaged layers created by a loose or cohered abrasive, an efficient gettering must be provided with a good defect draining mechanism to be functioning during later thermal treatment, since a way of transformation of defects–drains speaks about a functional resource of the getter, its capacity and the exhaustion rate of the gettering centre. These problems have not been studied so far. Though, the general regularities detected in the behaviour of dislocations and dislocation aggregates in semiconductors during a thermal treatment [Kontzevoy and Litvinov, et al. (1982); Mil'vidsky and Osvensky (1984)] make it possible to suppose that a cohered abrasive formed getter will possess a longer resource than that in a loose abrasive formed getter, since this type of getter creates a commensurable number of dislocations with positive and negative orientations of the Bűrgers vector edge component. During annealing, the locally high mechanical stresses in the damaged layer activate the displacements of dislocations and their surface–bound travells and also the annihilation of positive and negative dislocations; as a result of this, the concentration of gettering cenres decreases. In contrast to this, a priority of single–sign dislocations within a damaged layer (after the cohered abrasive high–temperature treatment) leads (alongside with the outmoves to the surface) to a polygonization, a formation of small–angle boundaries and a partial penetration of dislocations into the bulk of the crystal, i.e. to an increase of getter's thickness [Marousyak and Kosyatchenko (1989)]. In other words, in this case one may expect a more efficient gettering, due to an insignificant fall in the concentration and activity of the gettering centres in the layer damaged by a cohered abrasive. An actual picture of these processes is certainly more complex, for the kinetics of their running is affected by such factors as a chemical activity of the ambient within which an annealing is done, a spectrum and a concentration of the components of the impurity–defect composition in the semiconducting crystal itself, an availability of heterophase films on its surface, etc.

The above described considerations were proved by the results in [Bourmistrov and Pekarev, et al. (1978)] where it was studied how a getter formed by a cohered abrasive of the diamond synthetic wheel on a back side of the $p - Si(111)CZ$ ($\rho = 10\ Ohm \cdot cm$, boron–doped) wafers affects a density of defects near the front surface upon oxidation and epitaxy. After these operations, a selective chemical etching of these wafers has revealed that – independently of the thickness of the specimens – the density of vacancy–impurity clusters above a controlled getter–free polished side exceeds by one or two orders the density above the ground surface. With the wafers' thickness increasing from 200 to $400\, mcm$, the density of defects above the getter will increase, whereas above the controlled region it does not change practically, though a quantity of etching wells tends to grow. Upon epitaxy, a density of the stacking faults above the controlled surface also turned out to be one–two orders higher than that in the regions above the getter. An increase in thickness of the wafers brings an increase of stacking faults in both parts of the specimens.

These results are explained in [Bourmistrov and Pekarev, et al. (1978)] in the following way. If no getter, there arise two fluxes of point defects in the starting moment of oxidation. The first one is of vacancies conditioned by a formation of SiO_2 layer whose growth is accompanied with a vacancy absorption, since a specific volume of silicon dioxide is more than that of monocrystalline silicon. The second flux is of rapidly diffusing impurities (Cu, Au, Fe and the like) migrating surface–bound as this surface is their powerful drain. The vacancies when coming close to the surface of the $Si - SiO_2$ interface can establish the aggregates stabilized by the atoms of the rapidly diffusing impurities and capable of increasing their sizes up to those of the vacancy–impurity clusters revealed by a selective etching. In the layer adjacent to silicon dioxide there also appears an oversaturation of intrinsic interstitial atoms of silicon whose association and growth generate the oxygen–induced stacking faults being recognized by metallographical methods, together with vacancy clusters. When epitaxially growing layers on a front side of wafers, these two types of distortions serve as nuclei for producing stacking faults. The mechanically damaged getter on the back side of wafers causes their bending, due to an inhomogeneous in-depth distribution of internal elastic stresses. A gradient of these stresses makes the rapidly diffusing impurities and interstitial atoms of silicon travell towards the regions of crystalline lattice expansion, i.e. towards the getter. This process results in a drop of overabsorption across the interstitial atoms and that of the concentration of impurities in the vicinity of a front side of wafers, and this decreases the density of vacancy–impurity clusters, oxidation–induced and epitaxially–grown stacking faults.

The above considerations make it possible to describe on a qualitative basis a high–temperature gettering that provides a point defect mobility. However, as will be shown below, the gradient of static mechanical stresses is not a unique reason to suppress the defect formation when the crystals with damaged surface layers are processed thermally. Of not less importance may be the fields of dynamical elastic stresses arising during structural transformations and, in particular, during reconstructions in the gettering layer itself, for example, when annihilation or a surface–bound drift of dislocations occurs.

There have been studied the effects of the getter formed by a loose abrasive (a suspension of diamond micropowder, $20 - mcm$ grains) or by a cohered abrasive (a diamond synthetic paste) on the density of OSFs and that of the clusters of intrinsic interstitial atoms arising in the wafers of $p - Si(001)CZ$ ($\rho = 12$ $Ohm \cdot cm$, boron–doped) 460 mcm thick. Prior to oxidation, the wafer's side with the layer damaged by polishing was protected by a Si_3N_4 film $0.38 - 0.40 \, mcm$ thick; the film was deposited by a reactive spraying of silicon in the atmosphere of argon with nitrogen added. The wafers were oxidized in wet oxygen at $1273 \, K$. The length of oxidation varied from 30 to 150 min, with a $30 - min$ step. Upon completing a next oxidizing cycle, the density of OSFs and that of clusters were measured on the wafers, through the pictures of selectively etching in the Sirtl solution the region located not more than $2 - 2.5 \, mcm$ in depth from the SiO_2 layer The measuring accuracy of the defect density recorded by 15 – 20 scanning fields of the Jenatech Inspection microscope was not worse than 10 per cent. The wafers without gettering layers served as controlled specimens.

The original density of the microdefects measured prior to oxidation has shown that with a back side not mechanically ground the defect density on the front side in average was $4.144 \cdot 10^5 \, cm^{-2}$; upon a loose abrasive processing it was $4.108 \cdot 10^5 \, cm^{-2}$ and upon a cohered abrasive grinding $4.137 \cdot 10^5 \, cm^{-2}$. Though the differences revealed in the density of microdefects in the original specimens were less than 10 per cent, these differences, nevertheless, betray, first, the sensitivity of the microdefect structure in the bulk of silicon to external actions exerted upon a crystal surface, and, second, a difference in structural rearrangements accompanying a formation of damaged layers during a loose–abrasive and cohered–abrasive treatments. This difference was earlier noted in [Perevostchikov and Skoupov (1992); Karban' and Borzakov (1988); Perevostchikov and Skoupov (1987); Gorshkov and Perevostchikov, et al. (1989)]. These differences are conditioned, firstly, by the elastic wave fields becoming more intensive (because of loose–abrasive grinding of silicon and being responsible for microdefect dissolution) than when silicon is treated by abrasive wheels, with shifting stresses making the microdamages in the material more dominant.

How an oxidation length affects the defect density in the presurface getter–free layers of the wafer and in the layers with a getter of various types is shown in Fig. 3.1.

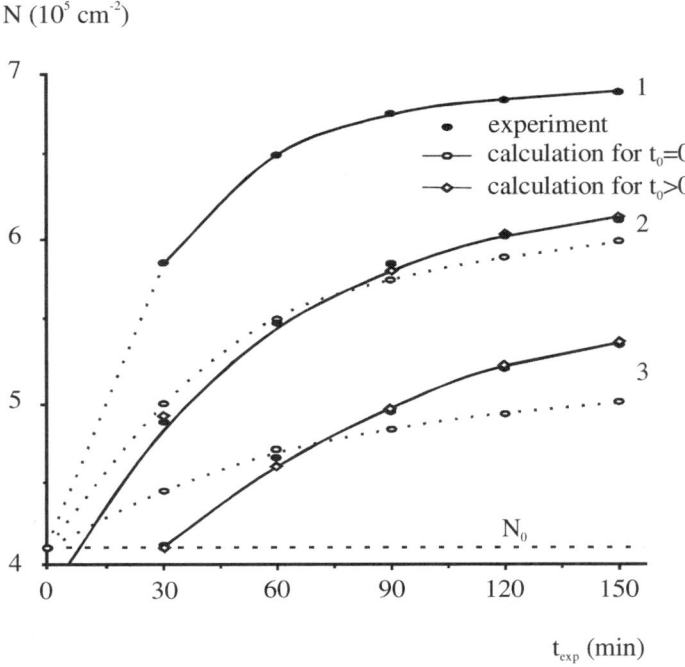

Fig. 3.1. Dependence of the microdefect density in the crystals of $p - Si\,(001)\,CZ$ ($\rho = 12\ Ohm \cdot cm$, boron–doped) on the oxidation length: **1** – getter–free; **2** – with a getter created by a loose–abrasive grinding; **3** – with a getter created by a cohered–abrasive grinding.

In it, it is seen that experimental values are well approximated by the dependencies of the type:

$$N(t) = N_0 + N_\infty \left[1 - \exp\left(-\frac{t - t_0}{\tau} \right) \right] \text{ for getter–introduced wafers}$$

$$N(t) = N_0 + N_\infty \left[1 - \exp\left(-\frac{t}{\tau} \right) \right] \text{ for getter–free wafers}$$

The numerical values of the variables for these expressions are given in Table 3.1.

As seen from Fig. 3.1, if the experimental points for the getter holding specimens are approximated by the curves with $t = 0$, then the inconsistency between the calculated and experimental data turns out to exceed the error of measuring the density of etching wells, especially in the case of the getter built by a cohered–abrasive grinding.

Table 3.1

The constants of approximating functions

Type of structure	Parameters			
	N_0 $(10^5\ cm^{-2})$	N_∞ $(10^5\ cm^{-2})$	τ (min)	t_0 (min)
Getter–free	4.1	2.8	31.0	0
With a getter (loose abrasive)	4.1	2.1	46.3	9
With a getter (cohered abrasive)	4.1	1.6	79.1	28

The constants in the approximating functions carry the following physical sense:

N_0 and N_∞ are respectively an original value for the defect density and its asymptotic complement at $t \rightarrow \infty$;

τ is the specific time determining a nucleating rate and a growth rate of defects;

t is a latent (incubational) period of bearing the oxidation–induced clusters and stacking faults.

That oxidation–induced defect density grows exponentially in response to an increase of the process length, and, hence, in response to the SiO_2 film thickness and elastic stresses in silicon – this dependence was proved by the results in [Shapovalov and Gryadoun, et al. (1995)]. The nature of deference in values of time constants in Table 3.1 can be understood, if elastic waves generated by high–temperature relaxation of structural distortions in the getter are supposed to play an active role in gettering during oxidation. As it is known [Lyubov (1985)], the speed V_g of the unconservative travell of edge dislocations (the structural analogs of Frank dislocation loops or oxidation–induced microdefects interacting with unbalanced intrinsic interstitial atoms when their concentration drops down due to annihilation with unbalanced vacancies) relates to the speed V_{0g} of overcrawling by dislocations (in the absence of annihilation) as

$$V_g = \frac{V_{0g}}{1 + \alpha C_V D_i t},$$

where

α is the coefficient describing an annihilation efficiency of point defects;

C_V is a concentration of vacancies;

D_i is a diffusion coefficient of interstitial atoms;

t is the current time.

Now, from the above expression we can find the most specific times for a growth of oxidation–induced stacking faults:

$$\tau = \tau_0 (1 + \alpha C_V D_i t) ,$$

where τ, τ_0 is the time for a getter–free and a getter–available specimens.

This formula shows that in the presence of unbalanced vacancies the flux of intrinsic interstitial atoms emerging in the vicinity of the $Si - SiO_2$ interface is slackening and the concentration and sizes of oxidation–induced defects are diminishing. Static elastic stresses acting within the $Si - SiO_2$ system are not able to produce vacancies in the amounts needed for efficient annihilation. This can be done by the fields of dynamical mechanical stresses of the elastic waves generated by relaxing structural distortions in the gettering layers [Alekhin (1983); Pavlov and Skoupov, et al. (1987)]. As exhibited in Table 3.1, most efficiently the annihilation of intrinsic interstitial atoms and vacancies runs in the specimens with the cohered abrasive getter, i.e. in such structures the relaxation of crystallographic distortions in the gettering layer will last longer, and is seemingly accompanied with elastic waves of a higher pressure, as matched against the amplitude of waves emitted by defects in the loose–abrasive getter. In accordance with the model of the "vacancy pump" [Alekhin (1983)] the unbalanced vacancies may be supposed to be formed on the boundary of the $Si - SiO_2$ interface as a result of a periodic "pumping" of interstitial atoms from the silicon lattice into the interstitial site or/and into the growing SiO_2 film. The frequency of this pumping process is similar to the frequency of elementary defect rearrangements in the getter, the greatest frequency of these rearrangements being close to the Debye frequency. In the case of pumping into the SiO_2 film, the silicon atoms appearing in the vacancy sites or in the micropores of the SiO_2 film must increase film density. This very effect has been detected by ellipsometrically measuring the refraction of SiO_2 layers by the ellipsometer. Measuring the getter–free controlled specimens $560 \pm 10 \, nm$ thick has given the average refraction value $n = 1.413 \pm 0.003$, with a selective variance across the entire surface in 15 test points being equal to $\sigma(n) = 0.069$; in the getter–available case the values were $n = 1.446 \pm 0.002$ and $\sigma(n) = 0.056$, respectively. In the specimens with the loose–abrasive and a cohered–abrasive getter the values of n and $\sigma(n)$ were found coincident, within the measuring accuracy.

The idea of unbalanced vacancies emerging near the $Si - SiO_2$ interface makes it possible to explain a misfit of latent periods t_0 for the structures with different getters. First, in the initial stage of the process, the annihilation of IP point defects makes low the rate of generating and growing of the oxidation–induced interstitial defects; this rate rises as soon as the sources of elastic waves in the gettering layers are getting exhaustive. Second, the unbalanced vacancies being gettered by these waves are also able to dissolve the original microdefects functioning as centres of generating the oxidation–induced clusters and stacking faults. An in-depth fall in vacancy concentration and an in-depth shift of the $Si - SiO_2$ interface, i.e. from the region "cleaned" from microdefects, brings a rise in the generation rate and growth rate of oxidation–induced stacking faults. If prior to oxidation, the structures with the getter protected by a Si_3O_4 film undergo a sufficiently long annealing in the inertial ambient at temperatures up to $1273\ K$, then the original defects will not become dissolved but the relaxation of distortions in the getter will activate, i.e. the elastic wave sources will become exhaustive. In the subsequent oxidation, the incubational period t_0 will substantially drop to the extent that it will vanish completely.

Through matching the data in Table 3.1, one may conclude that a cohered abrasive getter suppresses more efficiently the defect generating processes during oxidation, especially at large lengths of this operation, i.e. when thick SiO_2 layers are built up. For the case of thin SiO_2 films, say, used as a subgate dielectruc in the MOS-transistors, any of the above techniques for constructing a mechanical getter is possible to be used.

The efficiency of gettering by mechanically damaged layers can be increased through rational variants of grinding and optimal regimes and conditions in the later chemical etching of the getter, in order to reduce the residual elastic stresses in structures, The positive effect of such approach was demonstrated in [Kobazeva and Skvortzov (1986)] on the examples with the getters formed via treating the silicon with suspensions of micropowders ($10-$, $14-$ and $20-mcm$ grains) and further alkaline etching. It was found in [Kobazeva and Skvortzov (1986)] that a minimal density of dislocations and stacking faults in the epitaxial layers with a back–sided getter is gained, if the ratio of hard and liquid components of the abrasive suspension is $h:1=1:1.5$, the etching length for the 30 per cent caustic soda solution is $15-20\ s$ and temperature is $368-373\ K$. These very conditions of getter formation helped to obtain a maximum yield of fit transistor structures built on the basis of the above epitaxial compositions.

Therefore, a basic technological line for manufacturing silicon wafers with a mechanically–built getter involves the following operations [Labounov and Baranov, et al. (1983); Perevostchikov and Skoupov (1992)]:

a)   slicing an ingot;

b)   removing by chemical etching the layer damaged by slicing

c)   grinding and polishing the surface of slices to obtain the needed macrogeometrical parameters for the wafers;

d)   a chemicodynamical two–sided polishing of slices;

e)   a special abrasive treatment of the back side of slices, i.e. building up a gettering layer;

f)   etching the slices in an alkaline solution at high temperature, for smoothing a relief zone in the getter and partially removing its elastic stresses;

g)   oxidizing the slices or annealing in the inertial ambient, for activating the gettering processes;

h)   a finishing chemicomechanical or chemicodynamical polishing;

i)   protecting the getter chemically and by a thermoresistant film, for example, by a Si_3N_4 film.

A mechanically–formed getter makes it possible not only to substantially reduce the defectiveness in wafers and epitaxially–grown layers but also to strengthen their electrophysical characteristics and also the parameters of devices manufactured on the basis of these structures. In particular, gettering increases the lifetime of minor charge carriers, reduces the leakage currents in diodes and transistors, makes better the noise characteristics of devices, and, in the final end, increases a yield of fit devices. A presence of getter in device structures reduces the probability of the degradation induced by various destabilizing factors, say, such as radiation fields. This is conditioned by arising unbalanced IP defects which are captured by the getter and do not participate in the formation of the impurity–defect aggregates in the active regions of devices.

The gettering efficiency, as was demonstrated by numerous experiments, is determined not only by the gettering layer properties but also by the regimes and conditions of those thermotreatments which activate cleaning the structures from impurities and defects. For example, in [Nemtzev and Peckarev, et al. (1981); Labounov and Baranov, et al. (1983)] it was found that a mechanical getter diminishing the concentration of OSFs in silicon is, however, able either to increase or in no way to affect the lifetime of minor charge carriers – depending on the oxidation temperature and a crystal cooling rate. The positive effect, i.e. an increase of lifetime, is only observed in the course of slowly cooling the structures (at the rate not more than 1 degree/min). It is only in this case that the impurities manage to generate electrically inactive precipitates in the region of the getter. Whereas, during a fast cooling (hardening) the rapidly diffusing metallic impurities retain in the interstitial sites of the silicon lattice and build deep recombination centres within the prohibited zone; these very centres decrease the lifetime of minor charge carriers.

A mechanical getter is also efficient for cleaning from unwanted impurities and defects not only the silicon monocrystals but also other semiconductors (*GaAs,*

GaP); and also for the polycrystalline silicon slices applied in solar elements [Labounov and Baranov, et al. (1983)]. However, in contrast to silicon, many applications of the gettering technology to other semiconductors have been studied insufficiently and so studying this problem remains urgent for the contemporary microelectronics. In particular, one of such problems is a problem of changing a stehiometrical composition in the presurface layers in semiconducting compounds during a high–temperature treatment; during this treatment it is possible to detect the gettering properties in structurally–damaged regions. One of the possible solutions for this problem can be a irradiation–stimulated gettering, to be discussed below.

As it was shown, though the getters built by various abrasives are highly efficient, relatively cheap and adaptable to streamlined production, they, nevertheless, have the below drawbacks:

a)   the getter parameters are not highly reproductive; in particular, this is due to the getter's structural inhomogeneity;

b)   this kind of getter has an increased probability of brittle damage in the semiconducting structures;

c)   they can be uncontrollably polluted by impurities from the surrounding ambients and these impurities can be transported into a wafer;

d)   such getters possess spontaneous relaxation processes resulting in an increased homogeneity of distribution of gettering centres, etc.

And here comes the problem of how to regulate the quality of a mechanically–created getter, and, chiefly, control its structural homogeneity that mainly determines the gettering results, i.e. a spectrum and distribution profiles of the residual impurities and defects in the material to be cleaned.

As one of such methods for nondestructive and sufficiently expressive control of getter's parameters there may be recommended the roentgendefractometrical method [Perevostchikov and Skoupov (1992)]. Its idea lies in the following: a microscopic curvature C in the slices with one–sided damaged layers t thick is independenty measured by a roentgen two– or three–crystal spectrometer. On the basis of this experimental data it becomes possible to calculate the deformation (the stresses) in the getter in the direction being normal to the surface and estimate its efficient thickness:

$$\delta = \frac{d_{layer} - d_0}{d_0} \ .$$

For approximating a step distribution of defects in the getter, the below equations are valid for the measurable deformations and curvature:

$$\delta = \frac{\Delta d}{d} = \varepsilon_0 - \frac{2v\sigma}{E} = \varepsilon_0 - \frac{2v}{1-v}\left[\varepsilon_0 - \varepsilon_0\frac{t}{H} - xC\right]$$

$$C = \frac{6\,t\,\varepsilon_0}{H^3}(H - t) \ .$$

Here,

ε_0 is a deformation of the damaged layer;

τ is the stresses in the layer acting within a plane of the getter–wafer boundary;

x is a current structure thickness coordinate taken from the middle plane;

H is a structure thickness;

E and v are, respectively, an elasticity module and the Poisson coefficient for the wafer material.

Solving the given system of equations with respect to ε_0 and t for $x = \dfrac{H}{2}$ will give:

$$\varepsilon_0 = \frac{H^3 C}{6t(H-t)} \quad \text{or} \quad \sigma_0 = -\frac{E}{1-v}\left[\varepsilon_0 - \varepsilon_0 \frac{t}{H} - C\frac{H}{2}\right]$$

$$t = \frac{H}{2}\left[\frac{3(1-v)\delta - 4vCH}{(1-v)\delta - vCH}\right] \cdot \left\{1 - \left[1 - \frac{6HC(1-3v)[(1-v)\delta - vCH]}{[3(1-v)\delta - 4vHC]^2}\right]^{1/2}\right\}$$

These formulas make it possible to estimate deformation (stresses) across the getter's thickness with the error nor worse than ± 15 per cent for the cases when a damaged layer has a sufficiently sharp profile as, for example, when grinding or polishing is done by a large–grained abrasive. The getter's structural homogeneity is calculated through the values of variance $\varepsilon_0(\sigma_0)$ and t, when the wafer surface is scanned by the probe X-ray [Perevostchikov and Skoupov (1992)]. The formulas for calculation are sufficiently simplified if the condition $H \gg t$ is fulfilled.

A reproductive production of mechanically–damaged layers with a macroscopic homogeneous distribution of defects may be arranged with a use of anisotropic abrasive grinding or polishing. The results in [Karban' and Koi, et al. (1982); Perevostchikov and Skoupov (1992); Karban' and Borzakov (1988)], where the kinetics of defect formation in silicon during an abrasive processing of the surface was studied with a due account of its crystallographical anisotropy, make it possible to state that an oriented grinding and polishing mainly incorporate into the silicon's presurface layer the single–type distortions with an ordered spatial distribution. For example, these are the dislocations with a dominating orientation of the Bŭrgers vector edge component. One of the ways to diminish the arising macroscopic bending in the slices is depositing onto the ground side the coats of the associated thermal expansion coefficient. To fully eliminate the wafers' macrobending, the parameters of the coat must satisfy the below condition:

$$C_0 = \frac{6t(H-t)}{H^3} \Delta\alpha \cdot \Delta T \ ,$$

where

C_0 is an original curvature in the slice of thickness H ;

t is a thickness of the coat;

$\Delta\alpha$ and ΔT is, respectively, a deference between the thermal expansion coefficients for coat and wafer materials and the deference in temperatures for coat precipitation and gettering.

Earlier it was noted that coating a back side simultaneously passivate its active surface with a developed damaged structure thus preventing the penetration of pollutants into the wafer.

The above–described technique of building a gettering layer was experimentally implemented on the slices of $p - Si\,(111)\,CZ\,(\rho = 15\ Ohm \cdot cm$, boron–doped) 300 mcm thick. One side in the slices was ground by the micropowder suspension (7-mcm grains) along the $<110>$ or $<112>$ crystalographical directions and then a Si_3N_4 film 0.4 mcm thick was deposited onto it by the reactive silicon spraying in argon, with nitrogen added. Upon this, the structures were oxidized in the stream of wet oxygen at 1423 K during 30 min. On the basis of the selective etching pictures, the anisotropic $<110> - $oriented grinding was found to drop the densities of OSFs near the front side of slices from $3.6 \cdot 10^6\ cm^{-2}$ to $4.0 \cdot 10^3\ cm^{-2}$, and the $<112> - $oriented grinding diminishes the OSFs down to $4.2 \cdot 10^3\ cm^{-2}$. On the controlled specimens with a nonoriented grinding, the density of OSFs made up $7.3 \cdot 10^3\ cm^{-2}$. Upon gettering, in all three cases the variance of the density of OSFs was found to drop across the surface of the slices almost by three orders as matched against the nongettered specimens. The anisotropic treatment of the back side of slices may be supposed to create such a structure in the damaged layer that it will facilitate a rise of the amplitude of the wave field being formed from elementary excitations in each interaction of unbalanced IP defects with the edge regions of dislocations during a gettering annealing. Gettering efficiency may be increased if a polygonized dislocation structure is formed in the damaged layer, for example, with the help of low–temperature processing of slices by high hydrostatical pressure [Perevostchikov and Skoupov (1992)].

3.1.2
Impact–acoustic treatment

One of the techniques of forming structural distortions on a back side of the semiconducting wafers is a local ultrasonic impact [US pat. No. 4018826; UK pat.

No. 1483888; Schwuttke (1978)]. A discrete series of such impacts is generated by vibrating tungsten balls of 300 mcm in diameter. The balls, 800 – 1000 pieces in number, are poured into a front box of fluoroplastic on the bottom of which there is placed (with a back side up) a slice to be treated, with its back side up. A semiconducting slice may be also used as a cap of the box. The box gets connected to an acoustic line connected to a diffuser of a powerful loudspeaker ($60 - 100 \ W$) to which are supplied the rectangular periodic impulses coincident with a resonance oscillating frequency ($\approx 1.38 \ kHz$) of the slice to be treated. Through varying the power, the quantity and the duration of impulses, and the processing length as well, one can regulate a defectiveness level of the getter under formation.

As shown in [Schwuttke (1978)], an impact–acoustic treatment of silicon wafers mainly gives birth to the defects of two types in the presurface layer: shallow rectangular grooves and cone–like pits from which the shallow cracks (crow's feet) are running away. The grooves can turn to the clusters of up to $15 \ mcm^2$ in square, of density up to $10^8 \ cm^{-2}$ and $0.2 - 0.4 \ mcm$ in depth. The cone–like defects penetrate as deep as $10 - 20$ mcm and the cracks up to $50 \ mcm$. With the density of such defects exceeding $10^6 \ cm^{-2}$, a silicon slice may destruct. The specificity is that these two types of defects do not grow in depth during high–temperature treatment.

The getter built by impact–acoustic treatment of the back side of silicon slices will efficiently clean the bulk of background impurities during later high–temperature annealing, and this results in an increase by 1–1.5 orders of the lifetime of minor charge carriers, and the variance of values of this parameter will be significantly decreased. This getter helps to practically suppress stacking faults in silicon during oxidation or epitaxy. It should be emphasized here that in its essence the gettering by impact–acoustic treatment starts at the time of this treatment, i.e. at low temperatures. This occurs due to the acoustic waves that stimulate a diffusive redistribution of fast–diffusing metallic impurities and IP defects between the bulk of the crystal and the getter on its surface [Perevostchikov and Skoupov (1992); Pavlov and Skoupov, et al. (1987); Koulemin (1978)].

As compared to the traditional abrasive grinding or polishing, the impact–acoustic treatment can increase a reproductiveness of gettering layer parameters. Though, it is highly probable that its application can damage the wafers and, besides, this type of treatment is of low productivity, since it may be performed on the slices by the piece only.

3.1.3
Diffusive doping

The technique of gettering by diffusing layers is based upon unbalanced point defects being saturated by the misfit dislocations induced in the semiconducting slice by doping with the atom–size impurities which differ in sizes from the basic material atoms and in the concentrations being near to the dissolution limit at the given temperature. In this technique, the gettering anneal is, in its essence, made combined with a getter forming process.

A simplified model for creating the misfit dislocations during a diffusion process involves the following [Kontzevoy and Litvinov, et al. (1982)].

Suppose that into a wafer of thickness d there is introduced an impurity, with the radius r_{im} of this impurity being different from the radius of atoms of the basic substance r_0 and the concentration of this impurity on the wafer's surface being constant and equal to C_S. During diffusion, the difference in atomic sizes will create an elastic deformation whose value in the planes parallel to a diffusion front will be determined by the expression:

$$\varepsilon_x = \frac{\gamma C_S}{1-v}\left[erfc\left(\frac{d-Z}{2\sqrt{Dt}}\right) - \frac{2\sqrt{\pi Dt}}{\pi d}\right].$$

Here,

• the X-axis lies within the diffusion plane, with its normal to the front in the direction Z;

• $\gamma = \dfrac{r_{im} - r_B}{r_B}$ is the coefficient for compression/expansion of the lattice by the impurity atoms;

• D is the coefficient of impurity diffusion;

• v is the Poisson coefficient;

• t is a diffusion length.

As seen from the formula, according to the given model the maximum mechanical stresses for ε_x are obtained in the initial instant of time and their values become equal to

$$\sigma_{max} = \frac{E}{1-v}\gamma C_S.$$

Dislocations arise, if these stresses exceed the critical stress of the shift for the wafer's material at the diffusion temperature $\tau_{cr}(T_{dif})$, i.e. there is fulfilled the condition:

$$\tau = \sigma_{max}\cos\alpha\cos\varphi > \tau_{cr}(T_{dif}),$$

where

α is an angle between the sliding plane and the diffusion front;

φ is an angle between the sliding direction and the diffusion plane.

Generating the dislocations is supposed to start spontaneously at $t = 0$. The distribution of dislocations across the diffusive layer is described by the below ratio:

$$N_d = \frac{\gamma}{|\overline{b}|} \cdot \frac{dC}{dZ} \quad ,$$

where $|\overline{b}|$ is a module for the Bűrgers vector of the edge dislocation within the diffusion plane.

In the crystals of the diamond lattice the dislocations are most often formed within the $\{111\}-$ planes, and the shift vector is $< 110 > -$ oriented. The density of dislocations per a unit of the length of the diffusive layer N can be calculated as

$$N = \gamma \frac{C_S}{|\overline{b}|} \quad .$$

The dislocations generated in the initial moment are diffusing alongside with the diffusion process into the depth of the wafer to form a grid of misfit dislocations.

The model discussed describes in general a qualitative idea of the mechanism for generating misfit dislocations during diffusion and does not cover the entire variety of phenomena in semiconductors arising due to highly–concentrated impurities introduced into them. However, from the point of view of the gettering properties of the diffusive layers these phenomena (a stage–by–stage and a local formation of dislocations, a generation and a role of IP defects in forming a dislocation grid, an anisotropy of the latter, etc. [Kontzevoy and Litvinov, et al. (1982); Hu (1973); Ravi (1981)]) seem to be of minor importance and here are left beyond our discussion.

Table 3.2 holds the atomic radii of silicon and those of basic dopants and their compressing coefficients as well.

Table 3.2

Atomic radii of silicon in the basic dopants and their compression coefficients

Atom	silicon (Si)	boron (B)	phosphorus (P)	arsenic (As)	antimony (Sb)
Radius (nm)	0.117	0.088	0.110	0.118	0.136
γ (conv. u.)	0	-0.25	-0.06	0.009	0.16

As seen from this table, the atomic radii of boron, phosphorus and antimony are the most different from silicon. Thus, these impurities, if incorporated in definite concentrations, are most expected to create a grid of misfit dislocations that can serve as a getter for unbalanced point defects. This picture is observed in reality. Most popular in practice has become the technique of gettering by diffusive layers of phosphorus, and of boron less. Antimony being a mixture with a small diffusion coefficient turns out to be less effective.

Let us now consider some examples of gettering by diffusing layers [Laurence (1969); Labounov and Baranov, et al. (1983); Ravi (1981); Rozgonyi and Kyshner (1976); Baldi and Cerofolini, et al. (1978); Murarka (1976); Verkhovsky (1981); Bogatch and Gousev, et al. (1981); Yenisherlova and Marounina (1984)]. A gettering layer on a back side of wafers is built up by diffusing phosphorus according to the following technological scheme [Rozgonyi and Kyshner (1976); Bogatch and Gousev, et al. (1981)]. Prior to the first technological process, a low–temperature SiO_2 layer, a masking material for a further diffusing process, is deposited onto a front side of the wafer. Upon this, phosphorus is being diffused into the back side of the slice in the regimes and concentrations capable of creating a grid of misfit dislocations and a gradient of mechanical stresses at the level $10^{25}\ cm^{-2}$, i.e. in the diffusive layers the phosphorus concentration at temperatures $1323 - 1423\ K$ becomes not less than $10^{21}\ cm^{-2}$. The interaction between unbalanced point defects and edge components of the dislocation grid will result in a gettering of the nuclei of the stacking faults. As a source of phosphorus, $POCl_3$ is usually used. The diffusion at $1323 - 1423\ K$ during 1–7 hrs is considered to be an optimal regime for creating a grid of misfit dislocations. Under these conditions, at the depth exceeding the depth of the diffusion front by 1.5 – 2 mcm there arises a dislocation grid parallel to the surface. If the temperature of the process is lower than $1323\ K$, then a lattice does not appear and a further oxidation will produce a large number of stacking faults. A presence of a grid of misfit dislocations on the back side of the diode silicon structures (this grid suppresses a birth of the spatial and presurface OSFs) reduced by 2–3 orders the leakage currents in diodes and increased an output of fit devices, up to 90 per cent. It was also noticed that gettering the unwanted impurities by phosphorus layers may run also in the cases when no grid of misfit dislocations is formed, i.e. the clearing reactions are only stimulated by elastic stresses conditioned by the diffusing layer.

The gettering properties of the layers with a high density of boron – induced dislocations of the concentration of up to $5 \cdot 10^{20}\ cm^{-3}$ have turned out to be somewhat worse than those in the phosphorus–doped regions. Though, this case also provides better parameters in the devices, in particular, a drop of leakage currents in transistors when an isolating boron diffusion is done [Labounov and Baranov, et al. (1983)].

In contrast to boron, the efficiency of using phosphorus as a getter lies not only in the difference in the density of misfit dislocations but sooner it relates to a greater dissolution of impurities in the n+-layers than in the p+-layers of silicon; this is explained by a capture of the impurity atoms by the E-centres (P + V structures) available within the phosphorus–doped regions [Baldi and Cerofolini, et al. (1978); Murarka (1976)]. Therefore, in [Baldi and Cerofolini, et al. (1978)] it is supposed that if an impurity in the silicon lattice is in a dissolved state, then low temperatures will be enough to getter it by a phosphorus diffusing layer. In accordance with the model of gold dissolution in silicon, the coefficient of gold segregation between undoped and phosphorus–doped silicon may be written as

$$K^{Si(P)}{}_{Si} = 1 + \frac{N_r}{N_{Si}} \exp\left[\frac{E_{Au/Si} - E_{Au/Si(P)}}{RT} \right],$$

where

$K^{Si(P)}{}_{Si}$ is a segregation coefficient;

N_p, N_{Si} is an atomic concentration of phosphorus and silicon respectively;

R is a gas constant;

$E_{Au/Si}$, $E_{Au/Si(P)}$ is, respectively, the energies for activating the gold dissolution in an undoped and doped monocrystal.

Gettering via impurity segregation in a diffusive layer becomes efficient if $K^{Si(P)}{}_{Si} \geq 1$. With this accounted, the formula for the segregation coefficient easily produces the below condition:

$$\frac{E_{Au/Si} - E_{Au/Si(P)}}{R} \geq T.$$

From this it follows that gettering needs high concentrations of phosphorus (that create E-centres) and comparatively low temperatures (being determined by a diffusion of impurity into a gettering layer and by a capture of this impurity by fixed vacancies). The optimal temperature for the segregative annealing for gold gettering makes up $973 - 1073\ K$ [Baldi and Cerofolini, et al. (1978)].

Such model has turned out to be also valid for gettering other heavy metals in silicon, provided they are not connected with precipitates. If an impurity in the original crystals generates precipitates, then a segregative annealing must be preceded with a high–temperature treatment, to have them dissolved.

In a real situation, gettering the metallic impurities by diffusive phosphorus layers is seemingly conditioned by a simultaneous action of both mechanisms considered: by capturing on a misfit dislocation and forming the ionic phosphorus–impurity pairs. Gettering properties of such layers are strengthened if, upon diffusion, a Si_3O_4 or Al_2O_3 layer 0.2 – 0.4 mcm thick is deposited onto a back side of wafers. Such layers, first, suppress an autodoping process, i.e. an uncontrolled redistribution of phosphorus between a front and a back sides in the wafers, and, second, increase – due to an increased gradient of mechanical stresses – a drifting component in the impurity streams to the getter. However, it is worth noting here that the role of elastic stresses in gettering by diffusive layers is far not clear yet, as well as in the case with a mechanical getter.

As revealed by the by-layer neutron–activation analysis [Yenisherlova and Marounina (1984)], phosphorus diffusing to the back side of the silicon wafers with the concentrations at which a misfit dislocation grid is formed decreases significantly such impurities as copper, sodium and gold lying in the presurface layers near the front side. If no misfit dislocations are formed, then the concentration of impurities (except copper) near the front side undergoes slight changes. Alongside with this, gettering by diffusive layers can increase a concentration of metallic impurities in the bulk of slices. This phenomenon is thought [Yenisherlova and Marounina (1984)] to relate to a fall of the diffusing rate of the metallic impurities in the phosphorus–doped regions of silicon caused by the compressing stresses [Sprokel and Fairfield (1965)] in these regions. It should be also noted that possible sources of metallic pollutions in ready wafers were analyzed in [Yenisherlova and Marounina (1984)]. A chemical–mechanical polishing has been found to be a main source of pollutions; during this operation the metallic impurities are pinning to the presurface layers of silicon. Upon oxidation in the stream of dry oxygen, the concentration of some metals (sodium, copper, gold) may increase a concentration of the basic dopant in silicon, and to drop these metals becomes possible via gettering only.

3.1.4
Accelerated–ion implantation

This technique rests on the idea of creating structurally–damaged layers in semiconducting wafers through irradiating them by high ionic doses of energies up to $300 - 500 \ keV$ and further annealing them. Ions may be implanted both into a back and a front side of the slice – through special masks in definite local regions between the elements of the device to be made. Irradiation doses and densities of the ionic current are usually set so that during a postimplantation annealing one could build a stable defective structure like a large accumulation of dislocation loops or grids of dislocations. As of to–day, there is a rich convincing empirical knowledge and there have been developed theoretical models for defect formation during ionic implantation; this kind of implantation is thought to be one of the promising techniques in semiconductor doping; when trying to build an efficient getter this makes it easier to choose more efficient regimes and conditions for irradiation and further thermal treatments. Here, it should be certainly kept in mind that whereas during doping these irradiation–induced defects represent a negative aspect of implantation, then during gettering processes they perform a basic generating function.

Gettering the metallic impurities in silicon by ionically–damaged layers was seemingly first demonstrated in [Buck and Pickar, et al. (1972)]. Here, it was shown that upon irradiating the crystals by silicon ions of the energy $100 \ keV$, by the dose $10^{16} \ cm^{-2}$, and upon annealing at $1173 \ K$ in the back–scattered spectrums of helium ions one could detect in the getter an increase of such impurities as iron, cobalt, nickel and copper.

A gettering capability of structural distortions born by annealing depends on the spectrum and the concentration of irradiation–induced defects, and, consequently, on the type, mass, energy, dose and the density of the flux of the incorporated ions. A structure of the formed getter is also determined by a crystallographical orientation of the slice's irradiated surface, the temperature and duration of the postimplantation annealing and its ambient, and also by the technological history of the crystal, i.e. by a set of physicochemical treatments of the surface prior to irradiation. It was proved in [Nemtzev and Peckarev, et al. (1981); Labounov and Baranov, et al. (1983); Bogatch and Gousev, et al. (1981)] that silicon can be gettered by the ions of neutral (Si, N, O, S) and doped (B, P, As, Sb) impurities, the ions of inertial gases (Ar, Ne, Kr, Xe) and molecular ions ($AlCl_2^+$, $AlBr_2^+$, BF_2^+). Of great importance is also the temperature of the wafer during the irradiation period, since a growth of temperature brings a growth of annealing the simplest initial irradiation defects, and this decreases the probability of forming metastable aggregates able to establish a stable dislocation structure in the getter in the course of the later anneal.

The studies of the comparative efficiency of gettering by different ions of impurities in silicon have shown [Labounov and Baranov, et al. (1983); Verkhovsky (1981); Bogatch and Gousev, et al. (1981)] that under other equal conditions the efficiency grows in the sequence $Ar > O > P > Si > B > As$, in accordance with the magnitude of residual structural damages. The results of this study [Seidel and Meek, et al. (1975)] are presented in Table 3.3 where there is also the ratio C (relative units) between the concentration of the gold atoms being gettered by ionically–damaged layers and the concentration of gold being gettered by the diffusive phosphorus layers at $1273\ K$; also the coefficients of silicon lattice compression (expansion) by the incorporated atoms are compared.

Table 3.3 shows that the sizes of ions are not enough to explain the deference between the gettering properties of damaged layers. For this, one has to take into account the entire group of the above factors affecting the structure of the gettering layer. Also should be taken into consideration a possible chemical interaction of the incorporated impurity with the target atoms when this impurity undergoes implantation and annealing. The last sooner occurs in the case of oxygen and silicon when the incorporated ions together with the irradiation–induced defects create dispersive inclusions of SiO_X additionally distorting the lattice and thus increasing a gettering efficiency (Table 3.3).

The typical defects being observed upon annealing in the implanted layers are full and partial dislocation loops, dislocation segments and dipoles and point defect clusters as well. The impulse annealing makes small–size defects dominant; they grow in sizes as the length of the thermal treatment increases. As a rule, upon a high–dose irradiation even a long–time annealing is not sufficient to obtain the structures free from dislocations and dislocation loops.

Table 3.3

Comparative efficiency of gettering
Au and Si by different ions

Ion	Degree of lattice disordering (%)	C (conv. u.)	$\gamma = \dfrac{r_{ion} - r_{Si}}{r_{Si}}$
Argon (Ar)	100	1.20	+ 0.3
Oxygen (O)	60	1.15	– 0.44
Phosphorus (P)	82	0.67	– 0.06
Silicon (Si)	28	0.25	0.00
Boron (B)	10	0.15	– 0.25
Arsenic (As)	5 – 10	0.12	+ 0.01

If upon the implantation, a thermal treatment in the oxidation ambient or an additional diffusion of impurities is arranged, then in the irradiated layer there appear the stacking faults and impurity–defect precipitates. The studies of influence of the type of defects in the irradiated layer on their gettering capabilities have revealed [Labounov and Baranov, et al. (1983); Bogatch and Gousev, et al. (1981); Geipl and Tice (1977)] that the dislocations with the

Bûrgers vector $\overline{b} = \dfrac{a}{2} < 110 >$ become the most powerful drains for point

defects. With a less efficiency the point defects are gettered by the dislocation

loops with $\overline{b} = \dfrac{\alpha}{3} < 110 >$ – this is explained by a less amplitude of elastic

stresses around partial dislocations. Alongside with this, a gettering action is not only the property of the dislocation layer but the entire presurface zone within which the elastic stresses induced by postannealing defects are located. Upon annealing, the dislocation density increases with an increase of irradiation dose, and it seems that the gettering capability of ion–implanted layers should be expected to grow. However, in reality this capability is nonmonotonously dependent on the dose, because the latter's increase is usually accompanied with precipitation and cluster formation of the introduced impurity around the dislocations – this diminishes the elastic stresses and, as a result, weakens the gettering properties of the damaged layers [Geipl and Tice (1977)]. The irradiation dose needed for the efficient gettering can be reduced, if the ions are implanted through thin SiO_2 layers , for in this case a damaged layer will be formed both by the ions and the silicon and oxygen recoil atoms taken from the film [Labounov and Baranov, et al. (1983)]. On the example with gold in [Sigson and Cspredi, et al. (1976)], the efficiency of gettering by an irradiated $\{111\}$–surface was shown to be by one order higher than with the $\{001\}$–orientation – this is explained by a more higher concentration of irradiation–induced distortions being incorporated into the crystal when irradiating the planes of a large reticular density.

Gettering by ion–damaged layers on the back side of silicon slices increases by several orders the lifetime of minor charge carriers, diminishes the defect concentration in the further thermal operations (epitaxy, diffusion, oxidation), diminishes the leakage currents of the $p - n -$ junctions, increases breakthrough voltages and, finally, a yield of fit devices [Nemtzev and Peckarev, et al. (1981); Labounov and Baranov, et al. (1983); Rozgonyi and Kyshner (1976); Yenisherlova and Alekhin, et al. (1991)]. The efficiency of this gettering technique was well demonstrated by the results gained in [Yenisherlova and Alekhin, et al. (1991)] where there has been studied an influence of ion–implanted layers in the wafer on the structurally sensitive characteristics of epitaxially–grown silicon structures: the films 3 mcm thick of $n - Si\,(001)\,CZ$ ($\rho = 0.7\ Ohm \cdot cm$, phosphorus–doped) and on the wafers of $p - Si\,(001)\,CZ$ ($\rho = 0.01\ Ohm \cdot cm$, boron–doped). Before being grown epitaxially ($T = 1473\ K$), the wafers were irradiated both separately and in series by carbon and oxygen ions of the energy $300 keV$, by doses $6 \cdot 10^{14} - 1.8 \cdot 10^{15}\ cm^{-2}$. Upon the epitaxial process, the film parameters were examined by the method of volt–farad characteristics, translucent electronic microscopy and by metallography. Ion implantation into the wafer was found to make it possible to build epitaxial layers free of the defects peculiar to the controlled nonirradiated specimens. In all the specimens obtained by preliminary ionic irradiation, the charge carriers across the entire thickness of the films had a practically uniform concentration profile. In [Yenisherlova and Alekhin, et al. (1991)] these results are explained by the sequential implantation of carbon and oxygen ions that allows to obtain the most optimal dislocation structure for the getter. This structure, on the one hand, manifests its gettering properties (in particular, its ability to hamper the travelling of active trap–centres from the wafer into the growing film), and, on the other hand, this structure does not immediately disturb the surface of the interface, as it occurs when only carbon ions are implanted.

It should be noted here that even during relatively low–temperature annealing ($\approx 773K$) the gettering ability of the ion–implanted layers built at high irradiation doses is possible to be strengthened by additional inhomogeneous elastic stresses that can initiate the films or the deformation of structures in the devices of 3– or 4–point bending. Depending on the sign and magnitude of additional mechanical stresses, such getter makes it possible to regulate in the bulk of the crystal the density and profile of distributions of such stable structural distortions as dislocations [Pavlov and Skoupov, et al. (1987)]. That the dislocation structure is recombined (and, moreover, it occurs at large distances from the irradiated surface (the "long–range effect")) indicates that gettering by ion–implanted layers relates not only to the gradients of static mechanical stresses but is also stimulated by elastic waves arising during irradiation–induced distortions and their transformations in the later annealing process.

Earlier it was noted that gettering the undesired impurities and other defects by ionic irradiation can be exercised via incorporating the ions both from the back and front side of the slices, thus forming the getter between the active elements and sometimes in them directly. This way was employed, say, in creating electrically inactive dislocation loops in the regions of the base or emitter of bipolar transistors, through incorporating low–energy ions of inertial gases. These dislocation loops work as getters during the formation of the device and when it runs [Labounov and Baranov, et al. (1983)]. The front–sided irradiation and structures' annealing are also of practical interest, because this case can give rise to an effect of planar gettering; this effect, when combined with a long–range effect of the implanted layers, helps to clean numerous active regions and elements of devices distorted by technological processes [Litovtchenko and Romanyuk (1983); Litovtchenko and Romanyuk, et al. (1986); Stroukov and Khromov, et al. (1992)].

Building a front–sided getter on a semiconducting slice helps to diminish the mechanical stresses in the structures, reduce the gettering time and overlap the gettering operation and the basic device manufacturing technology. The planar gettering techniques are oriented on exploiting the unordered interphase boundaries (of the type $Si - SiO_2$, $Si - Si_3N_4$) [Litovtchenko and Romanyuk (1983)] and establishing the conditions for accelerated diffusion of the point defects from the surface into the region of the getter – this may be implemented in growing the thin films. In this case, a considerable square of the surface is cleaned and the conditions of growing the films epitaxially change drastically. The latter has been shown in [Litovtchenko and Romanyuk, et al. (1986)] where there have been studied the effects of the local ionically–implanted getter on the epitaxial growth of silicon films. In [Litovtchenko and Romanyuk, et al. (1986)], the crystals of $p - Si(111) CZ (\rho = 10\ Ohm \cdot cm$, boron–doped) were used as wafers. In these crystals, the local gettering regions were formed by implanting through masks a wide range of the Ar^+ ions of $E = 50\ keV$ and of a wide range of doses; the crystals were then annealed at $873 - 1373\ K$ in dry nitrogen. Then, on this surface the layers $1.5 - 12\ mcm$ thick, with specific resistance of $0.2 - 0.8\ Ohm \cdot cm$, were epitaxially grown by the chloride technique. The structure of the layers was later examined with the help of electronic microscopy, X-ray topography and metallography.

The implanted regions were found in [Litovtchenko and Romanyuk, et al. (1986)] to be subjected to considerable shift stresses within the inclined and parallel $\{111\}$–planes, which substantially affect the defect formation in the further annealing operations. Near the edges of the implanted zones there are observed the regions (up to 40 mcm in width) of highly concentrated defects, and this betrays an impact of the mechanical stresses near the interface boundary on the kinetics of annealing the radiative distortions. With the decrease of the ionic current density, the defect concentration increases. A growth of the irradiation dose was revealed to increase the size of point defects and drop the amount of small clusters. Irradiating the surface with mechanical disturbances (scratches, microspalls) and a later annealing lead to their healing and create a small–fragment defect structure. Upon growing the layers epitaxially, it was found that the film sections into which the wafer had been implanted contain a large number of stacking faults, lying in the $\{111\}$ planes being inclined and parallel to the surface, and numerous dislocation loops and tetragonal stacking faults. Their concentration depends on the irradiating conditions and film thickness. No stacking faults and dislocations have been detected in the wafers on the pure (nonimplanted) sections of the film. Upon oxidation, the number of stacking faults in the implanted sections has decreased, and in the pure sections the density of OSFs made up $1-5\,cm^{-2}$. With the film thickness increasing up to 10 mcm, there are observed shallow etching wells with the concentration of up to $10^5\,cm^{-2}$ on the surface of pure sections. Round the gettering regions; defect–free zones up to 50 mcm wide were detected. As was detected through electronic microscopy, gettering provides a defect–free growth of the films in the sections of size 5 x 5 mm. Electrophysical parameters of such pure sections turn out to be much better that those in the getter–free films.

According to [Litovtchenko and Romanyuk, et al. (1986)], the epitaxial film on the wafer with locally–implanted sections is grown with the following specificities:

a)   the density of the nucleating centres is substantially different on the irradiated and pure sections and this change the growth conditions for the films at initial stages; and namely, a homogeneus growth on pure sections is provided in the undersaturation conditions;

b)   the epitaxial film on the implanted sections contains highly–concentrated epitaxial stacking faults and dislocation loops (up to $10^6\,cm^{-2}$), and, as a result, these sections serve as getters for the impurities and IP defects; hence, there appears a planar gettering.

During the epitaxial growth, a gettering process itself is thought to run as follows. In the course of the preepitaxial annealing, the compressing and extending stresses help to clean the nonimplanted sections; this process also occurs at the expense of the flux of the point defect flux running perpendicular to the surface and into the depth of the crystal (into the extending region) and the planar flux to the arising dislocations in the gettering region. Initially, the epitaxial growth on the pure sections runs in the unbalanced conditions, due to the outflow of adatoms into the gettering regions. The nuclei are first formed on the implanted sections, and this results in a delay of crystallization in the pure sections; and meanwhile the oxidation–induced defects manage to dissorb from the surface. After the ionic dissorption of oxidites and gettering the impurity disturbances there starts a structurally–perfect growth of the film in the pure regions.

When a foreign atom gets to a pure film section it easily migrates to the gettering region, i.e. to the position where it is of minimal energy and near a dislocation or a stacking fault. A migration length can be as long as some millimeters, due to large coefficients of point defect diffusion across the film surface. The unbalanced IP defects, which during the cooling operation oversaturate the lattice and create the clusters, also manage – for the time when temperature is cooled down to a room temperature – to escape into the depth of the crystal and into the gettering regions. With the film getting thicker (more than 7 mcm), the perpendicular–to–surface flux damps significantly, and the unbalanced IP defects obtain a possibility to drain into the getter only. Therefore, around the gettering sections there are observed the cluster–free regions. It is worthy of mentioning here that under other equal conditions the planar gettering may turn out to be more efficient in the rate and the degree of the crystal cleaning, as compared against the bulk gettering, since the IP defects and impurities interact with getter presurface dislocations with much greater force, and, as shown in [Kosevitch (1981)], the energy of such interaction becomes significantly less dependent on the interdefect distance as we approach the surface.

The planar gettering manifests its efficiency not only in epitaxy but also during a high–temperature oxidation of silicon, with the ion–implanted getters heterogeneously spreading across the surfaces of the sections. This is shown in [Stroukov and Khromov, et al. (1992)] where the specificity of the planar gettering in the situation, when the $n-$ type silicon is locally irradiated by argon ions of $E = 40\,keV$ and by doses $\Phi = 10^{14} - 10^{16}\,cm^{-2}$ with a later oxidation at $1423\,K$ in dry oxygen, was studied through measuring the relaxation time of unbalanced capacity of the controlled MOS-structures and by electrographic techniques. According to the authors, the planar gettering is most efficient and long ranged, if the structure of the implanted layer represents itself a nonoriented polycrystalline phase (the dose of $\approx 10^{15}\,cm^{-2}$). An ordering arising in polycrystals, i.e. a texture, is accompanied with a drop of long–range effects of gettering. The gettering region can spread along the surface as far as 25 – 30 mm. Here, close to the boundary of the implanted area there is formed a layer less than 3 mm wide where there has been detected a visible degradation of the crystal's electrophysical parameters. Such layer was born by a specificity of elastic stresses and by a defective structure of the materials having the locally doped sections [Kontzevoy and Litvinov, et al. (1982); Lyubov (1985); Pavlov and Skoupov, et al. (1987)]. In exploiting a local ionic irradiation, searching for the ways to kill or suppress the negative effects of the transient layers on the semiconducting structures constitutes itself an urgent task for the technology of planar gettering.

For gettering purposes, ions can be implanted not only into silicon but also into other semiconducting materials, homogeneous and heterogeneous structures. For example, irradiating a back side of the $GaAs$ wafers and a succeeding annealing operation make it possible to obtain more perfect epitaxial layers with a uniform in-depth profile of charge carrier distribution and few crystallographical defects. It has been also revealed that upon the annealing at $1273\,K$, the oxygen ion implantation (with $E = 280$ and $350\,keV$ and by doses of 10^{16} and $(1-3)\cdot 10^{16}\,cm^{-2}$, respectively) makes better the structure and electrophysical parameters of the device silicon layers in the SOS-structures. The ions implanted are supposed to create not only a damaged layer of the getter but also interact with excessive aluminium atoms from the sapphire wafer preventing their penetration into the silicon film and thereby suppressing in it the uncontrollable changes of charge carrier concentrations [Labounov and Baranov, et al. (1983)]. It should be also said here that in contrast to silicon, the ionic gettering of other materials is so far under study, though it has real perspectives of application in manufacturing the devices.

3.1.5
Effects of laser irradiation

A substance being subjected to laser irradiation of sufficiently high power practically changes all its physicochemical characteristics; it becomes changed to the extent that a new aggregate composition appears as the boiling or evaporating temperature is reached. In semiconductors, absorbing laser light in accordance with a wave length can be defined by one or several mechanisms [Vendik and Gorin, et al. (1984)]: by fundamental absorption, impurity absorption, lattice absorption, by absorption by free charge carriers or by exciton absorption. During technological treatment, i.e. when performing a modified treatment of the materials by laser light, one usually employs a channel of fundamental absorption that is accompanied with impurity and exciton mechanisms. The photoelectrons arising in this case are unbalanced and during the time $10^{-10} - 10^{-9}$ they first drop to the bottom of the conductivity zone and then get through recombination centres to a valency zone. Here, a crystal lattice obtains the energy being a sum of energies – the energy of photoelectron relaxation in the conductivity zone and the energy of emission–free recombination (if no photoluminescence is considered). In essence, in the event of a powerful laser action all the photonic energy is converted to heat. With the falling photonic flux being of high density during the average efficient lifetime $10^{-9} - 10^{-8}\, s$ of photoelectrons in the conductivity zone, their concentration can become equal to $10^{21}\ cm^{-3}$, i.e. in this case a semiconductor metallization occurs. This case creates a supplementary channel for heating – absorbing the radiation by free charge carriers.

If the absorbing coefficient α for metals and semiconductors is known, then an in-depth variance of absorbed energy in the radiated material is possible to be found. If a irradiation power density on the surface is $q_0 A$, where q_0 is a laser power density and A is a portion of the power absorbed by the material, then the density of the power being absorbed in the depth X will vary as follows:

$$q_{satur} = q_0 A \exp(-\alpha X) ,$$

where

$$\alpha = \sqrt{2\omega\mu_0\sigma_{el}} \ ;$$

ω is an irradiation frequency;

σ_{el} is a material conductivity; and

μ_0 is an absolute magnetic permittivity.

The depth at which there occurs a maximum energy discharge is equal to the depth of the skin–layer $\delta = \dfrac{2}{\alpha}$ or $\delta = \sqrt{\dfrac{2}{\omega\mu_0\sigma_{el}}}$.

Since the energy portion absorbed by the metal is $A = \sqrt{\dfrac{8\omega\varepsilon_0}{\sigma_{el}}}$ (here ε_0 is a vacuum dielectric penetration), α and δ can be rewritten in the form

$$\alpha = \frac{8\pi}{\lambda_0 A} \quad \text{and} \quad \delta = \frac{\lambda_0 A}{4\pi} \ ,$$

where λ_0 is a wave length for laser irradiation in vacuum. In the point of the phase transfer the depth of the skin–layer may increase. For metals, the absorbing coefficient is usually equal to $\alpha = 10^6 - 10^7\, cm^{-1}$, and for semiconductors it is $10^4 - 10^5\, cm^{-1}$.

Irradiation yields a heating of the material surface; the surface temperature may be calculated as

$$T_{surf} = \frac{2Aq_0\sqrt{\alpha\tau}}{\chi\sqrt{\pi}} \ ,$$

where a and χ are the temperature conductivity and heat conductivity, respectively; and τ is a length of laser impulse action.

The laser power density corresponding to a start of material melting is determined from the condition $T_{surf} = T_{sq}$ and it will be equal to

$$q_{0cr} = \frac{T_{sq}\chi\sqrt{\pi}}{2A_{aver}\sqrt{a\tau}} \ .$$

Under laser irradiation, the material's presurface layers will be heated at the below rate:

$$\frac{dT}{d\chi} = -\frac{Aq_0}{\chi\sqrt{\pi}}\sqrt{\frac{a}{\tau}} \ ,$$

The lower is the material conductivity, the more will be the heating rate, since heat fails to fast drive away from the surface and this yields a gradient of temperatures:

$$\frac{dT}{d\chi} = -\frac{Aq_0}{\chi} \, erfc \, \frac{\chi}{2\sqrt{a\tau}}$$

that may achieve $10^5 - 10^6 \deg rees / cm$. The high heating rates and the thermal gradients give birth to strong mechanical stresses in materials. In general, to calculate such stresses is not so easy and usually this problem is numerically solved for some specific situation [Mil'vidsky and Osvensky (1984); Vendik and Gorin, et al. (1984)]. As a first approximation, the stresses may be estimated by the formula

$$\sigma \approx \frac{\alpha E}{1-v} \sqrt{a\tau} \cdot \frac{dT}{d\chi} \,.$$

All the above–listed factors exhibit that the laser irradiation provides high heating rates and thermal gradients, and also a selectivity and locally–made actions that induce an inhomogeneous dynamical field of mechanical stresses – all these processes do not only positively change the properties or parameters in a piece of work but can also generate various structural defects arising within a presurface zone of the initial energy discharge. In [Banishev and Novikova (1992)], when studying the silicon irradiated by millisecond laser impulses (with a use of aluminoyttrium garnet (AG) of $E = 3 \, J$ and $\lambda = 1.06 \, mcm$) the authors have found the following regularities in the defect generation:

•   the laser impulsive irradiation of the density $8.5 \, J / cm^2$ gives rise to local melting regions on the surface of crystal;

•   there is a threshold value of the energy density ($\approx 7.5 \, J / cm^2$), below which no defects are induced by irradiation and above which linear defects arise being oriented along the crystallographical directions;

•   the arising defects can be reversible, i.e. can vanish upon a cease of irradiation, or nonreversible;

•   the most probable type of defects generated by the laser over–threshold irradiation may be the dislocations being generated prior to a start of local melting; and a melting process itself is running on them.

That structural defects and relief on the surface of the silicon crystals were induced by impulsive irradiation from a ruby laser (the energy density varied from $20 \, to \, 130 \, J / cm^2$) was also noted in [Boushouev and Petrakov (1992)]. And in this case, the basic type of defects generated by impulsive thermal stresses were dislocations and their aggregates. Therefore, the laser irradiated structure in the presurface layers in its composition looks similar to the gettering layers obtained by mechanical treatment or ion implantation and annealing. Hence, the idea to exploit a laser irradiation for constructing the gettering layers in the semiconducting slices seems quite natural.

One of the first techniques of laser gettering was suggested in [US pat. No. 4131487]. The scheme of the apparatus is given in Fig. 3.2.

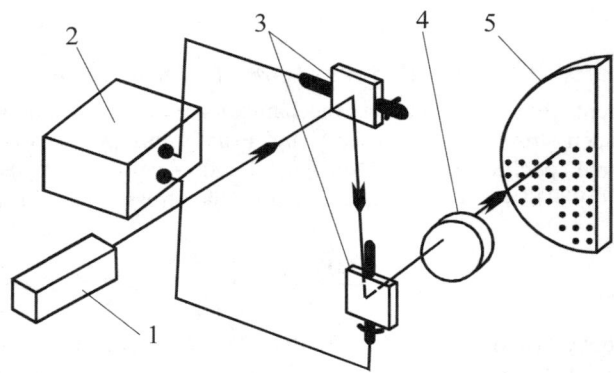

Fig. 3.2. The scheme of apparatus for creating a damaged layer by laser irradiation; **1** – a laser; **2** – a display generator; **3** – mirrors; **4** – a lens; **5** – a semiconducting slice.

The idea of this technique lies in the following. A focused ray of the impulsive laser is scanning a back side of the semiconducting slice and creates rows of craters (hollows) 5 – 15 mcm in depth and 40 – 60 mcm in diameter. A partial melting and recrystallization in unbalanced conditions make the morphology in a crater different from the morphology of the nonirradiated silicon. The degree of defectiveness of the slice is determined by the sizes and the number of craters. During the further thermal treatment ($T = 973 - 1573\,K$), near the irradiated regions there appears a grid of dislocations decorated by impurities. The silicon wafers are usually irradiated by the AG Nd-laser of the output power $8\,W$. The dislocation density and the depth in the damaged layer decrease with a drop of laser frequency (varying from 10 to 20 kHz), with a decrease of the beam diameter (60 – 45 mcm) and with an increase of the craters overlap (from 0 to 70 per cent of the square). In practice, heavy metals in silicon can be also gettered by a ruby laser ($\lambda = 0.684\,mcm$, of the energy density $1 - 10\,J/cm^2$, impulse length $10 - 15\,ms$) [Germany pat. No. 2829983].

Optimal irradiating regimes for the AG Nd-laser to form an efficient getter on silicon may be chosen on the basis of information concerning the structural transformations in the material being irradiated; the results of these transformations were classified in [Labounov and Baranov, et al. (1983)]. Here, it was revealed that with the radiation density being less than $7\ J/cm^2$ there occurs a melting in the local regions followed with epitaxial recrystallization. Here, the residual defect density in a damaged layer is small and no gettering occurs practically. With the irradiation density exceeding $7\ J/cm^2$, silicon is being evaporated (in parallel to melting) and a polycrystalline structure emerges in the central region, whereas in the periphery of the melted region there arise dislocations and stacking faults. With the energy density being less than $10\ J/cm^2$, the dislocation aggregates get unstable, and only the laser irradiation density exceeding $15\ J/cm^2$ gives rise to stable fan–like dislocation lines and clusters (Fig. 3.3). A further growth in the energy density increases a damaged layer depth and may cause a spreading of the dopants until a front side of slices,

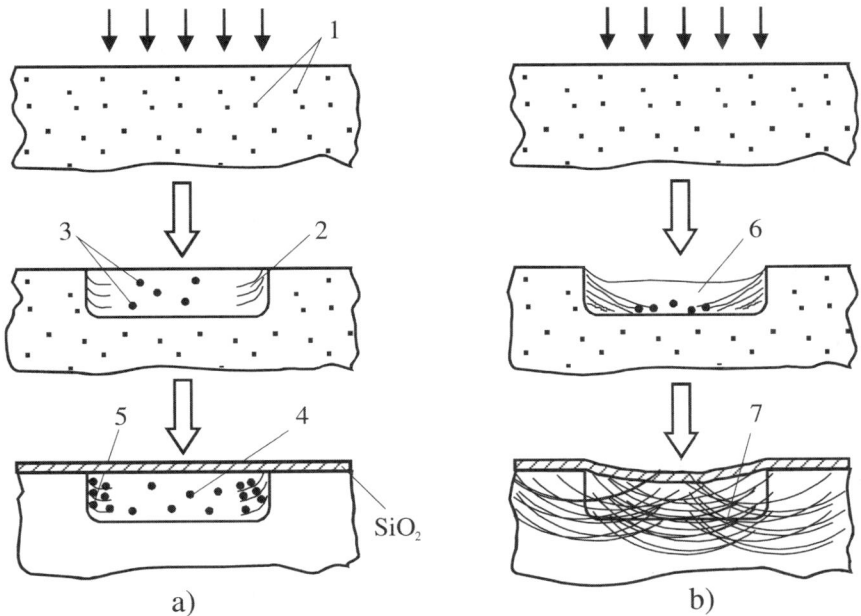

a)   b)

Fig. 3.3. The structure of the damaged layer in the presurface region of silicon during a laser irradiation with a density of up to **a** $15\ J/cm^2$ and **b** more than $20\ J/cm^2$; **1** – impurities; **2** – dislocations; **3** – swirl–defects; **4** – precipitates; **5** – precipitates on dislocations; **6** – a stressed region; **7** – a dislocation grid.

i.e. until the active regions of the semiconducting devices. That is why a laser irradiation is usually performed with the irradiation density being

$9-20\ J/cm^2$ – this yields a damaged layer of silicon $5-10\ mcm$ deep. In addition, the lifetime of minor charge carriers will increase significantly, the stacking faults will vanish practically and the yield of fit devices increases.

Cutting the slices by laser results in a gettering layer being formed along the perimeter of the chip; in the further thermal treatments this layer suppresses a conservative and unconservative growth of the edge defective region. This effect was detected in cutting the slices of silicon and gallium arsenide [Lourye and Sorokina, et al. (1991)]. The latter also betrays a possibility of applying the laser gettering to semiconducting compounds. This possibility was directly proved in [Bobitsky and Bertcha, et al. (1992)] where the effects of argon laser irradiation of $\approx 1\ W/cm^2$ and nitrogen laser (this laser was used to create a surface diffraction lattice) on intrinsic defect composition of silicon were studied. By the photoluminescent technique and by the destroyed full internal reflection, the laser irradiation was revealed to getter impurities into a irradiation release region. This gettering process, and this is important to note, was detected throughout the entire thickness of the specimens (up to 300 mcm). A surface phase lattice built through scanning by a powerful nitrogen laser also serves as an efficient getter; in this case, there are gettered into this region not only the impurities but the intrinsic structural defects as well.

The laser gettering techniques can be exploited in a combination with other technologies. For example, the laser irradiation of silicon, as shown in [Labounov and Baranov, et al. (1983)], may be combined with the annealing in the chlorine ambient. In this case, the laser irradiation will stimulate a gettering of not only alkaline and transient metals (as in a usual chlorine oxidation) but also gold and metals of platinum group as well. Thanks to this, a pollution of silicon by gold drops down to the concentrations close to a sensitivity limit of the neutron–activation analysis.

3.1.6
Gettering by porous silicon layers

According to its physicochemical properties, porous silicon (PS) is a unique material being obtained through electrochemically processing the monocrystalline silicon in water–alchohol solutions of phtorohydrogen acid [Labounov and Baranov, et al. (1978); Nikolaev and Nemirovsky (1989)]. As to its structure, porous silicon is a large set of zig–zag wells (pores) primarily oriented to the silicon wafer surface and separated by monocrystalline regions. In the cross–section, the layers of porous silicon involve a thin presurface region of amorphous silicon and a region of the porous material itself. Porous silicon is formed by chemicoelectrical reactions whose running needs the mobile carriers of positive charge (holes) in the presurface layer of the silicon anode and negatively–charged ions of fluorine in the reaction region. For the $p-$ type silicon, such conditions are met easily and for the $n-$ type with difficulties. Therefore, the pore generation kinetics for these materials is specific. The parameters of porous silicon layers (thickness, porousness or density in the material of the layer) are, mainly, linearly dependent on the $p-$type silicon anodizing regimes. The dependences of the basic parameters of $n-$type porous silicon differ not only from the same dependences for the $p-$type silicon but do not coincide in slightly and strongly doped crystals. Therefore, during the anodization of the $n-$type silicon there become very essential such aspects as a concentration of the donor in the original material, an illumination intensity for a reaction surface providing a hole generation process, and a concentration of hydrogen fluoride in electrolyte. At present, there are many technologies to build porous silicon layers on silicon of different types and values of conductivity; they are described in [Labounov and Baranov, et al. (1978); Nikolaev and Nemirovsky (1989)] and a wide choice of references is given in these reviews.

The analysis of the porous silicon structure shows that an anodization process slightly changes the crystalline lattice in the original silicon, in particular, a period of the porous silicon lattice increases not much as compared against the monocrystalline lattice period, though the layer is penetrated with a grid of pores whose sizes and configuration are determined by a type of conductivity and a level of doping the original material, the anodizing regimes, and by a presence of additional aspects stimulating the electrochemical reactions. Thanks to pores – microchannels, porous silicon has a substantially larger specific surface than the original monocrystalline silicon. This results in a higher chemical activity of porous silicon, and as a result of this, it reacts very well with alkali solutions at room temperature – this is very important for solving numerous technological problems employing porous silicon. From the technological point of view, very interesting is also a behaviour of porous silicon in response to high–temperature actions, and mostly a change in the specific surface of layers during annealing. The results of this study were suggested in [Nikolaev and Nemirovsky (1989); Bondarenko and Dorofeev, et al. (1985)] where a high–temperature annealing ($1273-1472\ K$) in the neutral ambient is shown to drop a specific surface of porous silicon more than by an order of magnitude. It is supposed that at the initial stage of porous silicon sintering there occurs a "shut-up" of pores–channels to form closed cavities being isolated from the crystal surface and each other. A later thermal treatment results in a by-vacancy growth of big pores due to a dissolution of smaller pores; near these pores the vacancy concentration is largely higher. For the pores lying near a free surface, together with their sizes increasing, a significant role also plays a competing process of dissolution; it occurs thanks to a withdrawal of vacancies from the pores to the external surface. This results in large, almost spherical, cavities being formed in the depth of the porous silicon layers; the cavities diminish the sizes as soon as they approach the surface of the structure.

The above information concerning porous silicon allows to consider it to be an efficient getter which has a sufficient and structurally–developed surface for draining the impurity atoms whose diffusion is accelerated by free vacancies induced by high–temperature treatment. In their essence, the layers of porous silicon, when treated thermally, should strengthen the impurity aggregation process in the presurface position; these aggregations are usually explained by a high density of unsaturated surface interatomic bonds [Labounov and Baranov, et al. (1983)]. This very technique of gettering by porous silicon layers was suggested in [US pat. No. 3929529], where a gettering process consists of four basic operations (Fig. 3.4):

- forming via anodizing in the hydrofluoric acid (HF) solution a porous silicon layer on a back side of wafers;
- performing a high–temperature gettering annealing;
- oxidizing the porous silicon; this operation may be overlapped with the second operation;
- removing in the hydrofluoric acid the porous silicon layer saturated with impurities.

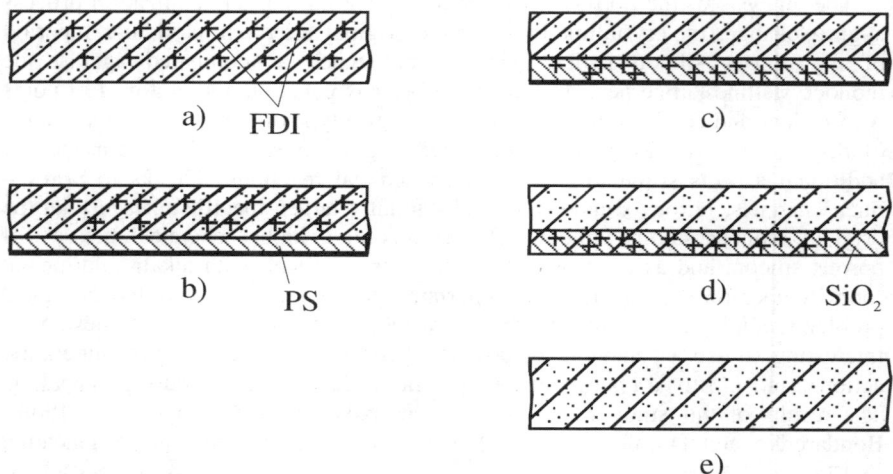

Fig. 3.4. Gettering the fast diffusing impurities (FDI) by a continuous layer of porous silicon: **a** original silicon film; **b** porous silicon forming; **c** high–temperature annealing; **d** porous silicon oxidizing; **e** removing the oxidized porous silicon.

Gettering by porous silicon layers can be done both prior to epitaxially growing the silicon films on a front side of the wafers and also upon completing this process, when a porous layer is built on a back side. In order to strengthen the gettering properties in porous silicon, into this layer, before annealing, may be diffused or implanted boron or phosphorus; these implants being of concentrations close to the dissolubility limit. Through the porous silicon, these implants are diffusing easily thus creating (in a monocrystalline wafer) heavily–doped layers of $n-$ and $p-$type conductivity, with an increased density of misfit dislocations. During the later annealing in the neutral ambient and oxidation, a gettering effect is gained both at the expense of a developed surface of porous silicon and due to impurity saturation by the misfit dislocations [Labounov and Baranov, et al. (1983)]. A gettering annealing may be performed via heating the structures by photonic irradiation of sufficient power – this allows to shorten an irradiation length [Bondarenko and Yakovtzeva, et al. (1984].

Porous silicon may be used to locally getter the regions near a front side of wafers, where during the subsequent technological operations there will be formed the active elements of the discrete devices or IC components. One of the variants of such technique [Goetsberger and Shockley (1960)] is shown in Fig. 3.5. Its idea lies in the following. Prior to growing a silicon film epitaxially, in the wafer there is built up a porous silicon layer, through a front–sided Si_3N_4 mask. The topology of this layer corresponds to the configuration of the isolating regions. Upon anodizing, a gettering annealing in the inertial ambient is arranged, to remove undesired impurities from the monocrystalline regions between porous silicon. Further, porous silicon is oxidized and the silicon epitaxial film is deposited; over the gettering sections in the wafer this film has a monocrystalline structure and a polycrystalline structure over the oxidized porous silicon. This is already a ready structure for constructing an integrated circuit. The specificity of this variant lies in constructing the gettering regions of porous silicon immediately close to the active elements of the devices and that makes it possible to substantially diminish the cleaning operation length and reduce the annealing temperature.

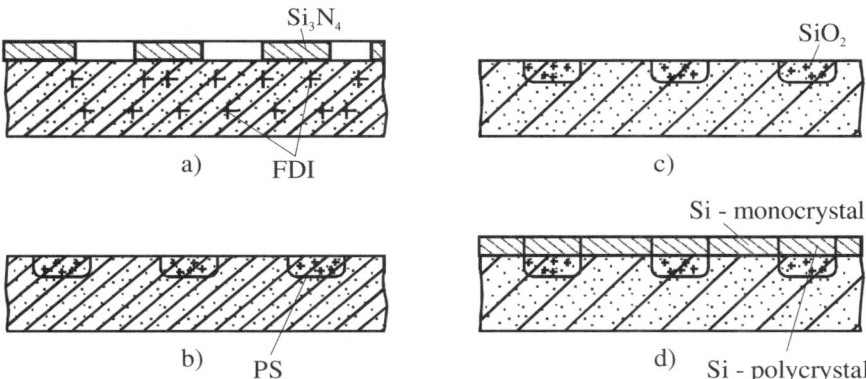

Fig. 3.5. The stages in the technological process for creating epitaxial layers through a use of local gettering regions of porous silicon: **a** creating a nitride mask; **b** forming porous silicon and annealing in the inertial ambient; **c** oxidizing porous silicon; **d** building up porous silicon epitaxially.

It is to be noted in conclusion that in exploiting the technology of gettering by porous silicon layers one should bear in mind a highly probable uncontrollable contamination by undesired external impurities which will be then diffusing into the front regions of the crystal [Nikolaev and Nemirovsky (1989)]. Therefore, when dealing with the structures with porous silicon layers one should thoroughly obey the electronic hygienic requirements at the technological places and try to keep these structures as short as possible between the porous silicon formation and the gettering anneal.

3.2
Gettering by heterophase layers

This subsection describes a group of gettering techniques exploiting the interactions between impurity atoms and the heterophase films deposited onto the silicon surface. As such gettering films there are used the metallic layers, impurity–silicate glasses, silicides of hard–melting metals, silicon nitride and polycrystalline silicon. High–temperature gettering by such layers is possible due to different solubility of impurities in the monocrystalline, chemical or elastic interaction of impurity atoms with the film material. These techniques provide a good cleaning of silicon crystals and other semiconducting materials from fast–diffusing impurities that bring into the prohibited zone the deep levels – the recombination centres responsible for low lifetime of the minor charge carriers [Hu (1973); Ravi (1981)].

3.2.1
Metallic films

Fast–diffusing impurities (FDI) are gettered by the layers of nickel, cobalt, gallium, aluminium, zinc, alloy of tin with lead [Labounov and Baranov, et al. (1983); Verkhovsky (1981)]. In one of the first papers on gettering impurities in silicon [Goetsberger and Shockley (1960)], it was detected that depositing a nickel film and its burn-in at $1273-1373\ K$ increase the lifetime of minor charge carriers. Depositing the nickel or zinc layers onto a surface of wafers with diode structures and subsequently annealing them in hydrogen atmosphere for 1h increased the number of diodes with "sharp" volt–ampere characteristics. It was supposed by the authors that at high temperatures there arises a liquid phase space in the silicon – metallic layer interface; in this space the fast–diffusing impurities accumulate. However, gettering by the molten metals is limited by some concentration level of the extractable impurities; this level relates to a final value of the coefficient for impurity distribution between a hard (silicon) and liquid metallic phase. Besides, the gettering temperature is good for metal atoms (nickel, cobalt, zinc) to be dissolved in silicon to form deep levels serving as centres for recombining charge carriers, i.e. constraining an upper limit of the possible increase of carriers' lifetime. This limit may be diminished through dropping the annealing temperature down to $973-1173\ K$ and raising the annealing length up to dozens of hours [Konakova and Shouman (1970)]. In this case, the solubility of getter's atoms in silicon will drop but the diffusion coefficients of fast diffusing impurities will change little. Such gettering regimes can induce precipitation of fast diffusing impurities, and this will make them electrically inactive.

The eutectic tin – lead alloy has turned out to be a good getter for copper from silicon [Labounov and Baranov, et al. (1983)]. The dissolubility of copper in this alloy at $1273\ K$ makes up $33\ at.\%$, and in monocrystalline silicon $10^{-4}\ at.\ \%$ only. The gettering efficiency was estimated on the silicon wafers especially saturated with copper through processing them in the copper nitrate solution and annealing in the argon atmosphere for 30 min at $1273\ K$. Such treatment makes it possible to have copper deposited on dislocations. Upon gettering by the $Sn - Pb$ eutectic alloy, the dislocation contrast recorded by the IR microscope (the dislocations were decorated by copper atoms) was diminishing, owing to copper being extracted into the melt.

Besides silicon, gettering by metallic layers may be also applied to other semiconducting materials. For example, a zinc melt was suggested [US pat. No. 4244753] to be used for gettering the undesired impurities from the wafers of semiconducting junctions of the type $A^{II}B^{VI}$; this helped to substantially better the electrophysical properties of such crystals as $ZnSe$, CdS, $ZnTe$. A gettering anneal for such materials was arranged at $873 - 1173\ K$ during $0.5 - 16\ hrs$.

3.2.2
Extrinsic silicate glasses

One of the traditional gettering techniques is cleaning the semiconducting wafers by layers of phosphorus – silicate (PS) or boron–silicate (BS) glass [Verkhovsky (1981)]; Goetsberger and Shockley (1960)]. The metallic impurities at high temperatures are supposed to be diffusing from the bulk surface–bound where they chemically interact with oxygen or phosphorus (or boron). A glass coat may be deposited at any stage of the technological line, and a subsequent annealing at $1273 - 1373\ K$ extracts the undesired impurities of heavy metals and helps to produce the material with minor charge carriers of high lifetime. In [Labounov and Baranov, et al. (1983)], depositing (prior to a formation of a subgate dielectric) a PS or BS glass on a back side of the silicon wafer reduced the concentration of copper, nickel or iron in silicon, made better the quality of silicon dioxide and reduced leakage currents in transistors. Gettering the diode silicon structures by a PS glass becomes more efficient when the annealing temperature increases; as a rule, this efficiency turns out to be higher than in the case with the BS glass. If a $p - n -$ junction is formed by a single–stage diffusion, then a gettering process simultaneously occurs. In two–stage diffusion processes or when some high–temperature treatments are available, gettering by PS or BS glass layers (deposited on a back side of the structure) is better to be overlapped with the last of high–temperature operations [Verkhovsky (1981)]. Extrinsic silicate glasses are employed to clean not only silicon but also other semiconductors. For example, in [Labounov and Baranov, et al. (1983)], depositing a PS layer 0.45 mcm from a gaseous phase into a back side of the gallium phosphide and a later annealing at $1173 - 1273 \; K$ were shown to cause a substantial drop in the efficiency in epitaxial films and an increase of the quantum output of light emitting diodes produced on such structures. It should be noted here that the techniques of gettering by glasses (phosphorus – , boron –, lead –, and vanadium–silicate glasses), in contast to gettering by metallic films, have wide capabilities, for the glass depositing technologies have been well developed and are sufficiently productive [Labounov and Baranov, et al. (1983); Ravi (1981)].

Alongside with this, gettering by extrinsic silicate glasses, as well as in the case of porous silicon, should be exercised very carefully, since during annealing such layers are able to capture the undesired impurities from the external technological ambient; these impurities getting into this ambient, say, from a quartz tooling [Meck and Seidel, et al. (1975)].

3.2.3
Layers of polysilicon and silicon nitride

At present, polycrystalline films find a wide application in LSIC technologies for circuit metallization, in the manufacture of the transistor gates, resistors, isolating elements in ICs, and the like. Semisilicon layers may be built up by several ways, from which most often employed is a high–temperature decomposition of silicon carrying vapours (tetrachloride silicon, silane, etc). For gettering purposes, the polysilicon layers are usually deposited onto a back side of the wafers. During annealing, the impurities are captured by the boundaries of silicon grains thus diminishing their free energy. A small size of grains and a sufficiently large thickness of polysilicon layers are able to build a getter of considerable capacity, and its thermostability approaching $1473 \; K$ makes it possible to use such getter practically in the course of the entire manufacturing line. A gettering ability of polysilicon layers can be controlled through varying the sedimentation regimes.

In [Labounov and Baranov, et al. (1983); US pat. No. 4053335] it was shown that a polysilicon sedimentation temperature is chosen with a due account of grains that start to grow when the temperature rises up to $1073 \; K$ and higher. This growth drops the getter's capacity. On the other hand, a fall in sedimentation temperature drops the gettering rate and gives rise to elastic stresses in the "polycrystalline–monocrystalline silicon" structure; this results in a structure warping. That is why the temperature within the range of $973 - 1073 \; K$ is considered optimal to build the polysilicon layers through pyrolyzing silane. The thickness of the layers is chosen from the condition of what square of the grains is needed and what length of the sedimentation is desired. Besides, it should be also considered that during a manufacture time the polysilicon can be oxidized and the silicon dioxide can be partially etched off. Usually, the films in polycrystalline silicon are made $0.5 - 5.0\ mcm$ in thickness.

Polysilicon is also obtained by electron evaporation. Here, both a polycrystalline structure and an amorphous structure (also having gettering properties) can be obtained in the film, depending on the wafer temperature and the sedimentation rate [Germany pat. No. 2738195]. The gettering process may be strengthened by additionally diffusing phosphorus into polysilicon [Japan pat. No. 51–34714]. In addition, the polysilicon layer is advisable to be protected by the film against the impurities from the external ambient; this film may be, for example, of silicon nitride or aluminium dioxide $0.05 - 0.20\ mcm$ thick [Labounov and Baranov, et al. (1983)].

During epitaxy or oxidation, a presence of a polycrystalline silicon layer on a back side of the wafer allows to significantly drop the concentration of stacking faults in the epitaxial films or near a front side of the oxidized wafers, and also to increase the lifetime in the minor charge carriers [Labounov and Baranov, et al. (1983); US pat. No. 4053335]. The polycrystalline silicon protected with a Si_3N_4 film was found most efficient.

The Si_3N_4 films themselves may serve as a good getter, thanks to inhomogeneous mechanical stresses being created by them themselves in the structure; these stresses stimulate a diffusive–drifting rearrangement of impurities in the semiconducting wafers. For example, in [Labounov and Baranov, et al. (1983); US pat. No. 3997368] in the beginning of the technological process the layers of pyrolytic Si_3N_4 $0.2 - 0.4\ mcm$ were depositing, at $1073\ K$, onto a back side of the dislocation–free silicon wafers $300 - 500\ mcm$ thick. Then, an annealing in nitrogen at $1273 - 1473\ K$ for 1 – 4 hrs took place; upon this, a SiO_2 layer 0.4 mcm thick was built on a back side of wafers by a thermal oxidation at $1323\ K$. After these operations, the wafers were going through all the stages of the technological line assigned for manufacturing the controlled diodes and bipolar epitaxial–diffusive transistors. It was found that Si_3N_4 films can reduce nearly as much as six orders the density of structural (technologically–induced) defects within the presurface regions. The stacking faults generated on the original growth microdefects in the material were less affected by the silicon nitride gettering, though in this case the impurities decorating the stacking faults during technological operations were noticed to drop their concentration. It is of interest that the local regions in the back–sided Si_3N_4 layers reduce the density of defects near a front side more readily than the continuous Si_3N_4 film. As was mentioned above, silicon nitride becomes more effective when used in combination with other gettering agents – mechanical or polycrystalline. Such combinations allow, in particular, to construct thinner Si_3N_4 films, and this makes the gettering process more technological.

3.2.4
Silicides of hard–melting metals

Comparatively not long ago, gettering the semiconductors was suggested to be done by the layers of silicides of such hard–melting metals as tungsten, tantalum or niobium [Vigdorovitch and Kryukov, et al. (1982); Gorelyenok and Kryukov, et al. (1994)]. In [Vigdorovitch and Kryukov, et al. (1982)], there have been studied the processes of generation and recombination of intrinsic interstitial silicon atoms relating to oxygen precipitation and gettering by a tungsten dissilicide layer. The getter was formed by the 50 nm tungsten film magnetronic spraying and later thermal treatment in the flowing hydrogen atmosphere at $1023 - 1473$ K for $1 - 240$ hrs, with a programmed cooling: the temperature dropping rate being 275 $K/\min$, upon 823 K more than 293 $K/\min$. Subjected to this study were the $n - Si$ wafers grown by CZ and FZ technique and irradiation–doped. The relaxation time τ_r for the unbalanced charge carriers was taken as a gettering criterion; this time was estimated through modulating the UHF-conductivity when the laser was excited by the $1.06 - mcm$ wave length. Oxygen concentration was estimated through IR-absorption spectrums within the band 1106 cm^{-1}, and the residual impurities by the neutron–activation analysis.

As was experimentally revealed in the specimens with highly concentrated oxygen [Vigdorovitch and Kryukov, et al. (1982)], interstitial precipitates start to be formed and the relaxation time τ_r starts to drop already at short–time thermal treatments. With the original content of oxygen being of the order $6.42 \cdot 10^{17}$ cm^{-3}, this process is not detected even by annealing for more than 24 hrs. Therefore, in the FZ and irradiation–doped specimens, the relaxation time τ_r was not found to be dependent on the annealing length. It was noticed that in the CZ silicon specimens the gettering efficiency is substantially higher (with original concentrations of oxygen being small) than in a more pure and relatively defect–free FZ material. The authors relate this to a stimulating action by the bulk generation of interstitial silicon atoms on the oxygen precipitates. The unbalanced Si_I is supposed to serve as a gettering process initiator, since it converts the (metallic) impurity atoms from a nodal state (lattice site state) into an interstitial state where they become more mobile and are captured by the getter. As for the unbalanced Si_I, it arises due to the formation and growth of oxygen precipitates Si_xO_{2x}. When used as getters, the silicides of hard–melting metals (they have a developed polycrystalline surface), first, provide an active capture of impurities into the film, and, second, suppress the emission of intrinsic interstitial silicon atoms into the bulk of wafers, because the growing silicides mainly generate vacancies.

Gettering impurities and defects in silicon, gallium arsenide and indium antimonide by hard–melting metal films or oxides was studied in [Gorelyenok and Kryukov, et al. (1994)]. Silicon wafers were coated with tungsten, tantalum, and niobium formed by magnetronic spraying. As getters on gallium arsenide there were used the SiO_2 films or two–layer structures (silicon – tungsten, silicon–chromium) and chromium 100 nm thick. The $InSb$ crystals were gettered by the own anodic oxide of $100 - nm$ thickness. In all cases, the thermally–treated specimens were cooled under a special program not allowing the additional defects to arise. The results gained in this research speak about an efficiency of copper gettering by metal desilicide layers on silicon. The dislocation density in gallium arsenide ($GaAs$) was detected to drop if the crystals with the $W Si_2$ films are annealed; in the films, the charge carriers become more mobile and the chromium atoms assume an electrically active state. In the indium antimonide films, the gettering process has increased a mobility and lifetime of charge carriers and dropped the dislocation density as well.

In [Gorelyenok and Kryukov, et al. (1994)] the authors concluded that the gettering annealing of semiconducting wafers with coats, including the films of hard–melting metals, becomes universal to some extent, for it is compatible with a traditional technology of the wafer production. Here, the coats are deposited onto wafers upon cutting, then annealing, grinding and polishing operations go.

3.3
Thermal treatments in gettering ambients

The techniques of high–temperature annealing in inertial or chemically active gaseous ambients are mainly oriented upon suppressing a defect formation process in silicon crystals during a formation on their surface the semiconducting or dielectric layers; these techniques are also used for evacuating the fast diffusing impurities into the surrounding ambient.

The oxidizing of silicon slices with a preliminary annealing in the argon stream in the same oven where these slices then undergo oxidation is described in [Murarka and Levinstein, et al. (1977)]. Instead of argon, other noble gases, hydrogen or their mixtures, may be also used. Nitrogen is not recommended for use, because of its interaction with silicon. However, if a preliminary annealing is done in the nitrogen stream with small additions of chloride hydrogen and oxygen, then a further oxidation does not practically yield any stacking faults [Takeshi and Toshiharu (1978)].

Small concentrations (0.1 − 6.0 %) of chlorine or its compounds in the oxidative ambient are known to reduce defects and charge states in silicon dioxide and in the $Si - SiO_2$ interface and increase breakthrough stresses and stability in MOS-structures [Labounov and Baranov,et al. (1983); Mil'vidsky and Osvensky (1984); Ravi (1981)]. Simultaneously, the properties in silicon presurface layers are bettered, in particular, the lifetime of minor charge carriers increases and the concentration of oxygen–induced stacking faults decreases. "Chlorine" oxidation is supposed to neutralize a negative effect in the impurities of alkaline, transient and heavy metals which, interacting with chlorine ions, either are removed from the silicon surface in the shape of volatile compounds or form electrically neutral aggregates exerting no effects on device characteristics. During the "chlorine" oxidation, the impurities of gold and platinum metals are gettered slightly – this is evidently explained by their chlorides being unstable at high temperatures. That the oxygen–induced stacking faults in the oxidative ambients with chlorine added are suppressed is explained by a birth of the aggregate of positively charged interstitial silicon atom and a negative chlorine ion; as a result of this aggregation, a less number of silicon atoms is diffusing into the depth of the wafer and generating the stacking faults [Murarka and Levinstein, et al. (1977); Edel'man (1980)]. High reactivity of chlorine and chlorine hydrogen restrain their application in technological processes. Less active chlorine–containing additions include trichlorethylane, tetrachloride nitrogen and trichlorethane [Labounov and Baranov, et al. (1983); Volkov and Zaitzev, et al. (1983)].

The trichlorethane–added oxidation was experimentally revealed to better also the parameters of MOS-structures, decrease the density and the sizes of oxygen–induced stacking faults [Labounov and Baranov, et al. (1983); Volkov and Zaitzev, et al. (1983)]. Besides, the stacking faults born by the preliminary oxidation were observed to shrink and vanish. This mechanism is explained by the vacancies generated on the silicon surface and further migrating to the stacking faults. In [Rozgonyi and Pearce (1977)], it was shown that the "chlorine" oxidation getters interstitial oxygen from the surface zone in the silicon wafer, also lessens the probability of generating the oxygen–induced stacking faults. The trichlorethylane vapours incorporated into the oxidation process made it possible to form the MDS-structures with the density of high–speed surface states being not higher than $2 \cdot 10^{10} \ cm^{-2} \cdot eV^{-1}$, of the capacity relaxation time more than $100 \ s$ and of the introduced charge in value being equal to $10^{-9} \ C \cdot cm^{-2}$ [Volkov and Zaitzev, et al. (1983)]. For gettering purposes, the silicon slices also undergo a high–temperature preoxidation in chemically active ambients whose components interact efficiently with metallic impurities. Such ambients usually represent themselves a complicated mixture of gases. For example, concentrations of copper, iron and gold in silicon are reduced via a short–time annealing at $1223 - 1373 \ K$ in the gaseous mixture of nitrogen oxide, hydrogen chlorine and nitrogen as a gas–carrier [Labounov and Baranov, et al. (1983)]. Impurities may be also extracted from silicon by annealing the slices at $1073 - 1273 \ K$ in the atmosphere of hydrogen and hexafluorine sulphur [Japan pat. No. 51-35345]. Such gettering annealing reduces a concentration of metallic impurities and this substantially increases the lifetime of minor charge carriers in silicon.

Of not less importance in perfecting the extrinsic defect composition in silicon crystals are the postoxidation thermal treatments; they reduce concentration and sizes or fully dissolve the clusters of intrinsic point defects and stacking faults. Anneals are performed in vacuum or inertial ambients. The best results in removing the oxygen–induced stacking faults were obtained by thermal treatment in the nitrogen stream for $2-100\ hrs$ at $1373-1573\ K$ [Labounov and Baranov, et al. (1983)]. This is perhaps explained by a thin nitride or oxinitride film which was formed on a silicon surface and seemingly strengthened a gettering effect for intrinsic interstitial atoms in silicon arising during a dissolution of stacking faults. It is worthy noting here that such annealing regimes may also dissolve the growth microdefects in the dislocation–free silicon. Though, if no additional getter is present on the surface, the above thermal treatment does not fully clear the crystals from the components of the growth extrinsic defects. After such treatment, second–phase inclusions and impurity precipitates still retain in the bulk [Milevscky and Vysotzkaya, et al. (1980)].

3.4
Internal getter

In the last years more and more attention is paid by researchers and technologists to the technique of cleaning the silicon structures from impurities and defects by the damaged regions built in the central part of the wafer, the so–called internal gettering [Nemtzev and Peckarev, et al. (1981); Labounov and Baranov, et al. (1983); Mil'vidsky and Osvensky (1984); Volkov and Zaitzev, et al. (1983); Craven and Korb (1981); Nagasawa and Matsushita, et al. (1980)]. In contrast to other gettering techniques, this technique needs neither additional structural defects to be incorporated nor special layers to be built up on wafers' surfaces. An internal gettering involves a precipitation of oxygen dissolved in the original bulk of the wafer and an extraction of oxygen from a presurface region in order to create a defect–free zone, for semiconducting devices to be then established in it. As known from [Volkov and Zaitzev, et al. (1983); Rozgonyi and Pearce (1977); Nemtzev and Pekarev, et al. (1983)], a long–time annealing of CZ wafers results in the precipitates of SiO_x $(x \approx 2)$, arising in the neutral atmosphere at $1373-1423\ K$ inside the slices, and the driven-off prismatic dislocations. These are the very precipitate–dislocation aggregates that serve as getters for impurities and unbalanced intrinsic point defects. These very precipitate–dislocation aggregates serve as getters for impurities and unbalanced IP defects. On the surface of the wafer treated by high temperatures there is usually formed a presurface layer free from precipitate–dislocation aggregates (excluding the specimens with a high original content of oxygen). Within the planes parallel to a free surface of wafers, the precipitate–dislocation aggregates are distributed in the getter in the way similar to the oxygen distribution profile in the original state; in particular, a swirl–like distribution of impurities may occur. With the length of thermal treatment increasing, the density and the sizes of precipitate–dislocation aggregates will grow, and the internal getter will be supplied with new dislocations gettering the point defects. This means that the internal getter will retain its efficiency throughout the entire technological route of device manufacturing. Though, from one operation to another this efficiency may somewhat drop or vary, as a result of the growing dislocation aggregates in the getter being mutually annihilated as dislocation loops during annealing become larger in sizes [Ravi (1974)].

In the thermally treated oxygen–carrying crystals the oxide precipitates are born at the rate of the form [Nemtzev and Pekarev, et al. (1983)]

$$J = n_k \varpi Z ,$$

where

n_k is a balanced concentration of critical nuclei;

ϖ is a frequency of atom jumps into a critical nucleus;

Z   is the Zel'dovich constant considering the probably that not all the critical nuclei are able to grow.

If a nucleus is spherical in shape, then, ignoring the impact of mechanical stresses, we shall have for n_k

$$n_k = n_1 \exp\left[-\frac{\Delta G_k(r_k)}{kT} \right] ,$$

where the free energy increment, if a nucleus appears, will be determined through a bulk and surface components as

$$\Delta G_k(r_k) = \frac{4}{3}\pi r_k^3 \Delta G_v + 4\pi r_k^2 \sigma .$$

From the condition of free energy minimum $\dfrac{\partial \Delta G_k(r_k)}{\partial r_k} = 0$ the critical nucleus radius is found to be equal to $r_k = -\dfrac{2\sigma}{\Delta G_v}.$

In the above formulas:

σ is a specific energy of the interphase boundary between a precipitate and a crystalline matrix;

ΔG_v is a bulk free energy of the precipitate;

n_1 is a concentration of possible places where precipitates are generated heterogeneously or the concentration of oxygen atoms in the event of a homogeneous nucleation.

A frequency of jumps of the oxygen atoms into a nucleus is found through a probability for such process and through its reverse variant, i.e. a diffusive departure of atoms from the precipitate. This may be expressed as

$$\omega = 4\pi r_k^2 n_1 \, p \, d \, v \cdot \exp\left(-\frac{E}{kT}\right) - 4\pi r_k^2 n_1 \frac{D}{d} =$$

$$= 4\pi r_k^2 n_1 d \left[p v \exp\left(-\frac{E}{kT}\right) - \frac{D}{d^2} \right],$$

where

d is an interatomic distance;

p is a probability of the oxygen jumping into a critical nucleus;

v is a frequency of oscillations of oxygen atoms;

$E = 2.4 \, eV$ is an activation energy of the oxygen diffusion in silicon;

D is a coefficient of oxygen diffusion.

If taken into account that $D = \dfrac{d^2}{2} p v \cdot \exp\left(-\dfrac{E}{kT}\right)$, then the last formula is easy to be converted into the form $\omega = 4\pi r_k^2 n_1 \cdot \dfrac{D}{d}$.

The increment in the specific free energy ΔG_v may be expressed through an enthalpy of oxygen dissolution in silicon $H = 6.67 \cdot 10^{10} \, erg \, / \, cm^3$, provided the crystal has undergone a heavy overcooling:

$$\Delta G_v = \Delta H_v \cdot \frac{T_E - T}{T_E} \, ,$$

where T_E is a temperature of the silicon saturation by oxygen, i.e. the temperature at which an actual concentration of oxygen is maximal. For example, for $N_O = 1.1 \cdot 10^{18} \, cm^{-3}$ we get $T_E = 1573 \, K$.

With a due account of all the ratios devised for the nucleation rate we find that

$$J = 4\pi n_1^2 \, Z \frac{D}{d} \left[\frac{2\sigma T_E}{\Delta H_v (T_E - T)} \right]^2 \exp\left[-\frac{16\pi \sigma^3 T_E^2}{3kT \Delta H_v^2 (T_E - T)} \right].$$

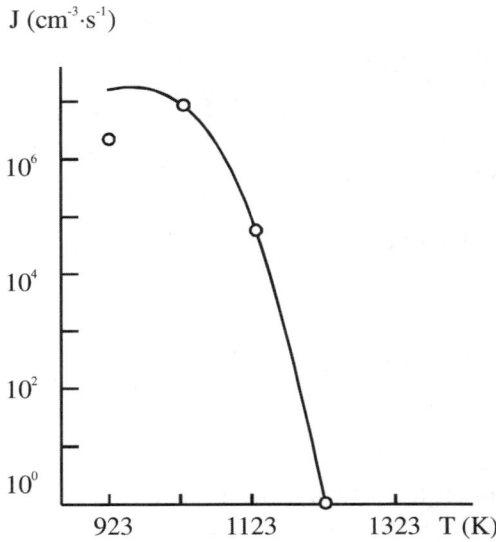

Fig. 3.6. A thermal dependence of the rate of homogeneous birth of oxide precipitates at temperatures lower than $1223\ K$.

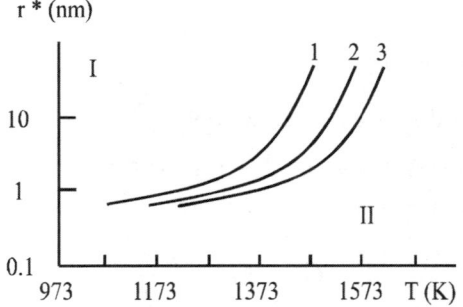

Fig. 3.7. A critical radius of the precipitate depending on the temperature for various relative concentrations of oxygen $(x10^{6})$: **1** – 10; **2** – 15; **3** – 20.
I – growth; **II** – dissolution.

The results calculated [Nemtzev and Pekarev, et al. (1983)] by this formula for the rate of oxygen precipitate nucleation depending on the temperature at $\sigma = 430\,ern\,/\text{cm}^2$ and $Z = 10^{-3}$ are presented in Fig. 3.6; and Fig. 3.7 shows how a radius of the critical nucleus depends on temperature when the content of oxygen in silicon is varying.

Figs. 3.6 and 3.7 demonstrate the greatest rate of precipitate nucleation to be obtained at $873 - 1073\ K$, and within this very range the critical sizes of nuclei are less dependent on the oxygen concentration in original crystals.

A presurface region of the wafer free from precipitate–dislocation aggregates can be created by oxygen extraction either prior to precipitation or during or after it.

Specific regimes for constructing an internal getter are usually chosen empirically, in accordance with the content of the impurities in the crystal and its technological history. The annealing schemes most often used in practice for the internal getter formation are given in Fig. 3.8.

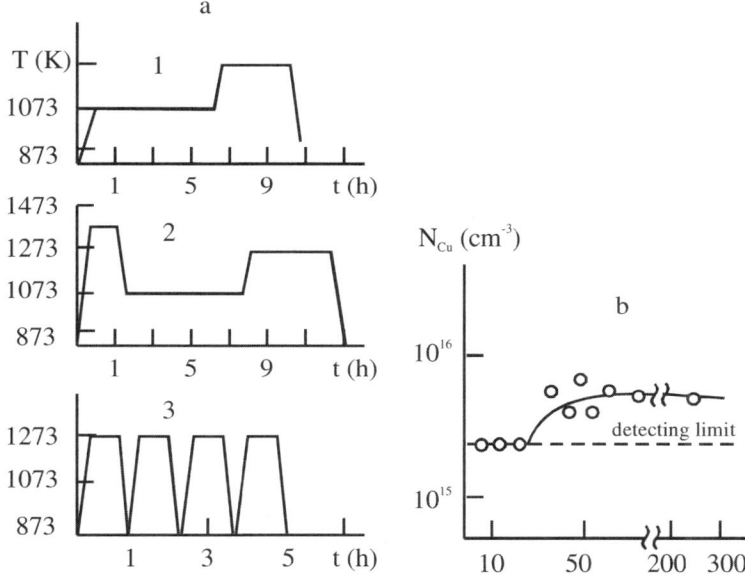

Fig. 3.8. a) Annealing regimes for experimentally studying the internal getter: **1** – two–stage; **2** – three–stage; **3** – oscillating; **b)** A profile of copper distribution in the wafer from the surface, upon the internal gettering (DL – a limit for detecting copper in silicon).

Removing the point defects and suppressing a formation of stacking faults by an internal getter increases a lifetime of the minor charge carriers in the cleaned presurface region. The major advantage of internal gettering lies in that that the oxidized precipitates at technological temperatures $T < T_E$ are in thermodynamical equilibrium with the crystalline matrix encircling them and with the oxygen dissolved in it [Nemtzev and Pekarev, et al. (1983)]. Besides, the getter lying symmetrically throughout a wafer's thickness practically causes no bending moments, i.e. it does not cause a warping in semiconducting structures.

In practice, an internal gettering is usually performed in three stages (Fig. 3.8 a): a high–temperature annealing, to extract oxygen from a presurface region; a low–temperature precipitation; and a high–temperature growing of precipitates and a gettering process itself. The last stage is usually arranged coincident with the thermal treatment of the structures. On the basis of the results in [Frolov and Zaitzevskaya, et al. (1986)], let us consider how an internal getter affects the parameters of silicon MDS-structures formed on the crystals of $p - Si\,(001)\,CZ\,(\rho = 10\ Ohm \cdot cm$, boron–doped). The gettering effects were measured through a density of generation current in the $n^+ - p -$ diodes with a controlled semisilicon gate. Subjected to measurements were the density of reverse currents being measured by gate of $n^+ - p -$ junctions in the regime of subgate region depletion. Since a gate square exceeded a junction square as much as two orders, subjected to measuring was actually a density of generation current in MDS-structure.

A batch of single–type wafers in [Frolov and Zaitzevskaya, et al. (1986)] was treated by the below four technologies:

    1)   a three–stage internal gettering: thermal treatments in nitrogen at $1373\ K$ (3 hrs), $1023\ K$ (24 hrs) and then a standard technological process;

    2)   constructing the precipitation centres without a free zone: annealing in nitrogen at $1023\ K$ (24 hrs) and then following a standard procedure. Here, a generation activity of gettering centres was supposed to be estimated;

    3)   constructing a free zone without establishing the precipitation centres: a 3-hr thermal treatment at $1373\ K$, prior to a standard technological process;

    4)   A controlled variant – a standard $n -$ channel MDS-LIC technology.

As was shown through measurements, the structures manufactured by the second technology demonstrated the greatest density of generation current. This case provides the greatest precipitation of oxygen but simultaneously a selective annealing also reveals a considerable concentration of stacking faults and dislocations, in contrast to the specimens produced by other technologies. The first technology has given a low current density – this is explained by a presence of the oxygen–free presurface zone and this coincides with the data obtained on the structures built by the third technology. As was revealed by the authors, an internal gettering is the most efficient, if the original specimens contain oxygen equal to $N_0 \geq 10^{18}\ cm^{-3}$. With the oxygen concentration being less than $(8-9) \cdot 10^{17}\ cm^{-3}$, the current density in the first technology turns out to be higher than in the controlled technology, i.e. in this case the gettering is not full. Thus, it was found by the authors that with the oxygen in the original specimens being equal to $N_0 > 7 \cdot 10^{17}\ cm^{-3}$ the efficiency of internal gettering will be maximal. With the oxygen concentration less than $(8-9) \cdot 10^{17}\ cm^{-3}$, the current density in the first variant turns out to be higher than in the controlled variant, i.e. in this case the gettering is not full. Hence, the authors concluded that a homogeneous precipitation of oxide inclusions needs the concentration $N_0 > 7 \cdot 10^{17} \, cm^{-3}$. Alongside with this, an average–temperature annealing at $1023 \, K$ creates a comfortable condition for heterogeneously nucleating the precipitates on all possible defective centres near the surface of structures.

An expediency of reducing oxygen in the silicon presurface layers via annealing in inertial ambient at temperatures $1423 - 1473 \, K$ is shown in [Frolov and Zaitzevskaya, et al. (1986)]. Here, an internal gettering is shown to have at least two factors differently affecting the generation centres being created within an active region in the wafers. The first factor is the internal gettering itself whose efficiency is determined, except the technological regimes, by the concentration of interstitial oxygen. Therefore, to make this gettering technique applied industrially, ingots and slices should be preliminarily sorted, in accordance with the original concentration of oxygen. The second factor is precipitating oxygen on the available defects in slices and in the interfaces during the average–temperature anneals; this precipitation may affect the leakage currents in the structures with an internal getter. A high–temperature preliminary anneal of the wafers in the nonoxidizable ambients is recommended to be used in the $n -$ channel MOS-technology, for this makes it possible to suppress a negative impact of the oxygen surface precipitation on the parameters of transistor structures.

A quality of an internal getter and a precepitation rate are dependent on the original microdefects held in silicon wafers. In [Prokofyeva and Sokolov, et al. (1991)], the high–rate precipitation materials also included the monocrystals or slices of the microdefect density exceeding $5 \cdot 10^4 \, cm^{-2}$, and the low–rate precipitation materials had the density less than $4 \cdot 10^4 \, cm^{-2}$. A dynamics of internal gettering was studied on the slices of $n - Si(001) CZ$ ($\rho = 4.5 \, Ohm \cdot cm$, phosphorus–doped) and $p - Si(001) CZ$ ($\rho = 12 \, Ohm \cdot cm$, boron–doped) having different precipitation levels, undoped and doped by $Zr(Hf)$. It was detected that in the course of thermal treatments simulating a technological SMOS IC manufacturing process the high–rate precipitation slices were forming a getter retaining its properties during the subsequent operations. According to authors [Prokofyeva and Sokolov, et al. (1991)], no preliminary low–temperature treatment is recommended on such slices, since it kills a defect–free zone. However, for the low–rate precipitation slices such thermal treatment is needed, to nucleate a new phase. It is shown by them that the gettering efficiency is better to be increased by a zirconium doping, in particular, to get oxygen homogeneously spreaded in the material. In this way one obtains a homogeneity in the internal getter. On the basis of the results obtained [Prokofyeva and Sokolov, et al. (1991)], a technological line for manufacturing the original wafers was suggested to be supplemented by the following operations: upon chemically removing the layers structurally damaged by abrasives one should perform such operations as a controlled thermal treatment, to generate the precipitates in the bulk of slices; an estimation of precipitate density on the spall of twin–slices; a formation of a batch of slices of low and high precipitations; a sorting of slices with respect to their precipitation levels and further in the technological line.

In [Yenisherlova and Mil'vidsky, et al. (1991)] there has been studied an interrelation between annealing regimes for a getter formation and the structural specificities of gettering centres; the transformation of these centres during subsequent anneals imitating the standard technological treatments was studied as well. An optimal density of defects in the bulk of slices was found to obtain a maximal gettering effect. Also it was shown that a use of gettering annealing for a chlorine ambient in the first stage increases many times the thickness of the cleaned presurface region. As noted in [Yenisherlova and Mil'vidsky, et al. (1991)], one of the aspects describing a gettering capability of structural defects arising in the bulk of slices is a presence of extended and compressed regions of microdeformation in the vicinity of these defects. Oxygen containing precipitates are capable of capturing the point defects only during dissolution, whereas during their growth or transformations (resulting in phase transformations) they themselves emit into the bulk of the crystal their intrinsic interstitial atoms of silicon. Therefore, the dislocation loops, and among them mostly the full dislocation loops, are supposed to be major drains for fast–diffusing impurities.

The effects of the three–stage gettering annealing on the density of the surface states on the boundaries in the $Si - SiO_2$ interface were studied in [Vasilyeva and Kolotov, et al. (1992)]. Subjected to studying were the MOS-structures formed on the wafers where the density of growth microdefects approached $10^6 \ cm^{-2}$ or they were defect–free. It has been found that the internal gettering with subsequent anneals of slices in dry oxygen ($1373 \ K$, 4 hrs) and nitrogen ($923 \ K$, 16 hrs; $1273 \ K$, 4 hrs) results in a defect–free zone on the specimens with microdefects 31.2 mcm wide, and without microdefects 24 mcm wide. For these specimens, the average density of distortions in the getter made up $4.7 \cdot 10^6$ and $1.7 \cdot 10^6 \ cm^{-2}$, respectively. Without gettering, the density of the surface states on the boundary of the $Si - SiO_2$ interface was $(1-6) \cdot 10^{11} \ eV^{-1} \cdot cm^{-2}$, and upon gettering it fell down to $(0.4 - 0.8) \cdot 10^{11} \ eV^{-1} \cdot cm^{-2}$; besides, the most vivid fall was detected in the specimens with a minimum of original growth microdefects. It is specific that in the structures with a high concentration of growth microdefects there has been observed an increase in the density of surface states upon the second, low–temperature, stage of the gettering annealing, as matched against the original values. This event the authors related to a more active precipitation in these specimens running heterogeneously along with the microdefects and resulting in the increased concentrations of defective bonds on the boundary of the $Si - SiO_2$ interface; these concentrations arise due to deformations in the silicon lattice near the growing inclusions of oxide. After a full gettering cycle is completed, the density of the surface states is falling, and the parameters of the interface boundary become better. The effects of internal gettering on the properties of the $Si - SiO_2$ interface in MOS- and MDS-structures have been studied through depositing at $1153\ K$ onto oxidized slices a Si_3N_4 film 75 nm thick from monosilane with ammonia [Vasilyeva and Kolotov, et al. (1992)]. The SiO_2 films were 100 nm thick and grown thermally at $1173\ K$ in wet oxygen with trichlorethane added. A three–stage gettering annealing of the $Si - SiO_2$ and $Si - SiO_2 - Si_3N_4$ structures was performed in a nitrogen ambient: 1-st stage – at $1373\ K$ (4 hrs); 2-nd stage – $923\ K$ (16 hrs); 3-rd stage – $1272\ K$ (4 hrs).

Then, aluminium contacts were formed on the surface, by thermal spraying and further by photolithography. Volt–farad high–frequency characteristics were measured in the MOS- and MDS-structures at $T = 78$ and $300\ K$ through which a density of surface states (N_{SS}) was then estimated. The energy distribution of N_{SS} across the forbidden zone of silicon was recorded by DLTS spectrums. The structure of specimens was also analysed by metallographical and roentgenotopographical methods.

Depositing a Si_3N_4 film onto a SiO_2 layer was found to increase the surface state density in the $Si - SiO_2$ interface, by an order of magnitude. But this effect is reversive, and upon removing Si_3N_4 the values of N_{SS} are restored to the original value. The Si_3N_4 film also changes a nature of the energy distribution of densities of N_{SS} in the upper half of the prohibited zone of silicon; for MDS-structures this distribution is maximal for energies $0.3\ eV$ from level E_c, as opposed to a sufficiently smooth U-like distribution observable in MDS-structures. Coating the $Si - SiO_2$ structures with a Si_3N_4 film induces some additional mechanical stresses resulting in additional broken bonds which create maximum N_{SS} when the energy is of the order $E_c - 0.3\,eV$. As was revealed by roentgenotopographical studies, the specimens with a two–layer dielectric have higher stresses than in the single–layer SiO_2 films or after removing a Si_3N_4 coat.

Metallographic analysis of spalls in the slices (they underwent an entire cycle of internal gettering) has revealed in the specimens a presence of a defect–free presurface zone in all the cases. The idea that such zones arise due to the oxygen being extracted from silicon into the external environment [Nemtzev and Pekarev, et al. (1983); Frolov and Zaitzevskaya, et al. (1986); Prokofyeva and Sokolov, et al. (1991); Yenisherlova and Mil'vidsky, et al. (1991); Vasilyeva and Sokolov, et al. (1991)] was not supported in [Vasilyeva and Kolotov, et al. (1992)], since all the stages of gettering annealing were performed on the specimens whose surface was coated with SiO_2 or $SiO_2 - Si_3N_4$. Dielectric films hamper an escape of oxygen from the silicon presurface zone and so in the situation under question the internal getter forming mechanism differs from those described above; so far it remains unknown. Upon gettering, the density of N_{SS} for both types of structures were also found to fall but this change affected the MDS-structures more than the MOS-structures. Upon gettering, a shape of energy spectrums of N_{SS} retained. The X-ray topograms helped to reveal that the gettering annealing results in a relaxation of mechanical stresses which in the MDS-structures remain, nevertheless, higher than in the MOS-structures.

The residual impurities (their presence was recorded through the lifetime in unbalanced charge carriers) were most intensively gettered in the specimens with SiO_2 layers being in thickness more or less than some critical value; this value is determined by a set of conditions for a structure surface, in particular, by a composition of the ambient, a preliminary treatment, etc [Amal'skaya and Bagraev, et al. (1992)]. If a SiO_2 layer is thin, the gettering process will produce unbalanced intrinsic interstitial atoms of silicon, and if thick, then unbalanced vacancies. The interval of intermediate thicknesses in the SiO_2 layer characterized by an intensive annihilation of vacancies and intrinsic interstitial atoms near the surface of slices does not make it possible to obtain an effective gettering.

As was noted above, it is practically impossible to construct – by traditional thermocyclic treatments – such a presurface region which could be fully free from precipitates and have sufficiently abrupt (step–by–step) transfer in the interface with the internal getter. One of the solutions of this problem may be an irradiating of the front side in wafers by high–energy light ions, for example, by protons or $\alpha-$particles of the energies of some MeV [Skoupov (1996); Kiselyev and Obolensky, et al. (1999)]. In the initial segment of their track, such ions loose their energy because of ionization effects and mainly generate the mobile IP defects and elastic waves; this, as a whole, provides a preliminary (prior to gettering anneals) cleaning the presurface region (from units to dozens of micrometers in thickness) from original microdefects and nuclei of precipitates (aggregates of point defects). In the end of the run, where the elastic losses of the energy in the introduced particles start to dominate, there are formed the irradiation–induced defect aggregates which are either annealed or become the centres of generating second–phase particles during a formation of the internal getter. A monochromatic nature of the original irradiation and a low density in the streams of accelerated particles (to avoid an overlapping of tracks in the initial stage of deceleration) provide a structural homogeneity in the defect–free zone and a sharpness in the interface–getter boundary. A positive effect of the preliminary irradiation by $\alpha-$particles, with $E = 4.5 \, MeV$ by doses $10^{11} - 10^{12} \, cm^{-2}$, on the silicon–gold structures with an internal getter was shown in [Kiselyev and Obolensky, et al. (1999)]. This kind of treatment helped to increase breakthrough stresses between metallic contacts both across the surface and the thickness of structures and also to better volt–ampere characteristics and make the $Au - Si$ structures more resistant against protonic irradiation. The pregettering irradiation also allows to smooth the heterogeneity in the distribution of densities of growth and technological microdefects in the slice cross–sections parallel to surface, and in this way to make the internal getter structure more homogeneous. On the basis of the results obtained in studying the effects of high hydrostatic compression (in the impulse and static regimes) on a composition of structurally damaged layers in semiconducting crystals upon abrasive–chemical treatments [Perevostchikov and Skoupov (1992)], it becomes possible to say with a great portion of assurance that for the case of internal gettering this kind of treatment is able to produce a positive effect. In hydrostatically–compressed slices with an internal getter the functioning of dislocation sources (oxide precipitates), becomes activated, and accordingly it increases a capacity of the getter with respect to unbalanced point defects. This direction of studies is still in its very beginning.

3.5
Comparative analysis of gettering technique efficiency

The classification of gettering techniques with respect to the type of the defects being removed and the position these operations hold in a general technological cycle of manufacturing the $Si -$ based devices is presented in Table 3.4. In it, a technique of internal gettering is seen to provide, for silicon in any case, a cleaning of the presurface regions with device's components throughout the entire technological line.

Table 3.4

Comparative data for the gettering techniques in the silicon technology

Gettering technique	The defects to be gettered	A place of the gettering operation in the entire technological treatment
Mechanical treatment of the back side in the wafers	SFs, clusters, metallic impurities	In the start of the technological treatment
Laser irradiation of the back side in the wafers	Same	Same
Ionic implantation	Same	Same
Oxidation with the chlorine containing gases added	Same and oxygen	At high–temperature stages – in the beginning or at definite stages
Diffusion of phosphorus (boron) directed to the back side	SFs, PD clusters, fast–diffusing impurities	In the beginning of the technological treatment
Depositing of heterophase films	Same	Same
Anneals in various ambients	SFs, impurities of metals	Stage-by-stage, in accordance with the technological line
Internal gettering	Dislocations, SFs, clusters, impurities including oxygen	During the entire technological treatment

However, the difficulties caused by the application of this technique, especially when applied to large–diameter monocrystals with impurities and defects spreading inhomogeneously across the entire bulk hamper its utilization in industry.

The efficiency of the most popular gettering ways was studied by a deep–level relaxation spectroscopy in [Bobrova and Galkin, et al. (1991)]. This method makes it possible to directly detect the deep–level centres related to the atoms of fast–diffusing metals, aggregates of impurities or intrinsic point defects. Also studied was the correlation between deep levels in the active regions of the controlled structures, their generation–recombination parameters and the defective regions which were revealed by the sectional X-ray topography and metallographically. Subjected to study were the controlled transistor structures manufactured on the wafers of $n - Si(001)CZ$ ($\rho = 20\ Ohm \cdot cm$, phosphorus–doped) with oxygen concentration equal to $9 \cdot 10^{17}\ cm^{-3}$. Experiments were performed on the three types of getters.

The first getter, an internal one, was formed by a three–stage thermal treatment: oxidation at $1473\ K$ (2 hrs) in dry oxygen, with a 2 per cent addition of HCl; annealing in argon at $1023\ K$ (5–6 hrs) and $1273\ K$ (6 hrs). A getter of this type is used in manufacturing large ICs and CCD–structures.

The second getter, diffusive, was established by phosphorus diffusing to a back side of wafers in the concentrations very close to the solubility limit ($\approx 1\cdot 10^{21}\ cm^{-3}$). The "drive-in" operation was conducted in the regimes providing a formation of the misfit dislocation grid in the vicinity of the wafer's back side.

The third getter was an ion–implanted layer on the wafer's back side. This getter was established by implanting silicon ions of the energy $100\ keV$ and dose $3.1\cdot 10^{14}\ cm^{-2}$, with subsequently annealing in dry oxygen and removing a SiO_2 layer.

The data obtained on structures with the above three types of getters was compared against the results measured in the controlled specimens.

In the controlled specimens, there have been detected three levels of electronic traps in the forbidden zone of silicon:

1-st – with the activation energy $E_c = 0.54\ eV$, the capture cross–section $\sigma = 10^{-15} - 10^{-14}\ cm^2$ and the concentration $N_1 = 8\cdot 10^{12}\ cm^{-3}$;

2-nd – with the activation energy $E_c = 0.22\ eV$, $\sigma = (0.4-1)10^{-18}\ cm^{-2}$ and $N_2 = 4.8\cdot 10^{11}\ cm^{-3}$;

3-rd – with the activation energy $E_c = 0.15\ eV$, $\sigma = (1-4)\cdot 10^{-18}\ cm^{-2}$ and $N_3 = 2.4\cdot 10^{11}\ cm^{-3}$.

The spectrums of the gettered structures showed an ion–implanted getter to stably reduce the concentration of centres N_1 with the level $E_c = 0.54\ eV$ by about three times, and with the level $E_c = 0.22\ eV$ by 2 times, and the peak with the level $E_c = 0.15\ eV$ was practically vanishing. A diffusive and an internal getters provided practically a complete vanishing of all peaks in the spectrums, i.e. the deep level concentrations N_1, N_2 and N_3 were less than the sensitivity limit $5\cdot 10^{10}\ cm^{-3}$ of the technique.

The values of lifetime in the minor charge carriers on the controlled structures were largely scattering from the average value_ $\approx 2\cdot 10^{-5}\ s$. In the specimens with an ionically–damaged getter this lifetime made up $6\cdot 10^{-5}\ s$, and the dispersion was reduced to some extent. The same tendency has been also detected in the other two getters; though some separate structures with an internal getter had the lifetime of minor carriers equal to $\approx 10^{-3}\ s$. The level $E_c = 0.54\ eV$ is thought to be determined by gold getting into silicon from quartz attachments during oxidation; the level $E_c = 0.22\ eV$ is close to the divacancy ($E_c = 0.23\,eV$); and the level $E_c = 0.15\ eV$ was not identified finally.

According to X-ray topograms, the thermal treatments exert no impact on the defectiveness of structures from the controlled batch [Bobrova and Galkin, et al. (1991)]. On the structures with an internal getter, a defect–free presurface zone was found to be 40 – 60 mcm thick and a region adjacent to it had a high density of defects (up to $10^9\ cm^{-3}$). All these specimens have demonstrated elastic stresses across the entire bulk of the crystals. The structures with a diffusive getter showed a field of high stresses spreading nearly up to a wafer's front surface of the slices being 380 mcm thick. The getter created by silicon ion implantation with a later oxidation represented itself a region of local lowly–concentrated defects creating around themselves the elastic distortions of the crystalline lattice. Metallographic studies of spalls of the irradiated wafers have revealed the dislocation loops and stacking faults available in their bulk. The authors [Bobrova and Galkin, et al. (1991)] emphasize that similar defects were observed by them earlier in the boron–irradiated silicon, but were not found after implanting heavy ions of argon and crypton when amorpous layers were emerging.

Thus, as shown in [Bobrova and Galkin, et al. (1991)] all three types of getters are able to diminish the concentration of centres bringing deep levels into a prohibited zone of silicon; and this results in a rise of lifetime of minor charge carriers. The best, and nearly similar, results are given by the diffusive and internal getters. Cleaning the crystals from impurities and defects by these getters relates to a specificity of the elastic stress fields being established by them across the entire bulk of silicon.

Alongside with this, a diffusive getter is functionally efficient only at initial stages of the technological line and during not so lengthy high–temperature operations, because the misfit dislocations functioning as point defect drains are of limited capacity. In this respect, an internal gettering is, certainly, of the highest priority, provided all the technological problems of its reproduction are solved. It is evident that in practice a combination of different gettering techniques is most expedient, they will be mutually complementary and rise the efficiency of killing undesired impurities and extended defects in the layers. Nowadays such combinations are already exploited: an abrasive treatment and an internal gettering; an abrasive treatment and a silicon nitride sedimentation; a phosphorus diffusion and a semisilicon depositing; an internal gettering and a thermal treatment in special ambients; etc [Labounov and Baranov, et al. (1983); Ravi (1981)].

3.6
Mechanisms and models of gettering processes

Gettering may be defined as a set of elementary interactions of impurity atoms with unbalanced IP defects or with extended regions, i.e. the drains located inside, on the surface or beyond a semiconducting structure. This interacting process results in a fall of point defect concentration within the bulk under cleaning. According to [Nemtzev and Peckarev, et al. (1981)], the gettering mechanisms may be classified as they are schematically represented in Fig. 3.9.

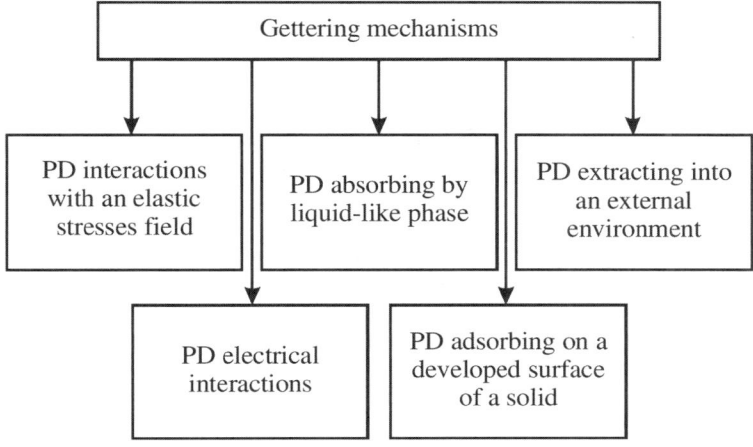

Fig. 3.9. Gettering mechanisms.

However, it should be noted that while implementing some definite technique of gettering several mechanisms from Fig. 3.9 can work simultaneously. Therefore, as each gettering mechanism was considered in detail in [Kontzevoy and Litvinov, et al. (1982); Mil'vidsky and Osvensky (1984); Lyubov (1985); Hu (1973); Ravi (1981); Bullough and Newman (1970)] they are discussed below in brief only.

3.6.1
Interaction of point defects with a field of elastic stresses

Practically any point defect creates in its ambient its own field of elastic stresses that deforms a surrounding matrix. In a continuous approximation, a point defect may be supposed to be some elastic spherical inclusion of radius $r_i = r_0 (1 + \delta)$ placed into a spherical cavity of radius r_0 in the matrix. If a relative numerical difference between an inclusion and a cavity is $\delta > 0$, then near this inclusion in the matrix there will act radial compressing stresses, and at $\delta < 0$ extending stresses. Thanks to this, a point defect interacts in the crystal with any field of

mechanical stresses which can relate to external or internal sources. If an external (with respect to a detected point defect) elastic field change inhomogeneously across the entire bulk of the crystal, then the point defects for $\delta > 0$ will strive to occupy the regions in which a crystalline lattice undergoes an extension deformation and the defects for $\delta < 0$ a compressing deformation. When analysing the gettering kinetics, one should distinguish an interaction of point defects with inhomogeneus macro– and microscopic fields of elastic stresses. The first are induced by external forces and embrace the entire crystal or its regions whose sizes largely exceed the interatomic distances. As an example of this are the macrostresses in the slices induced by their inhomogeneous heating or in the course of depositing the films with other structural and thermomechanical characteristics. Microscopic fields of elastic stresses arise around extended structural defects like dislocations, stacking faults, second–phase insertions, etc. By its nature, the interaction of these distortions with point defects may be referred to one of the following [Bullough and Newman (1970)] types:
*   dimensional interaction (of the first and second order) being determined by the nonzero value of δ ;
*   modular interaction determined by the difference between efficient elastic inclusion constants and the matrix;
*   electrical interaction being determined by charge states in the interacting defects;
*   photonic interaction induced by the difference of oscillating frequencies in the vicinity of interacting defects.
For semiconductors, the dimensional and electrical types of interactions between defects [Lannoo and Bourgoin (1981)] are most essential, primarily, from the point of view of their long–range impact.

With a use of the results obtained in [Bourmistrov and Pekarev, et al. (1978)], let us consider first how the interaction between point defects and the inhomogeneous macroscopic field of elastic stresses affects a gettering process in semiconducting slices. Here, a t – depth layer is supposed to have been constructed by the abrasive treatment of a back side in the H – thickness slices. In this case, a slice obtains a field of elastic stresses being inhomogeneus in its thickness; this field can cause a bending in the slice. During gettering annealing, an inhomogeneus distribution of stresses will yield the drifting streams (I) of point defects, as a result of nonzero gradients of defect chemical potentials in the bulk of slices:

$$I = \frac{C_0 D}{\Omega k T} \, grad \, \mu,$$ (3.1)

where

C_0 is a balanced concentration of point defects in a nondeformed slice;

D is a diffusion coefficient;

Ω is an atomic volume; and

μ is a chemical potential of defects.

As a first approximation it is possible to consider that $grad\,\mu = \dfrac{\Delta\mu}{H-t}$ and to express a change of the chemical potential through a sum of compressing σ_c and expanding stresses σ_e acting in the vicinity of the slice surface under gettering: $\Delta\mu = \Omega(\sigma_c + \sigma_e)$.

We shall calculate the stresses with the help of the equation by S.P. Timoshenko [Ustinov and Zakharov (1977)] describing an elastic–stressed state in the slice with an inhomogeneously–distributed in-depth deformation $\varepsilon(z)$. For a circled slice, this equation in the right coordinate system with the XOY plane coincident with a middle plane of the slice will be of the form:

$$\sigma_x(z)=\sigma_y(z)=\sigma(z)=-\frac{E}{1-v}\left[\varepsilon(z)-\frac{1}{H}\int_{-h}^{h}\varepsilon(z)\,dz-\frac{12z}{H^3}\int_{-h}^{h}z\,\varepsilon(z)\,dz\right]$$

where $h=H/2$; E is an elasticity module; and v is the Poisson coefficient.

Suppose that

$$\varepsilon(z)=\begin{cases}\varepsilon>0 \ at \ h-t\le Z\le h\\ \varepsilon=0 \ at \ -h\le Z\le h-t\end{cases}.$$

In this case, the equation for the stresses varying with respect to the slice thickness will assume the form:

$$\sigma(z)=-\frac{E}{1-v}\left[\varepsilon(z)-\frac{\varepsilon t}{H}-\frac{6z\varepsilon t(H-t)}{H^3}\right]=$$

$$-\frac{E}{1-v}\left[\varepsilon(z)-\frac{\varepsilon t}{H}-\frac{z}{R}\right],$$

(3.2)

where $R=\dfrac{6\varepsilon t(H-t)}{H^3}$ is a slice curvature radius.

From the last equation we find maximum compressing stresses for $z=-H/2$ and expanding stresses for $z=\dfrac{H}{2}-t$:

$$\sigma_c=-\frac{E}{1-v}\left[-\frac{t}{H}+\frac{H}{2R}\right]\cdot\varepsilon$$

$$\sigma_e = \frac{E}{1-v} \left[\frac{t}{H} + \frac{H-2t}{2R} \right] \cdot \varepsilon \; .$$

Using these formulas we obtain a final expression for estimating a magnitude of the drifting stream when point defects are gettered by a damaged layer:

$$I = \frac{E}{1-v} \cdot \frac{C_0 D}{kTD} = \frac{6E\varepsilon}{1-v} \cdot \frac{C_0 D}{kT} \cdot \frac{t(H-t)}{H^3} \tag{3.3}$$

From (3.3) the stream of point defects is seen to grow the more, the more becomes the magnitude of the deformation in the damaged layer; the thickness of the damaged layer is less than that of the entire slice. In the way similar to (3.3) it becomes also possible to analytically describe the drifting streams of point defects when gettering is performed by foreign films.

The expressions (3.1) and (3.3) help us to understand how exactly the streams of alternate–sign point defects and different dilating power δ are redistributed in an inhomogeneous field of elastic stresses. And actually, a presence of mechanical stresses $\sigma(z)$ exerts impact upon the diffusion coefficients of point defects through changing the activating energy of the migrating process U_m by the magnitude $\Delta V \sigma(z)$, where ΔV is an activating volume for performing an elementary act of the process. As a first approximation we may take $\Delta V = 4\pi r^2 \delta$. Then, with $U_m > \Delta V \sigma(z)$ the expression (3.3) may be rewritten as

$$I = \frac{E}{1-v} \cdot \frac{C_0 D}{kTR} \left[1 + \frac{4\pi r^2 \delta \sigma(z)}{kT} \right] ,$$

where D is a diffusion coefficient for an unstressed crystal.

From this formula, with accounting a sign of stresses at various depths (3.2), it follows that for $\delta > 0$ the drifting streams of point defects will ascend in the crystal regions of extending stresses ($\delta > 0$) and descend in the compressing regions ($\delta < 0$). For the defects with a negative sign of dilating power, the effect will be reversive and this has been proved by the gettering experiments described in the previous subsections.

Gettering the point defects from a wafer is evidently started in the zone immediately adjacent to a getter and having maximal elastic stresses. With an increase of length of the gettering anneal and a decrease in the point defect concentrations in this zone, there arises a gradient of point defect concentrations throughout the entire depth of the slice, and alongside with drifting streams there will emerge diffusive fluxes of impurities and of unbalanced IP defects; these streams are running to the getter or to the opposite (free) surface of the slice. Having got into the getter the point defects come into interaction with the dislocations thus unconservatively rearranging the defects (if they are IP defects) or creating impurity aggregates and precipitates. Most intensively these processes are running when point defects are interacting with edge full dislocations or their aggregates. In the event of high oversaturations (higher than a solubility limit), impurity discharges can emerge both on partial and even screw dislocations, and also they can arise during homogeneous precipitation. In general, a picture of phenomena running in a thermally treated getter is extremely complicated, and it has not been studied so far. Liable to studying are only its separate elementary fragments, with a use of the theories describing the interactions between point defects and dislocation aggregates and a decomposition of hard solutions within a field of elastic stresses of extended defects [Lyubov (1985); Bullough and Newman (1970); Fistoul' (1977)]. Alongside with this, a need in such investigations is evident, for their results could be applied to estimating a power and a functional ability of getters and also in choosing efficient regimes and conditions of gettering operations.

3.6.2
Electrical interaction of point defects

In semiconductors, the majority of the structural defects, the point defects being also in their number, constitute themselves the charged centres and create acceptor or donor levels in the prohibited zone. Therefore, via inhomogeneously distributing one type of impurities throughout the entire bulk of the crystal it becomes possible to regulate the spatial distribution of other charged point defects, i.e. to getter by specially–doped layers. As was noted above, this fashion of interaction is of a long–range nature [Lannoo and Bourgoin (1981)]. In subsection 3.1.3 this mechanism was said to occur in gettering the gold atoms by the diffusing layers of phosphorus where a high concentration of E-centres was established. Capturing the gold atoms by the vacancies of such centres increases the coefficient of impurity segregation between a phosphorus–doped and undoped regions in the silicon crystal. The segregation coefficient increases with a growth of phosphorus concentration and a fall of temperature, naturally to the values when a diffusive mobility of gold still retains. This model has been proved valid not only for gold but also for other heavy metals [Nemtzev and Peckarev, et al. (1981)]. It is worthy of mentioning here that alongside with the electrical interaction there also works in this case the elastic interaction between the impurities under gettering and the misfit dislocations induced by high–level doping.

3.6.3
Absorbing the point defects by a liquid–like phase

It is known [Nemtzev and Peckarev, et al. (1981); Mil'vidsky and Osvensky (1984); Fistoul' (1977)] that for the majority of impurities in semiconductors the balanced coefficients of distribution between a hard and liquid phases are less than unity. For example, in silicon we have for copper $K = 4 \cdot 10^{-4}$, for gold 10^{-15} and for iron 10^{-5}. Therefore, if in some region of the crystal one could establish a liquid phase, then the remaining, monocrystallic, part of the crystal could be cleaned by $K < 1$. In the capacity of such liquid–like phase there may be used any largely–defected (up to amorphousness even) layer of the material; for example, the abrasively–damaged layers, the films of polycrystalline and amorphous silicon, etc. In its essence, the impurity extracting mechanism here will initiate high temperatures, elastic and electrical fields being induced by the getter, and a solubility of impurities in it is determined by a composition of amorphous inclusions.

3.6.4
Extracting the point defects into an external environment

During high–temperature annealing of semiconductors in a neutral ambient or high vacuum, both atoms of the basic substance and impurities are being evaporated. This gives rise to concentration of vacancies and decreases a concentration of interstitial atoms in the presurface layers of the crystals, in contrast to their thermodynamical balanced composition in the entire bulk of the material. Therefore, this creates in-depth running diffusing fluxes of vacancies and those of interstitial atoms ascending surface–bound – this process diminishes the concentration, the sizes of clusters, the stacking faults, carries the dislocations surface–bound or shifts them within the bulk of the material, depending on the orientation of the Bűrgers vector and the concentration ratios of various IP defects being saturated by these fluxes. In the course of this process, the volatile components of the material (first of all, impurities) are evaporated and a presurface defect–free zone is established. The extracting effect may be additionally reinforced by evaporating atoms chemically interacting with the molecules or atoms of the external environment, with the chemically active components being incorporated into them. As an example of this process there may serve a "chlorine" oxidation of silicon (subsection 3.3) where the presurface vacancy concentration exceeds a spatial thermodynamically balanced concentration [Nemtzev and Peckarev, et al. (1981); Labounov and Baranov, et al. (1983)].

3.6.5
Adsorbing the point defects on a developed surface in a solid body

A solid body surface is known to easily adsorb point defects, since this process decreases free energy. If an efficient square in the surface of the crystal is enlarged in some definite places, then point defects could be then extracted to create an oversaturation or undersaturation of IP defects in the vicinity of such regions. This results in a diffusive rearrangement of vacancies and intrinsic interstitial atoms followed by gettering the impurities and aggregates of IP defects. This mechanism seemingly works during gettering by silicon porous layers and by a structurally–damaged surface zone. Near a developed surface, the IPD forming energy is substantially lower than that in the bulk of the material, and this determines a higher sensitivity of the material's presurface layers to various external actions [Alekhin (1983)]. This regularity can be exploited as a basis for developing the low–temperature techniques of gettering submicronic semiconducting layers. It is worth noting here that even in constructing a developed relief surface, for example, abrasively and chemically or by silicon anodizing, impurities may be gettered into the zone where mechanical, chemical or electromechanical reactions occur [Perevostchickov and Skoupov (1992)].

3.6.6
Models for diffusive redistribution of point defects in gettering

One of the first models discribing a gettering as a process of redistribution of point defects between the bulk of the semiconducting wafer and the getter built on one of its sides was suggested in [Gousev and Bogatch, et al. (1984)]. Its essence lies in the following. There is a semiconducting slice of thickness H, where the original concentration of point defects is described by the function $N_{pd}(x,0)$; here x is a coordinate along the normal to the slice surface. The origin of coordinates $x = 0$ coincides with a front side and $x = H$ with a back side of the slice on which a getter of thickness $h \ll H$ has been built up. The task is to estimate the point defect distribution in the slice subjected to a gettering anneal and to calculate a cleaning level in the front side δ_N of the slice. The cleaning level δ_N is equal to the ratio of the original surface concentration of point defects $N_{pd}(x = 0, t = 0)$ to the postgettering concentration $N_{pd}(x = 0, t)$. In the model [Gousev and Bogatch, et al. (1984)] it is supposed that:

a)   the point defect concentrations on both sides of the slice–getter interface boundary are described by the segregation coefficient $K = \dfrac{N^c{}_{pd}}{N_{pd}(H)}$ which is a phenomenological parameter;

b)   a product of the segregation coefficient by a getter thickness is not time–dependent, i.e. $K \cdot h = cons \tan t$;

c)   in the getter a concentration of the gettering centres and the point defects under gettering does not depend on a coordinate, and a real inhomogeneous distribution can be reduced to a homogeneous one through averaging the concentrations within a gettering layer;

d) During gettering, there are no thermal gradients and mechanical stresses in the slice, and a total number of point defects in the slice–getter system retains invariable.

The original diffusion equation is of the form

$$\frac{\partial N_{pd}(x,t)}{\partial t} = D_{pd}\frac{\partial^2 N_{pd}(x,t)}{\partial x^2} , \qquad (3.4)$$

where D_{pd} is a diffusion coefficient for a given type of point defects.

Boundary conditions for this equation will be:

$$- D_{pd}\frac{\partial N_{pd}(x,t)}{\partial x} = 0 \text{ for } x = 0 ;$$

$$- D_{pd}\frac{\partial N_{pd}(x,t)}{\partial x} = k \cdot h\frac{\partial N_{pd}(H,t)}{\partial t} \text{ for } x = H .$$

Original conditions:

$$N_{pd}(x,t) = N_{pd}(x,0) \text{ for } t = 0, 0 \le x \le H$$

$$N_{pd}(x,t) = \frac{N^c{}_{pd0}}{k} \text{ for } t = 0, x = H .$$

Solving the equation (3.4) by splitting the variables in the way similar to the calculations done in [Bevzouck (1971)], the concentration of point defects in [Gousev and Bogatch, et al. (1984)] was expressed as

$$N_{pd}(y,t) = N_{pd}(1,0) + 2\sum_{m=1}^{\infty} M_m\left[1 - \frac{\cos(\gamma_m \cdot y)}{\cos(\gamma_m)}\exp(-\gamma_m{}^2\theta_t)\right] ,$$

where

$$M_m = \frac{\gamma_m \int N_{pd}(y,0)\cos(Y_m \cdot y)\,dy - N_{pd}(1,0)}{\sin\gamma_m[1+\theta_g^{-1}+\theta_g \cdot Y_m^2]} \; ;$$

$y = \dfrac{x}{H}$ is a standardized coordinate;

γ_m is a root of the equation $tg(\gamma_m) = -\gamma_m \theta_g$;

$\Theta_g = \dfrac{kh}{H}$ is a gettering criterion, and $\Theta_t = \dfrac{D_{pd}\cdot t}{H^2}$ is time.

If an original PD concentration in the slice does not depend on the coordinate, i.e. $N_{pd}(x,0) = N_{pd0}$, then a solution is simplified:

$$N_{pd}(x,t) = N_{pd}(H,O) +$$

$$+2\big[N_{pd} - N_{pd}(H,O)\big]\sum_{m=1}^{\infty}\frac{1-\cos(\gamma_m y)/\cos\gamma_m \, \exp\!\big(-\gamma^2_m\Theta_t\big)}{1+\Theta^{-1}_g + \Theta_g \gamma^2_m} \; .$$

In most practically important cases we have $K \gg 1$ and $N^c_{pd}(t=0) \cong N_{pd0}$. Therefore, the concentration $N_{pd}(H,0)$ is largely less than N_{pd0}. With this taken into account, the cleaning level $\delta_N = \dfrac{N_{pd(0,0)}}{N_{pd}(0,t)}$ may be written as

$$\delta_N = \left[2\sum_{m=1}^{\infty}\frac{1-\dfrac{\exp(-\gamma^2_m\Theta_t)}{\cos\gamma_m}}{1+\Theta^{-1}_g + \Theta_g\gamma^2_m}\right]^{-1} \; .$$

The time dependence of the cleaning level calculated through the formula (3.5) in [Gousev and Bogatch, et al. (1984)] shows that the major specificity of gettering by a limited layer lies in the cleaning level becoming saturated during long–time annealing. This fact demonstrates that a final distribution of point defects in the slice is homogeneous, and therefore for a long–time gettering the formula for cleaning a front side may be simplified. A constant quantity of point defects in the slice–getter system prior to and upon annealing is obtained if the below condition is obeyed:

$$HN_{pd0} + hH^c_{pd0} = HN_{pd} + khN_{pd} \; ,$$

from which it follows that

$$\delta_N = \frac{(H + kh) N_{pd0}}{HN_{pd0} + hN^c_{pd0}} \; .$$

Since we usually have $N^c_{pd0} \approx N_{pd0}$ and $H \gg h$, then

$$\delta_N \approx 1 + \theta_g \; . \tag{3.6}$$

The cleaning level δ_N calculated for deferent times for different gettering criteria is given in Fig. 3.10. A broken curve shows a dependence of minimally

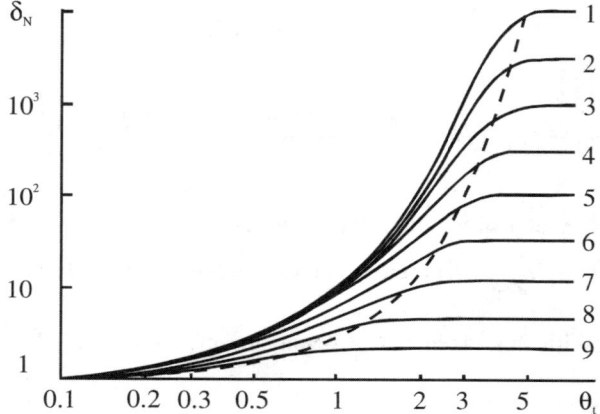

Fig. 3.10. Time dependence of the cleaning level for gettering by a limited layer at various values of the gettering criterion θ_g : **1** – 10^4 ; **2** – $3 \cdot 10^3$; **3** – 10^3 ; **4** – $3 \cdot 10^3$; **5** – 10^2 ; **6** – 30 ; **7** – 10 ; **8** – 3 ; **9** – 1.

needed cleaning time on the value of the gettering criterion. In Fig. 3.11 the minimally needed time for gettering by a limited layer is shown dependent on the value of the gettering criterion. A comparison of limiting cleaning levels determined in Fig. 3.10 and determined from the simplified formula (3.6) demonstrates a possibility of easily estimating the effects of PD redistribution.

From Fig. 3.11 it is possible to find that the gettering time minimally needed to obtain a cleaning level on the slice front side makes up 0.9 of its maximally possible time for the given conditions. Though for this one should be aware of what impurity determines a gettering process in the slices of the given type. In [Gousev and Bogatch, et al. (1984)], the fast–diffusing impurities were experimentally gettered from the slices of the commercial $n - Si\,(001)\,CZ$ ($\rho = 4.5\ Ohm \cdot cm$, phosphorus–doped) $350\ mcm$ thick, by the getter formed by the ionically–implanted phosphorus of the energy $100\ keV$ and dose $10^{16}\ cm^{-2}$. A gettering anneal of different lengths was conducted in the nitrogen

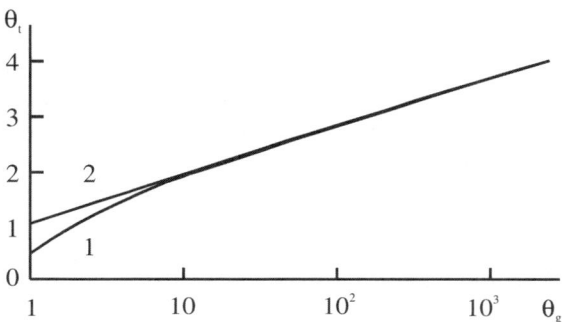

Fig. 3.11. The dependence of minimally needed time for gettering by a layer on the gettering criterion: **1** – a calculated dependence; **2** – an approximation.

ambient at $1173\ K$. Here, the getter was protected with a SiO_2 coating. The cleaning level of the slices was checked by the lifetime in the minor charge carriers: $\delta_N = \tau / \tau_0$, where τ_0, τ is the pre– and postgettering lifetime. The experimentally detected dependence of δ_N on the annealing time t helped to find a stabilization (saturation) segment and the time t_1 at which this occurs, i.e. the getter is being saturated with impurities. Further, with a use of the quasibalanced model the gettering criterion $\delta_N \approx \theta_g \approx 10$ was found through (3.6), and on the basis of Figs. 3.10 and 3.11 there was found the time $\theta_1 = 1.8$ corresponding to the given θ_g. Through the value of θ_t for the known t_1 and H there was calculated the diffusion coefficient for the impurity constraining the gettering process: $D_{pd} \geq \dfrac{1.8\,H^2}{t_1} = 9.2 \cdot 10^{-7}\ cm\ \cdot s^{-1}$. Having analysed the known information concerning how possible impurities in commercial silicon affect the lifetime in the minor charge carriers and also their diffusion coefficients, the authors have concluded that under the conditions of their experiments the impurity constraining the gettering annealing is iron. How the minimally needed time for gettering is dependent on the gettering criterion (Fig. 3.11) may be approximated exactly enough by the formula $\theta_{t\ \min} = 1 + 0.9 \lg \theta_g$, from which

$$ t_{\min} = \frac{H^2}{D}(1 + 0.9 \lg \theta_g). \qquad (3.7) $$

The expressions (3.6) and (3.7) may be used to quantitatively estimate the gettering regimes for fast–diffusing impurities.

The model for the impurity redistribution during gettering by a layer was further developed in [Dougaev and Ivanotchko, et al. (1987)] where the edge conditions for a diffusive exchange by impurities between a getter and a slice bulk were specified. The authors [Dougaev and Ivanotchko, et al. (1987)] have suggested a notion of "the time of catching impurities on the drain" (τ) in the getter and have additionally estimated the role of the transient region between the getter and the slice. The authors suggested the below formula for the concentration of front–sided impurities during a gettering anneal by a damaged back–sided layer:

$$ N(x,t) = N_0 \exp\left(-\frac{\overline{\lambda}\, t}{H}\right). $$

Here N_0 is an original concentration of impurities; $\overline{\lambda} = \dfrac{\lambda h}{h = \lambda \tau}$, where λ is a phenomenological parameter dependent on the getter–wafer interface and determined as $\lambda = D_c / L_{th}$ (D_c is an impurity diffusion coefficient in the transient L_{th}-thickness region). The model suggested has helped to explain a number of experimental facts, though in it a physical aspect of the processes was left beyond analysis.

In [Marousyak and Kosyatchenko (1989)] there is suggested a model in which (in contrast to the models in [Gousev and Bogatch, et al. (1984); Dougaev and Ivanotchko, et al. (1987)] a getter thickness is not thought small and a coefficient of impurity saturation by a damaged layer is introduced. Trough this, the authors managed to find a better approximation of the experimental profiles for impurity distribution in the vicinity of the gettering layer and also to prove that a slice cleaning level depends on a getter's thickness. In calculations, the numerical values of the coefficient, describing the impurity saturation by the getter, were made various, without specifying a nature of drains.

The models suggested have analysed a change of the impurity concentrations on a front side of slices during gettering irrespective of the doping level of the material under cleaning, mainly in the regions adjacent to a front side. In its essence, any semiconducting device represents itself a combination of topologically assigned planar–doped regions in which an impurity concentration, say, in emitting regions, may become very high. In this case, the device's active zones themselves start to function as getters for fast–diffusing impurities whose parameters, as a result, may be drastically worsened [Litovtchenko and Romanyuk, et al. (1986)]. To have this negative effect reduced, one uses a gettering a gettering of fast–diffusing impurities by damaged back–sided layers on the structures. Though it is clear here that capturing the undesired impurities in the doped regions will make the gettering process different from gettering in slightly–doped slices. The specificities of gettering the silicon structures with planar–doped regions has been first analysed in [Gousev and Bogatch, et al. (1990)] where the redistribution of fast–diffusing impurities from a planar–doped region into a gettering layer was simulated by a mathematical model suggested; the model also included the microscopic parameters of the definite impurities and capturing centres (traps). The analytical ratios obtained here make it possible to choose gettering regimes and getter's parameters, and this increases a structural perfection in planar–doped regions. It is shown that if the traps in the getter and in the planar–doped regions are of one type, then the cleaning maximum in the active regions of slices will not depend on the temperature of the gettering anneal. Theoretically and experimentally it was shown [Gousev and Bogatch, et al. (1990)] that among recombination–active fast–diffusing impurities in silicon, the planar–doped regions are most difficult to be cleaned from iron atoms, since a gettering process in this case includes not only a diffusive evacuation of the impurity through the bulk of the slice but also an emission of diffusing through the wafer bulk into the getter but also the emission of impurity atoms from the traps of the planar–doped regions. To raise the efficiency of terminal operations in gettering, the techniques of a forceful activation make the final operations of gettering more efficient, the atoms in the planar–doped region must be forcefully activated in their rejection from the traps into the interstitial site, being further diffused into the getter. For this purpose, during gettering the planar–doped regions are suggested to be subjected to an impulsive photonic annealing or irradiation. Since an efficiently of rejection of impurity atoms from the traps into the planar–doped regions depends on the energy characteristics of these regions, the authors [Gousev and Bogatch, et al. (1990)] think it necessary to incorporate special impurities into the getter; these impurities together fast–diffusing impurities form chemical compounds whose dissociation temperature exceeds the temperature of the gettering anneal. Such conditions can be met by nitrides and oxides of metals.

For the intensive cleaning of planar–doped regions from fast–diffusing impurities, the problem of how far lies the getter from these regions is important; because the locally high mechanical stresses induced by dislocation aggregates in the damaged layer can drastically decrease the energy barrier for emitting the impurity atoms from the traps. Such possibility is suggested by the internal gettering. A model of internal gettering in silicon was suggested in [Gaidoukov and Kozhevnikov (1995)]. The authors have studied a mechanism of long–range interaction in the "impurity–gettering centre" structure. In this model, a spherical oxygen precipitate works as a gettering centre; this precipitate interacts with the unbalanced IP defects only through a dipole mechanism. According to the authors, this is determined by the precipitate elastic field having a purely shifting deformation. The dipole mechanism works, if the point defects are unified into the simplest aggregates like the IP defects–impurity or impurity–impurity aggregates. The calculations done exhibit that the two defects with a spherically symmetric field of stresses are united to form a dipole with a tetragonal distortion of the lattice. It was found that independently of the sign the precipitate elastic field has this field (in its vicinity) is always enriched with PD–dipole aggregates, i.e. the oxygen inclusions created in the internal getter serve as impurity atom condensing centres. If an aggregate on the precipitate surface is being decomposed (at this an impurity atom releases), then for sustaining the equilibrium one needs a diffusive–drifting supply of new dipole aggregates. Thus, for the situation when mobile aggregates of the type "impurity atom – IP defects" are formed far from the oxide precipitate and when these aggregates decompose near its surface, this model is able to explain an internal gettering mechanism which is not limited by oversaturation and works as some "drifting pump" designed to drop the dopant concentrations in the bulk of the crystal.

4 Physical foundations for low–temperature gettering techniques

Physical fundamentals of low–temperature gettering in semiconductors are discussed. It is experimentally shown how the gettering processes are affected by the abrasive and abrasive–chemical treatment of crystal surface, structural changes and characteristics in semiconductors, semiconducting structures, epitaxial compositions and in active components of SOS–structures induced by hydrostatic compression, ultrasonic irradiation and accelerated ions irradiation.

Developing the production of functionally complicated ICs of submicronic parameters closely relates to a continuous perfection of the original monocrystal growing technologies, wafers production, deposition of superthin layers of conductors and dielectrics, high resolution lithography, ionic doping and the like. At this, the number of liquid processings for the surfaces of structures and high–temperature operations are brought to a minimum. To locally and selectively activate the chemical and physical reactions that form the active and passive elements in microcircuits, there are employed a coherent and incoherent photonic irradiation, electronic and ionic beams, electrical and magnetic fields.

Contemporary technologies for superhigh–speed ICs need a high–quality original silicon and monocrystalline device layers, defect–free layers in dielectrics, metals, polymerical resistors and in lithographic masks. In the IC crystals up to $10 \times 10 \ mm^2$ in size, a layer defectiveness must not exceed $0.1 \ def / cm^2$. With the device layers getting thinner, there sharply emerges a danger of contaminating them by uncontrollable process–induced impurities. This worsens the electrophysical characteristics of SHS ICs, and, in particular, increases noise levels against a background of low working currents and voltages at which the highly–integrated circuits and submicronic elements run. That is why purity and structural perfection in SHS IC device layers are severely demanded; these parameters are firstly determined by the extent to which the original wafers were contaminated. The dynamics of these growing demands for an admissible contamination of silicon wafers looks as follows [Mil'vidsky (2000)]: year 1995 – $5 \times 10^{10} \ cm^{-2}$, year 1998 – $2.5 \times 10^{10} \ cm^{-2}$, 2000 – $1 \times 10^{10} \ cm^{-2}$, 2004 – $5 \times 10^{9} \ cm^{-2}$. In this connection, gettering the impurities and defects in thin presurface layers of semiconducting structures is becoming a very crucial problem.

It is evident that such techniques must be low–temperature and must not bring a thermal erosion to the surface. These techniques could be exploited not only at the initial stages in the device manufacturing line but in the finishing operations as well, and also for treating the ready work pieces, in order to optimize and improve their characteristics.

This chapter discusses the physical fundamentals for developing low–temperature gettering techniques; these techniques are based on generating unbalanced IP defects and elastic waves that provide a mobility of impurities and a recombination of extended defects during a technological process. A number of specific ways to reduce a residual defectiveness in the semiconductor presurface layers is also suggested.

4.1
Gettering effects during an abrasive and chemical treatment of crystal surfaces

That abrasive and chemical treatment of the semiconductor surface induces unbalanced IP defects and elastic waves has been proved experimentally and theoretically [Gerasimenko and Dvouretchensky (1975); Maximyuk and Bourbelo, et al. (1977)]. Therefore, it has sense to describe the basic research in this field only briefly; the results gained in it make it possible to comprehend an action of the gettering mechanism during these treatments.

The vacancy aggregates (the analogs of VV–centres during an ionic irradiation) arising in the silicon layers structurally damaged by grinding and polishing have been recorded with the help of the electronic paramagnetic resonance and internal friction [Gerasimenko and Dvouretchensky, et al. (1972); Gerasimenko and Dvouretchensky (1975); Maximyuk and Bourbelo, et al. (1977)]. According to these authors, excessive vacancies and their aggregates are born by interacting dislocations induced due to a microplastic damage of the material in the local regions of its contact with the abrasive particles. Through the internal friction technique it has become possible to observe A − centres in p − and n − type silicon; these centres exhibit a relaxational maximum of internal friction of activation energy $0.38 \, eV$ and frequency factor $2 \times 10^{13} \, s^{-1}$ that corresponds to the reorientation of the vacancy–oxygen aggregate. This maximum vanishes, if a layer 30 mcm thick is etched off; this manifests a sufficiently deep penetration of unbalanced vacancies into the bulk of the crystal during a grinding process.

The greatest portion of information concerning the generation and migration of unbalanced IP defects to a considerable depth during the abrasive and chemical action on the semiconducting crystals was obtained in the course of studying the long–range effect [Perevostchikov and Skoupov (1992)]. This effect has been spotted in monocrystals and epitaxial structures of silicon and gallium arsenide having different types and values of conductivity after these structures were subjected to one-side grinding by a loose and cohered abrasives of various grain compositions and to chemicodynamical and chemicomechanical polishing. This effect is betrayed by changes occurring in such crystal parameters as residual mechanical stresses, photoconductivity, photoluminescence, microhardness, concentrated growth and process–induced microdefects; these features usually become evident near the side being reversal to that already processed [Fig. 4.1] [Perevostchikov and Skoupov (1994)].

External action

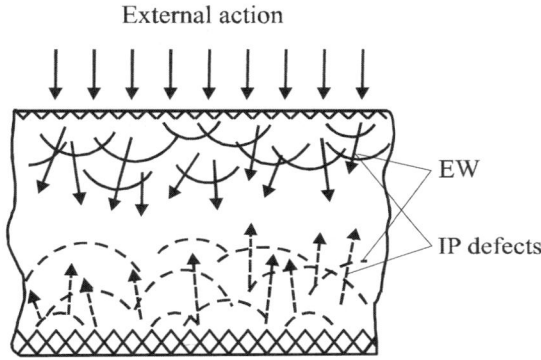

Fig. 4.1. A birth of a direct (continuous arrows) and a reversible (broken arrows) long–range effects.

In more details, the experimental observations of this effect are described in [Maximyuk and Bourbelo, et al. (1977)].

In contemporary literature, this kind of effect is thought to be born by the following three reasons: a) the elastic stresses created by a structurally damaged layer and a transforming state of the extrinsic defect composition; b) a recombination of the original imperfection of crystals under the action of the IP defects; and c) the elastic waves generated within the zone of the material microfailures induced by mechanical or chemical treatment of the surface. If the process–induced defects are highly concentrated, then the role of elastic stresses becomes significant; this takes place in deep abrasive operations (cutting, grinding). In the case of relatively soft actions, for example, polishing, when the thickness of the damaged layers and the concentration of defects in them become small, the importance of this aspect for the long–range effect may be neglected, and taken into account should be only the changes in the extrinsic defect composition caused by the fluxes of IP defects and elastic waves.

In mechanically processed crystals, during a plastic deformation of presurface layers the mechanisms for generating IP defects may be classified in accordance with the importance they play in creating a damaged layer; these mechanisms may be: a dislocation mechanism, a dynamical surface "source–drain" mechanism and a wave mechanism.

During a motion of dislocations, point defects may be induced, first, at the expense of the annihilation of the dislocation segments with the Bűrgers edge component (these segments are located in the neighbouring parallel sliding planes); and, second, by a dynamic instability of steps (terraces) on the dislocation line (at the edge of the extraplane) [Damask and Dienes (1963); Friedel (1964)]. The second process is real when the dislocations travell at the speeds close to that of sound. As for the first process that is most probable during a mechanical treatment, it may occur, according to [Indenbom and Orlov (1977)], in the course of the annihilation of closely lying dislocations induced by various sources or by a single source like the Frank–Read source giving birth to stepped dislocations. When a source is running, at the expense of steps there may be created dipoles which, at a small distance between them, will annihilate emitting the IP defects. One more essential mechanism for generating the IP defects relates to the consequences of passing by the dislocation through a forest of dislocations; as a result of such travell, on the moving dislocation there will appear the steps lying in the nearest planes. To escape from these consequences, the Bűrgers vector of the tree being passed must not be parallel to the sliding plane of the travelling dislocation. Thanks to their overcrawling, the steps formed are "selfexploded" to induce point defects [Friedel (1964)].

If the sources of IP defects are the moving dislocations with steps, then the PD concentration may be found as

$$N_{ipd} = \frac{LN_d\varepsilon}{b^2} ,$$ (4.1)

where

L is a maximal distance at which a dislocation is displaced or a loop is expanded from the Frank–Read source;

N_d is an original density of dislocations;

b is the Bűrgers vector modulus; and

ε is a value of the plastic deformation.

If the point defects arise during the annihilation of dislocation segments, then the concentration is described as

$$N_{ipd} = \frac{l\varepsilon}{L^2 b^2} ,$$ (4.2)

where l is a length of the annihilating segments approximately equal to the size of the dislocation source.

Using the information about the damaged layers [Perevostchikov and Skoupov (1992)] and, during a grinding process, taking for silicon $L = 10^{-3}\, cm$, $N_{ipd} = 10^3\, cm^{-2}$, $E = 10^{-3}$, $l \approx L$ and $b = 3 \times 10^{-8}\, cm$ we from (4.1) obtain $N_{ipd} \approx 10^{12}\, cm^{-3}$ and from (4.2) $N_{ipd} \approx 10^{15}\, cm^{-3}$. The last magnitude is close to the concentration of IP defects in silicon at T_{melt} [Alekhin (1983)]. It is to be noted here that if the point defects are also generated in a step–by–step fashion, then their concentration will be high, provided the effect of postoperational accumulation of defects is taken into account [Perevostchikov and Skoupov (1992)] and $N_{ipd} = 10^6 - 10^7\ cm^{-2}$ in (4.1) is assumed to be a density of dislocations that retained in the presurface layer from the previous rough abrasive treatment.

Another group of processes, whose activation caused by a mechanical treatment of the surface makes the unbalanced IP defects be injected into the bulk of the material, relates to dynamical oscillating elastic stresses within the zone where the crystal is contacting with abrasive particles [Alekhin (1983); Perevostchikov and Skoupov (1992); Perevostchikov and Skoupov (1997)]. The first channel for creating the IP defects is a transformation of the point defect aggregates induced in the presurface layer by the motions and interactions of dislocations. Within the field of spatial alternate–sign elastic stresses induced by the abrasive particles travelling across the crystal surface, the aggregates of vacancies and interstitial atoms, being different in sign and dilatation power, can decompose or grow in size. Therefore, the vacancy aggregates will decompose within the zone of extending stresses (deformations) and will grow in the places of lattice contraction. The aggregates of interstitial atoms behave in an opposite fashion. Their decomposition results in free IP defects, one portion of whom going into the bulk and another into the external and internal drains.

The second channel for generating IP defects is governed by the oscillating stresses that exert action on the surface and establish the local regions where the presurface layers become oversaturated and undersaturated with vacancies and interstitial atoms. This mechanism, called "a vacancy pump" in [Alekhin (1983)], implies the IP defects being pumped from the surface (where the vacancy and interstitial atom generating energies are approximately two times as less than in the bulk) into the regions where the external contact pressure of abrasive particles gives rise to undersaturated point defects of the given type. For example, the vacancies are being "sucked in" (ingested) by an expending region of the lattice. During the next phase of the oscillating pressure, i.e. the compression phase, such regions become oversaturated with vacancies; these vacancies are accumulated into aggregates that migrate into the bulk of the crystal. Most probably such aggregates are heterogeneously generated on impurities or second–phase inclusions. The vacancy aggregates and intrinsic interstitial atoms vary within the counterphases of external pressure oscillations, and a difference in diffusion coefficients of these defects determines a disagreement in depths between these aggregates and accumulations.

At the depths exceeding the diffusion lengths of point defects, the unbalanced IP defects may be generated by elastic waves spreading from the crystal surface being treated mechanically [Perevostchikov and Skoupov (1992)]. How the elastic waves affect the extrinsic defect composition in the mechanically treated semiconductors is considered below in more details.

As was already noted, the extrinsic defect composition of crystal is being changed throughout the entire cycle of mechanical and chemical treatment of the surface, including finishing operations as well. In [Batavin and Drouzhkov, et al. (1980)], it was found through positron annihilation that in the silicon crystal layers the concentrations of IP defects may be changed by chemicomechanical treatment as much as $100 - 150$ mcm in thickness. In the further thermotreatments, in particular, during an epitaxial process these defects get together into aggregates and clusters of impurity atoms whose concentration makes up $10^{15} - 10^{16}$ cm^{-3} and their average size is about 1 nm.

Upon treating the zero–dislocation silicon wafers chemicomechanically at $281 - 295$ K, the presurface layers are found to hold the defects that, as was revealed by selective etching, look like hillocks $1 - 3$ mcm in size and are of the density equal to $4 \times 10^4 - 3 \times 10^6$ cm^{-2}. Through the raster–electronic microscopy in the mode of induced currents, these defects were found to correspond to the regions of the amplified recombination of charge carriers. The defects revealed by raster–electronic microscopy were of the following two types: the first possessing an anisotropic and the second a spherically–symmetric field of elastic stresses. The sizes and nature of the contrast in defect images betray coherent and semicoherent second–phase inclusions [Petroff and Kock (1975)]. A distribution of defects across the wafer surface correlates with a growth distribution of impurities in the ingot cross–section.

A use of by–layer selective etching made it possible to obtain the in-depth distribution profiles for second–phase particles in the $n-$ and $p-$ silicon wafers, with various specific resistance being dependent on the conditions of chemicomechanical treatment. For example, for the wafers of $\rho -$ silicon (boron– doped, $\rho = 10$ $Ohm \cdot cm$) subjected to chemicomecanical treatment with an intensively cooled polishing agent the defect concentration will drop monotonously with depth. If a cassette with its wafers is cooled, then the defects are maximally concentrated far from the surface, with the sizes of discharges diminishing simultaneously. In both cases, the defects were observed available deeper than 120 mcm. In contrast to this, in the wafers of $n - Si(001)CZ$ ($\rho = 4.5$ $Ohm \cdot cm$, phosphorus–doped) the defect concentration drops to zero at the depths of up to 90 mcm. In the crystals of $p - Si$ CZ ($\rho = 1$ $Ohm \cdot cm$, boron–doped), no second–phase particles have been found, though the selective etching reveals within the layer of up to 10 mcm the unidentified defects like etching wells. In the wafers of Si sapphire–doped ($\rho = 0.01$ $Ohm \cdot cm$), no defects are revealed. If the polishing temperature for the wafers of $p - Si$ CZ ($\rho = 10$ $Ohm \cdot cm$, boron– doped) increases, the concentration and sizes of second–phase particles diminish; and upon chemicomechanical treatment there will emerge no defects, as well as in the specimens obtained by polish–free spalling. An increase of pressure makes the second–phase defects less probable. Practically no defects arise in the vicinity of such extended distortions like microcracks retained from the previous abrasive treatments. The second–phase discharges induced by chemicomechanical treatment serve as additional centres for defect formation during further high–temperature oxidation, diffusion and epitaxial build-up.

The results obtained by different authors are explained by silicon impurity atoms displacing from the surface into the depth of the crystal under the action of elastic waves arising in the zone where suspension hard–phase particles are getting into contact with the surface. At low temperatures, the impurities in the presurface region become oversaturated and, as a result, the hard solution decomposes to form second–phase particles. Since a rising concentration of the dopant increases the absorption of elastic waves, a defective layer, when treating the $n - Si\,(001)\,CZ$ ($\rho = 4.5\ Ohm \cdot cm$, phosphorus–doped), will be thinner than in the event of the $p - Si\,(001)\,CZ$ ($\rho = 10\ Ohm \cdot cm$, boron–doped). Similarly is explained the fact that when treating the $p - Si\,(001)\,CZ$ ($\rho = 1\ Ohm \cdot cm$, boron–doped) and the $Si\ CZ$ ($\rho = 0.01\ Ohm \cdot cm$, sapphire–doped) the defects are observed within a thin layer or vanish at all.

How layers are damaged during the chemicomechanical treatment of silicon by intrinsic interstitial atoms being formed and diffusing at large depths – this mechanism is seemingly the following. The local pressure being periodically changed in a given point of the surface and generated during the chemicomechanical process by hard–phase particles about $1\ mcm$ in diameter may approach $2.5\,GPa$ and regulate an interstitial mass-transfer of the material from the contacting zones [Indenbom and Orlov (1977)]. According to estimates in [Alekhin (1983)], within the wafer presurface layer (if there is a contact compression) the oversaturation with intrinsic interstitial atoms will be

$$\frac{C_p}{C_o} = \exp\left[\frac{p\Omega}{kT}\right] \approx 4.6 \times 10^5 \ .$$

To suppress the defect generating process induced by excessive interstitial atoms periodically emerging in the surface local regions at the instant of a force-down of abrasive particles, it is necessary to increase a rate of removing off the material up to the values at which the rate will exceed the rate of atom diffusion into the bulk of the wafer. However, even under this condition the residual process–induced defects at large depths are impossible to be eliminated.

The model suggested is supported by the results in [Alekhin and Litvinov, et al. (1984)] describing the effects of chemicomechanical treatment on the density of OSFs. The chemically and mechanically treated silicon wafers obtained a high density of OSFs when being oxidized in dry oxygen at $1423\ K$, though, this density was successfully dropped by a preliminary annealing at $623\ K$ in the same ambient. It was observed that a longer preliminary low–temperature annealing diminishes the concentration and increases the sizes of stacking faults;

at the thermal treatments longer than 2 hrs these faults become more than $10\ mcm$ in size. These results are in good agreement with the well–known idea of defect formation during the thermal treatment of silicon crystals that contain microdefects, i.e. the centres for generating the OSFs [Leroy (1979); Ravi (1981); Shelpakova and Yudelevitch, et al. (1984)], and prove the idea that similar defects are induced by the chemicomechanical treatment.

A final operation in preparing semiconductor wafer surfaces, and sometimes an intermediary operation [Perevostchikov and Skoupov (1992); Perevostchikov and Skoupov (1997)], is etching the wafers chemically; most often, this is a chemicodynamical polishing. This kind of treatment, alongside with possible contaminants from etchers [Shelpakova and Yudelevitch, et al. (1984)], can also induce unbalanced IP defects in the crystal [Fistoul' and Petrovsky, et al. (1979); Fistoul' and Sinder (1981); Fistoul' and Tchernova, et al. (1980); Italyantsev (1991)]. Actually, if the crystal is balanced with the ambient (an etcher), then above its surface there occurs a balanced concentration of the removing material equal to its solubility in the etcher, whereas in the crystal itself there will be, accordingly, a balanced concentration of vacancies. Here, the fluxes of the removing and condensing atoms (or injecting and "evaporating" vacancies) will be equal. If an off–etching substance is drained from the reaction surface or its solubility becomes very large, then the above fluxes will be not equal. The material's atoms will be continuously drained from the reaction surface and on it there will emerge unbalanced vacancies that will be diffusing into the depth of the crystal. As the concentration of vacancies within the presurface layer exceeds the balanced concentration, their oversaturated solution will then dissolve to form a porous structure. With the external conditions (the etching rate, the intermixing of the etcher and the temperature) remaining unchanged, the vacancy profile in the presurface region will, upon some time, become stationary. With the impurities available in the crystal, the vacancy distribution profile gets decorated with them; this is observed so [Fistoul' and Tchernova, et al. (1980)] in the case where the sodium atoms migrate to the surface of the silicon crystals being etched.

The atomic mechanism of generating unbalanced vacancies when crystals are etched chemically is suggested in [Italyantsev (1991)]. In many ways, this mechanism is similar to the classic Schottky model according to which the vacancies emerge as a result of the two sequential events: a) a formation of vacancies within the first presurface monolayer of the crystal (V_s); this formation is stimulated by a continuous travell of atoms from the surface into the etcher; and b) a thermal diffusion of vacancies into the bulk of the crystal. Though, the mechanism suggested is said to be crucially different from the Schottky model. First, the etching process induces the thermodynamically unbalanced vacancies, because in this process the chemical dissolution may play a role of some "pump" that is pumping the vacancies from the surface into the bulk of the crystal. Second, the kinetics of the chemically stimulated generation of vacancies is affected not only by temperature but by the rate of the forceful removal of atoms from the crystal surface as well.

The basic requirements to the etching process needed to ensure a stable generation of unbalanced vacancies are formulated in [Italyantsev (1991)]. These requirements come to being from the following. A chemical etching process may be thought to involve the two situations differing from the point of view of the etching mechanism and the IP defect generation. During the first situation, the material is removed at the expense of a new phase being created by the chemical interaction of the etcher components with the film crystals; this phase is then dissolved in other components of the same etching solution. In this case, despite the material having been actually removed, a chemical etching of the crystal material itself may not take place and the nature of IP defects, being induced in this case, will be determined by the types of hard-phase reactions to build a new phase, in the shape of a surface film. For example, when etching the silicon in liquid etchers consisting of the silicon oxidizer and dissolver SiO_2 (SiO_x), most probably there will be generated the intrinsic interstitial atoms rather than vacancies, in analogy to the thermal oxidation [Hu (1974)]. This is proved experimentally, too [Perevostchikov and Skoupov (1992)]. Another situation considered in [Italyantsev (1991)] describes the etching process producing no intermediate films and so the silicon atoms are travelling from the surface into the molecules of the reaction products in the etcher, i.e. there occurs an immediate etching.

Besides the immediate etching, to have the unbalanced vacancies generated it is necessary to have a portion of atoms being discharged by the normal mechanism, since in the case of the tangent dissolution the locational structure of the silicon atoms in the surface is selfproduced and no surface vacancies (V_s) appear. For nonsingular, i.e. for the atomically rough, surfaces of the crystal, the mechanism of normal etching (as well as that of growth) is basic. However, the surface vacancies induced on such edges of the crystal will slightly change the free energy of the surface that itself is sufficiently high as a result of the diffusive structure of the microrelief. Therefore, in spite of the normal etching running for the nonsingular edges naturally, it will not induce a considerable unbalance of surface vacancies that could provide a travell of vacancies into the bulk of the crystal. In this connection, according to [Italyantsev (1991)], the crystal can be oversaturated with vacancies with the help of etching the singular (atomically smooth) and vicinal (in their orientation being near to singular and flat–stepped or terraced) edges. For such edges, during an etching process there occurs a dominating tangent–oriented separation of atoms from the atomic steps, and in usual conditions the normal etching has a slight contribution. This contribution may be increased, if some agent is incorporated into the etcher whose atoms (molecules) migrate to the atomic steps faster than the atoms (molecules) of the chemically active agent; in this way the channel of tangent–oriented etching becomes partially blocked (Fig. 4.2) [Perevostchikov and Skoupov (1994)].

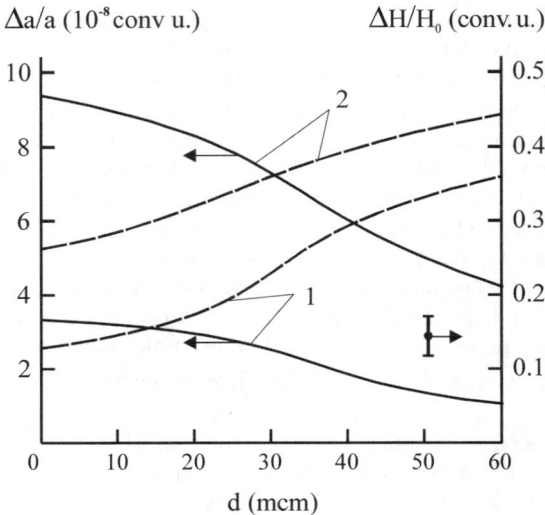

Fig. 4.2. Dependence of deformation (continuous curves) and relative changes of microhardness (broken curves) in silicon wafers (**1**) and gallium arsenide (**2**) on the sizes of abrasive particles, when treating the side reversible to that under study.

If a vicinal edge deviates from a singular one at sufficiently small angle θ so that the distance (L_{st}) between the steps exceeds more than two times the diffusing run (L_{ads}) of the etcher adsorbed atom (Fig. 4.3), then for some portion of etcher atoms $\dfrac{L_{st} - 2L_{ads}}{L_{st}}$ no drain to the atomic steps will be possible and the probability of atoms removed by the normal mechanism must grow.

For the surface under etching, the average value of L_{st} is $\dfrac{a}{\theta}$, and, according to [Barton and Kabrera, et al. (1984)], L_{ads} is calculated from $a \times \exp \dfrac{E_{des} - E_m}{2kT}$, where a is a height of the atomic step; E_{des} and E_m are the activation energies, respectively, for the desorption of the etcher atom from the flat region in the surface and for its migration across the surface. Thus, it may be expected that a proportion of the normal etching will increase at

$$\theta < \theta_0 = 0.5\exp\left[-\frac{E_{des} - E_m}{2kT}\right].$$

In real crystals, θ is estimated with a use of the density of dislocations having a swirl component and other defects creating on their surface additional atomic steps.

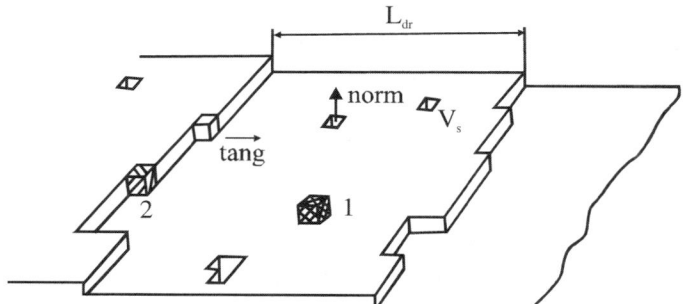

Fig. 4.3. The elements in the structure of the crystal vicinal edge during chemical etching: **1** an adsorbed atom of the chemical etcher; **2** an atom of the added agent in the etcher blocking the tangent etching.

The magnitude θ_0 depends on the surface diffusion of etcher atoms and thus in each specific case it has to be determined separately. In [Italyantsev (1991)], θ_0 was estimated for the case when E_{des} and E_m correspond to the intrinsic atoms of the (111) $-$ oriented crystal with face–centred lattice: $E_{des} - E_m \approx 2\varphi$ [Barton and Kabrera, et al. (1984)], where φ is an atomic energy of electronic bonding. For $\varphi = 0.5\ eV$ and $kT = 0.1eV$ we get $\theta_0 \approx 0.01^0$ that is obtainable by contemporary technologies. At high temperatures, an atomically–smooth surface of the crystal can change its structure and assume a nonsingular state. This temperature is determined by the nature of the crystal and is an upper limit for the etching process to be performed to incorporate excessive vacancies into the crystal.

Since during etching, the real (C_s) and balanced ($C_s^{\,0}$) concentrations of vacancies on the surface interrelate as $C_s > C_s^{\,0}$, the oversaturation of the surface (V_s) with vacancies will – under the law of acting masses – make the spatial layers oversaturated with unbalanced vacancies. One portion of unbalanced vacancies will travell into the bulk of the crystal where the enthalpy of their generation is higher than on the surface; this travell is caused by the configurational entropy of the vacancy solution in the bulk exceeding the surface entropy. The probability that during its migration time τ_s the vacancy will move into the lower atomic layers along the surface is calculated as

$$P = 1 - \left(1 - \frac{t_s}{t_v}\right)^{\tau s / ts} ,$$

where

t_v and t_s are the periods of diffusive jumps of vacancies in the bulk and on the surface of the crystal;

$1 - \dfrac{t_s}{t_v}$ is the probability that the vacancy will not "fall down" into the bulk in one of the surface lattice sites it is passing through;

$\left(1 - \dfrac{t_s}{t_v}\right)^{\tau s / ts}$ is the probability that no transition $V_s \rightarrow V$ will occur in any surface lattice sites;

P is the probability that a transition $V_s \rightarrow V$ will occur at least in one of the crystal sites τ_s / t_s visited by this vacancy.

The scheme of how the vacancy is travelling across the surface (V_s) and how the transition $V_s \rightarrow V$ is performed are shown in Fig. 4.4.

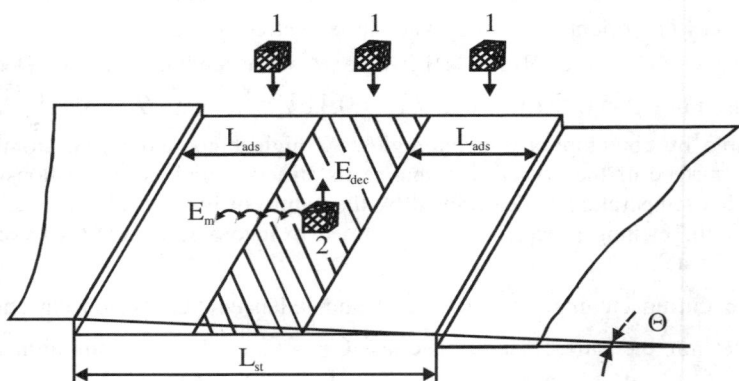

Fig. 4.4. A simplified scheme for a crystal vinicial edge disoriented by the small angle θ with respect to a singular plane. The shaded region is the region where the etcher atoms are adsorbed from the gaseous phase **1** from which the adatoms **2** are not able to approach the atomic steps.

A usual value of the ratio t_s / t_v is $10^{-4} - 10^{-2}$ [Gegouzin and Kaganovsky (1984)]. Therefore, the probability may be written as $P = 1 - \exp\left(-\dfrac{\tau_s}{t_v}\right)$.

Hence, at $\dfrac{\tau_s}{t_v} \gg 1$ the kinetic probability of the surface–to–bulk transition by vacancies will be close to unity. In other words, the kinetics of the transition $V_s \to V$ does not limit the vacancy injection into the bulk of the crystal, and the flux of vacancies into the bulk of the crystal is determined by a birth rate of V_s, i.e. by the rate at which atoms are removed during etching by the normal mechanism. If $\dfrac{\tau_s}{t_v} \ll 1$, then P will be small and saturating the presurface layers by unbalanced vacancies will be constrained by the transition $V_s \to V$. Such situation is possible to occur, say, when the surface is very much filled with atoms of some sorbent that will fast occupy (for times much less than t_v) the position V_s as soon as this position is created by the etching process.

Here comes a need in restraining an etching rate and this limitation arises because of two reasons. First, the monolayer etch-off time is an upper limit for the lifetime of V_s, and to fulfil the condition $\tau_s \gg t_v$ one should have the monolayer etch-off time exceeding $t_v = \dfrac{a^2}{D_v}$. If the etching-off process is not accompanied with the competing growth processes, then the microscopic etch-off time may be expressed through the macroscopic etching rate as $\dfrac{a}{V_{etch}}$. Then, the first limitation will be $V_{etch} \ll \dfrac{D_v}{a}$, where D_v is a bulk diffusion coefficient of vacancies, and a is a magnitude of the lattice period order.

The second limitation for the macroscopic etching rate is determined by the need to have this rate less than the diffusing rate of the vacancy front, since only in this case we may speak about the silicon presurface layers being oversaturated with vacancies. This condition may be written as $V_{etch} \ll \sqrt{\dfrac{D_v}{t}}$, where t is a characteristic time of etching. The quantitative estimates made in [Italyantsev (1991)] for neutral vacancies in $p-$ silicon have showed that the first of the above conditions is practically always fulfilled at usual etching rates

$10^{-3} - 10^{-1} \ mcm / s$, provided the temperature is some times higher than in the room.

The second condition ($t \approx 10^3$) within the given range of etching rates is fulfilled at temperatures $773 - 973 \ K$.

Thus, in order to have vacancies injected into the crystal during a chemical etching, it is necessary to obey the requirements assigned in [Italyantsev (1991)]. First, an etching process must be immediate, without inducing the new intermediates phases with their later dissolution. Second, the crystal surface must be precisely oriented with respect to the singular edges, to have the distance between atomic steps exceeding the assigned value. Third, the restraints for etching rates and temperature regimes must be obeyed, i.e. the low temperature is regulated by the ratio of the etching rate and vacancy diffusion, and the maximal temperature is determined by a spontaneous transformation of the atomically–smooth surface to the atomically–rough one.

The vacancies injected into the presurface layers can quasichemically react with the crystal defects of different nature and stimulate such phenomena as a diffusive impurity redistribution, a dissolution of incorporated defects, a change in the spectrum of the defects producing deep levels, a formation of micropores [Fistoul' and Sinder (1981); Fistoul' and Tchernova, et al. (1980)]. As shown in [Vyatkin and Italyantsev, et al. (1986); Italyantsev and Mityukchlyev, et al. (1988)], by generating the IP defects (chemically stimulated by gaseous etching at temperatures higher than room temperatures) – mostly vacancies – it becomes possible to efficiently regulate the properties of semiconductor monocrystals and the structures with ion–doped or dielectric layers. According to the authors of this research, this way creates a very wide possibility for complicated semiconductors, since for them a sequence of the chemically stimulated incorporations of point defects into various sublattices of the crystal may be strongly varied – thus creating a needed spectrum of defects for producing a material of the desired properties.

The last conclusion is supported by the results in [Kasatkin and Perevostchikov, et al. (1993)], where the effects of chemicomechanical and chemicodynamical polishing on the deep centres in $n - GaAs$ were studied. By the stationary capacitance spectroscopy, it was shown that in the crystals of low concentration of original defects, upon polishing them chemicomechanically by the alkali suspension of silicon dioxide, there arise electronic traps $E1$, $E2$, $E3$ of the activation energies 0.69, 0.54, $0.3 \ eV$, respectively. The concentration of these traps in the presurface layers reaches $10^{16} \ cm^{-3}$. The dominating traps ($E1$) were bonded with arsenic atoms in the interstitial sites. Chemicodynamical polishing diminishes the concentrations of the above defects as much as some orders and makes their in-depth distribution profiles more sharp.

In [Karzanov and Perevostchikov, et al. (1994)], through measuring the Hall effect it was found that the $n - GaAs$ crystals undergoing a liquid etching in the solution $H_2SO_4 : H_2O_2 : H_2O = 3 : 1 : 1$ at the rate $3 - 5\,mcm/\min$ change the concentration of charge carriers, their mobility and specific electrical resistance. In this research it was revealed that in etching off a front side of the wafers in a by–layer fashion (the electrical measurements were done on this side) and retaining a damaged layer (through polishing by diamond paste) the specific resistance of the material will drop near the back side, and the concentration and mobility of charge carriers will increase. A two-sided etching increases the mobility of charge carriers and simultaneously drops their concentration and the resistance of specimens. In the same way there will be changed the electrical parameters during etching a back side of the crystals only. The results suggested in [Fistoul' and Tchernova, et al. (1980)] are explained on the basis of the ideas that the extrinsic defect composition of the $GaAs$ crystals is transformed by elastic waves and unbalanced IP defects, vacancies mainly; being induced by the etching process.

In [Perevostchikov and Skoupov (1987); Perevostchikov and Skoupov (1987a)], the effects of the liquid chemical etching in the solutions $HNO_3 - HF$ of various compositions on the properties of the silicon crystal presurface layers were studied through measuring the microhardness and X-ray three-crystal diffractometry. The microhardness was found to increase with an increase of the etching rate; and the rates exceeding $5\,mcm/\min$ there appears a presurface hardened layer of the hardness of up to $15.5\,GPa$. Low etching rates brought the microhardness equal to $11\,GPa$. According to X-ray data, a rise of etching rates increased an average amplitude of residual deformation and simultaneously dropped the variance of the lattice period values for the presurface layers. By comparing the signs of the bend of atomic planes and the relative complement of the lattice period, it was concluded that chiefly positive centres of delatation are being accumulated in the silicon presurface layers and these centres are responsible for extending elastic stresses in the lower layers.

In [Perevostchikov and Skoupov (1987)], the experimental results were suggested to be explained in the below way: the silicon presurface layer is supposed to be saturated with atomic hydrogen generated by the dissolution of silicon in the $HNO_3 - HF$ solution and being highly mobile. In addition, structural distortions are caused by the elastic waves emerging on the reaction surface and by the unbalanced IP defects, mostly, the intrinsic interstitial atoms of silicon [Italyantsev (1991); Hu (1974)]. As was noted in [Perevostchikov and Skoupov (1987a)], a cumulative action of these phenomena leads to the long-range effects of chemical etching; in particular, these effects change the crystal microhardness on the side reversible to that being etched off. In this case the microhardness was also observed to increase with a rise of the etching rate, and especially when the region to be etched off has highly concentrated structural defects, for example, induced by diamond disk cutting and surface polishing operations. Since the crystals in [Perevostchikov and Skoupov (1987)] were rather defects, the long–range effect seemed mainly created by the elastic waves born by the defect transformations near the etching side; these waves stimulate a birth of unbalanced vacancies near the reversible side [Alekhin (1983); Perevostchikov and Skoupov (1992)]. The excessive vacancies efficiently dissolve the original growth and technological microdefects, i.e. the aggregates of intrinsic interstitial atoms that serve as potential sources for microplastic deformation in silicon when the indentor is forced. It is this very action that increases the microhardness.

That the elastic waves in silicon crystals are induced by one–sided chemical etching has been experimentally proved in [Gorshkov and Perevostchikov, et al. (1989)]. The elastic waves were recorded by the acoustoemission impulses from the previously chemically polished surface of zero-dislocation $p - Si$ (111) CZ crystals ($\rho = 20\ Ohm \cdot cm$) $0.4 - 0.42\ mm$ thick whose back side was etched in the $HNO_3 - HF$ solutions. Subjected to study were the specimens that prior to etching were grinded or polished by various abrasives, of grains from 5 to $40\ mcm$. Acoustic emission was recorded within the frequency range from 0.01 to $1.5\ MHz$. As was experimentally shown, in the course of etching a rise of concentration of the mechanical defects in the presurface layer of the specimen side increases the intensity and length of acoustic irradiation; here, the integral dependence of the signals is the more, the higher is the etching rate. Etching at slow rates $1 - 5\ mcm/\min$ generally makes the acoustoemission descrete, in contrast to a comparatively weak continuous component. The latter becomes more intensive with a growth of the crystal etching rate (Fig. 4.6).

In [Gorshkov and Perevostchikov, et al. (1989)], the elastic waves are thought to be born, first, by relaxation of local fields of elastic stresses [Boiko and Garber, et al. (1973)] that accompanies the dissolution of crystal and drift of structural defects to the surface, and also by the defects being deformed by forces acting during the etching process from the nearing surface. The second reason of their birth is the interaction arising between unbalanced IP defects injected by the reaction surface and the structural distortions available in the crystal. It is worth mentioning here that in the course of etching process the acoustic effect seems to be a common property in crystals, to be more exact, that of solids, irrespective of their structure. This is also proved by the results in [Khokonov and Shokarov (1988)] where the acoustoemission was observed to encounter when salol was dissolved in benzin and chloride sodium in distilled water.

The above said makes it possible to state that abrasive and chemical treatments of semiconductor surfaces induce unbalanced IP defects and elastic waves whose presence is a necessary condition for activating the low–temperature changes of extrinsic defect composition at considerable distances from the generation zone. lHowever, this condition may turn out to be insufficient to implement a gettering process, since fresh aggregates and accumulations can be formed by excessive vacancies and intrinsic interstitial atoms, hereby increasing a defectiveness of the material [Perevostchikov and Skoupov (1992); Perevostchikov and Skoupov (1987)].

material [Perevostchikov and Skoupov (1992); Perevostchikov and Skoupov (1987)].

From practical point of view, of greater interest are the ways for suppressing the defects induced by growth and technological microdefects – the clusters of point defects in the wafers of dislocation monocrystals. The conditions under which any external action upon the crystal can drop the concentration or completely eliminate the point defect clusters from its bulk or at least from device presurface layers are fully considered in [Italyantsev and Mordkovitch (1983)]. Since the conclusions made in this study are general enough and cover various technological actions, it is expedient to discuss this study in detail.

In the model suggested in [Italyantsev and Mordkovitch (1983)] for describing the transformations in the sizes of point defect clusters, it is supposed that the clusters and two–component hard solutions of mobile vacancies and interstitial atoms exchange particles at the expense of two processes – the emission of particles from the surface of point defect clusters and the condensation onto their surface. If the particles constituting the cluster and the IP defects condensing onto its surface are of one and the same type, then the size of point defect clusters will grow. Otherwise, the condensing IP defects and the cluster particles will be recombined and this will make the cluster less in size. Emission of particles from the surface of the cluster under the action of phonons always decreases the sizes of point defect clusters and increases the concentration of mobile point defects dissolved in the crystal. With these two processes taken into account, the equation, describing a change of efficient cluster radius ρ in time, will be of the form [Italyantsev and Mordkovitch (1983)]:

$$\frac{d\rho}{dt} = \pm \frac{\omega\varepsilon}{\lambda}\left[D_I(\overline{N}_I - N) - D_v(\overline{N}_v - N)\right], \tag{4.3}$$

where

λ, ε are the parameters dependent on cluster geometry;

ω is an elementary volume of the particle in a PD cluster;

$\overline{N}_{I,v}$ is the kinetically balanced concentrations of interstitial atoms and vacancies in the hard solution of IP defects of the crystal;

$N_{I,v}$ and $D_{I,v}$ are real concentrations of mobile IP defects in the crystal and their diffusion coefficients, respectively. The sign $\pm$ in the equations stands for the cases of vacancy-type or inclusion–type clusters.

The parameter ε takes either the value $\dfrac{S}{4\pi\rho^2}$, if the cluster is of the volume–type, or $\dfrac{S}{2\pi\rho b}$, if the defect is planar, where S is a square of the point defect cluster surface interacting with the solution of IP defects; b is the Bŭrgers vector modulus for a dislocation loop. For volume clusters, the parameter λ is equal to the defect efficient radius ρ corresponding to the radius of the sphere whose volume is equal to the volume of the point defect cluster. For the case of two-dimension point defect clusters, say, the oxidation–induced stacking faults or growth $A-$ clusters, it is assumed that $\lambda = 1$, where 1 is the magnitude equal in its order to the lattice constant.

The equation (4.3) shows that the nature of changes in PD clusters depends on the ratio of real and kinetically balanced concentrations of the Frenkel pairs in the crystal. In [Italyantsev and Mordkovitch (1983)] it is shown that the kinetically balanced concentration of point defects in the clusters, for the case of a two–component hard solution, always exceeds the critically balanced concentration of the same type of clusters interacting with a single–component of IP defects. It implies that near the cluster being balanced with a two–component solution there is a concentration gradient of IP defects of the same type that are held in the point defect clusters, whereas the cluster of the balanced size $\rho = \overline{\rho}$ ($\overline{\rho}$ carries an idea of the critical radius of the cluster determined from the condition $\dfrac{d\rho}{dt} = 0$) in a single–component solution does not excite a homogeneous distribution of mobile IP defects near its surface. A slight bias of IP defect concentration from the kinetic balance leads to the situation that the point defect clusters, initially being of size $\rho > \overline{\rho}$, will grow, and the clusters with $\rho < \overline{\rho}$ will be dissolved. Though, as shown in [Italyantsev and Mordkovitch (1983)], the sign of transformation of the point defect cluster (PDC) sizes, in contrast to the case with a single–component solution, is here determined not only by the sign of the bias of this or that component from the kinetically balanced value but also by the ratio of absolute magnitudes of these biases.

From (4.3) it follows that if the inequality

$$D_I \, \Delta N_I > D_v \, \Delta N_v \,, \tag{4.4}$$

where $\Delta N_I = N_I - \overline{N}_I$ and $\Delta N_v = N_v - \overline{N}_v$, is valid, then the introduced clusters (PDC_I) may be expected to expand and the PDC_v will diminish in size. In the opposite case:

$$D_I \, \Delta N_I < D_v \, \Delta N_v \,, \tag{4.5}$$

the sizes of PDC_I must decrease and PDC_v must grow.

As seen from (4.4) and (4.5), for a two-component hard solution of IP defects there are possible four variants of biasing from the kinetic balance each being accompanied with various types of cluster transformations. What evolutional changes the point defect clusters undergo in each of these four cases is described in [Italyantsev and Mordkovitch (1983)]; here, the results were also matched against the experimental data touching the transformations in the growth $A-$ type microdefects and oxidation–induced stacking faults in silicon crystals subjected to various external actions. For one of the pairs, $\overline{N}_v$ and $\overline{N}_I$, in the capacity of approximated values of the kinetically balanced concentrations there may be chosen the concentrations of the related IP defects arising in the entire bulk of the crystal during a long high–temperature annealing in the inertial ambient.

Hence, the following situations are analysed.

a) $\Delta N_I > 0$, $\Delta N_v \leq 0$. According to (4.4), such inequality must increase the sizes of PDC_I in silicon and decrease the sizes of PDC_v. Such ratio of concentrations encounters in oxidizing the silicon thermally; this causes a growth of $A-$ clusters and generates the incorporated oxidation–induced stacking faults [Ravi (1981)].

b) $\Delta N_I \leq 0$, $\Delta N_v > 0$. With this condition implemented according to (4.5), PDC_I must dissolve and the sizes of vacancy–type clusters must grow. As was emphasized in [Italyantsev and Mordkovitch (1983)], such situation occurs during a high-temperature annealing of silicon in the inertial ambient when Schottky–generated vacancies arise in its presurface layer. That the $A-$ clusters and the oxidation–induced stacking faults were dissolving in the presurface layers in silicon crystals during annealing in inertial gases or vacuum was experimentally observed in [Sugita and Shimizu, et al. (1974)].

c) $\Delta N_I > 0$, $\Delta N_v > 0$. In this situation, the PD clusters will be transformed in accordance with the ratio between ΔN_I and ΔN_v. Such situation emerges in crystals irradiated by high–energy light particles or $\gamma-$ quantums.

Since the rate of irradiation–induced vacancies and that of interstitial atoms is the same, the ratio between ΔN_I and ΔN_v (if $\Delta N_I >> \overline{N}_I$ and $\Delta N_v >> \overline{N}_v$) will be then determined by a difference in the rates of Frenkel pairs discharging onto the trap–centres. In irradiating the thick specimens, i.e. when $L >> \sqrt{Dt_0}$ (L is a thickness and t_0 is an irradiation length), the vacancies and interstitial atoms will be vanishing mainly thanks to their recombination, and as a result of this, the condition $\Delta N_I \approx \Delta N_v$ will be preserved within the entire bulk of the crystal. In this case, for silicon with $D_I > D_v$ the inequality (4.4) will be valid, and this will correspond to a growth of PDC_I and a dissolution of PDC_v. For thin specimens (with $L << \sqrt{Dt_0}$), a crystal surface itself will serve as a basic drain for IP defects. In the ideal case, a surface must be a more efficient drain for interstitial atoms, because $D_I > D_v$. That is why for thin specimens there is possible the situation when

$\Delta N_I < \dfrac{D_v}{D_I} \cdot \Delta N_v$, and in this case the incorporated clusters will dissolve and

PDC_v will grow.

d) $\Delta N_I < 0$, $\Delta N_v < 0$. In this case, similarly to **c)** , the PD clusters will behave in accordance with the concentration ratios ΔN_I and ΔN_v, and this process itself may be implemented via subjecting the crystal to a high–temperature high–speed heating when the IP defect concentration fails to approach a kinetically balanced value corresponding the current temperature. Under such conditions, one should expect a dissolution of both vacancy–type and incorporated clusters. As was experimentally shown in [Italyantsev and Mordkovitch (1983)], in silicon crystals treated by the impulse thermal process there arise – from some definite heating rate – the oxidation–induced stacking faults and the growth $A-$ clusters start to dissolve.

The ideas suggested in [Italyantsev and Mordkovitch (1983)] may be used to analyse and foretell the behaviour of PD clusters in crystals during various technological actions, including low–temperature, in order to choose the operations providing the most efficient fall of cluster concentrations. Besides, the suggested model, upon analyzing the states of clusters (microdefects) after some action upon the crystal, allows to specify a reason of PDC transformations and refer this reason to one of the four situations considered above. For example, this model helps to explain the fact (registered in [Perevostchikov and Skoupov (1987)]) that the growth and technological microdefects dissolve near the silicon wafer surface being reversible to the side grinded by the loose abrasive. When grinding one of the crystal sides, there evidently arise a vacancy oversaturation that diminishes the sizes and concentrations of clusters of interstitial atoms, i.e. the above situation **b** ($\Delta N_I \le 0$, $\Delta N_v > 0$) is being implemented.

Vacancies are generated in a structurally damaged layer by the above described mechanisms during a grinding process and in the course of later defect relaxation when the specimens are maintained in normal conditions [Perevostchikov and Skoupov (1992); Kapoustin and Kolokol'nikov, et al. (1992)]. However, the most portion of unbalanced vacancies seemingly emerges near the side opposite to that being grinded, under the action exerted by the alternate–sign field of elastic waves [Alekhin (1983); Perevostchikov and Skoupov (1992)].

On the basis of the experimental data given in this section, it is possible to suggest the following basic technological procedure for low–temperature gettering to be exploited at the very first stages of semiconductor wafer production.

a) Cutting an ingot into slices and a deep two–sided chemicodynamical polishing to remove a structurally–damaged layer. At this stage, prior to chemicodynamical annealing, the Si and Ge slices are appropriate to be annealed in the inertial ambient or vacuum at $0.4 - 0.5 T_m$ [Perevostchikov and Skoupov (1992)]; and as for semiconducting compounds like $A_3 B_5$ such annealing, but at $0.3 - 0.4\,T_m$, brings positive results upon chemicodynamical polishing [Perevostchikov and Skoupov, et al. (1985b)]. Residual defectiveness may be drastically reduced by treating the wafers thermally under elastic and stressed conditions [Bourago and Perevostchikov, et al. (1986)]. For this, upon slicing it is necessary to measure a curvature of slices and to bend them symmetrically to the axis so that the original convex side during annealing becomes concave but with the same curvature radius as it was upon cutting. Upon such treatment, one must choose a front (working) side of the slice, i.e. the side on which all active elements of LSIC will be formed. For this, the criteria formulated in [Perevostchikov and Skoupov (1992)] may be employed.

b) A one–sided grinding of slices from the back side by the loose middle–grain abrasive ($14 - 28\ mcm$).

c) A chemicomechanical and a superfinishing polishing of the front side.

d) A chemicodynamical two–sided etching at the depth equal to a total thickness of the relief–policrystaline and crack zones in layer structurally damaged by grinding near the back side of the slices.

In implementing this wafer producing procedure, a gettering process is arranged at the first two and the last stages of the treatment, respectively, preceding and terminating the stage of potential formation and accumulation of PD clusters near the front side of wafers, in the course of chemicomecanical polishing [Alekhin and Litvinov, et al. (1984)]. This namely drops a defectiveness in Si and $GaAs$ slices near the side under chemicomechanical polishing if a structurally–damaged layer is available on the opposite side. This, a so–called "reversible long–range effect" (Fig. 4.5) becomes strengthened if the abrasive particles used for grinding an opposite side of wafers grow in their sizes (Fig. 4.6); but this effect will gradually weaken if the specimens are long stored between grinding and chemicomechanical polishing operations.

A qualitative scheme for the emerging direct and reverse long–range effect during a one–sided excitation of crystals is shown in Fig. 4.5.

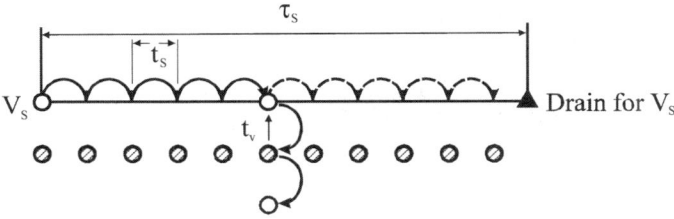

Fig. 4.5. The travelling trajectory of the surface vacancy V_s. The broken arrows show a potential trajectory of V_s, with the possibility of their injection into the crystal bulk left beyond the consideration.

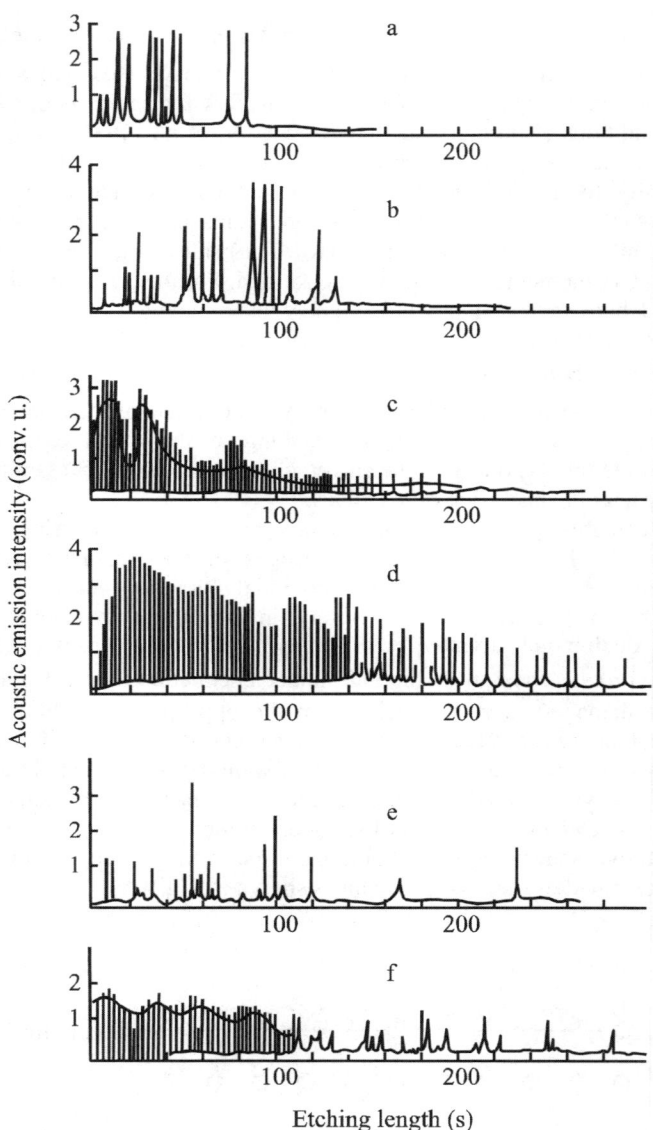

Fig. 4.6. A change in the intensity and amplitude of acoustic emission depending on the length of etching the Si wafer front side **a, b** by the polishing diamond paste ($5 - mcm$ grains) and by grinding it by **c, d** diamond disk ($40 - mcm$ grains) and **e, f** diamond disk ($28 - mcm$ grains). The etching ($mcm / \min$) was: **a, c** – 5; **b, d, f** – 10; **e** – 1.

Fig. 4.6 shows the deformation curves ($\Delta a / a$) and the microhardness relative increment growth curves ($\Delta H / H_0$, where H_0 stands for microhardness prior to the treatment) for the presurface layer jupon polishing, depending on the diameter (d) of the abrasive particles used for grinding and polishing a back side of the Si and $GaAs$ wafers $350-360$ mcm and $370-390$ mcm thick.

It is seen here that with a rise of defectiveness near the opposite side (i.e. with a rise of d) the microhardness will grow monotonically and the residual deformation of the presurface layer will drop upon polishing. Most probably, such behaviour of $\Delta a / a$ and $\Delta H / H_0$ is determined by a drop in the concentration of PD clusters (primarily of intrinsic interstitial atoms of the material under treatment which during polishing appear and accumulate near the surface). A relaxation of this effect is largely specific rather for $GaAs$ than for Si wafers; this seemingly relates to larger concentrations of growth structural distortions functioning as drains for point defects and elastic wave scattering centres in binary semiconductors [Bourago and Perevostchikov, et al. (1986); Perevostchikov and Skoupov (1994)].

Upon the finishing chemicodynamical polishing, partially etching off the back–sided damaged layer will retain its gettering properties in later stages of high–temperature treatment of structures. It is important here that in such treatment an absence of relief–polycrystalline and crack zones will significantly decrease the probability of slice bending and deformation [Perevostchikov and Skoupov (1994)]. A residual defectiveness in wafers may be additionally reduced through exploiting in this technology such operations as an abrasive placticization of two sides after slicing and a back side after polishing [Perevostchikov and Skoupov, et al. (1987)]. Besides, in the latter case the microroughness in the surface relief caused by the brittle microdamage of the material can be efficiently reduced if during a grinding operation the presurface layers are plasticized, for example, through illuminating the layers by the intensive visual spectrum flux; such illumination results in a photomechanical effect in semiconductors [Westwood (1972); Iyer and Kuczynswi (1971); Perevostchikov and Skoupov, et al. (1985)]. The layers adjacent to a front side can be structurally perfected via a side–by–side etching of slices [Perevostchikov and Skoupov, et al. (1991)]. In this event, the defects concentrated near the side undergoing no etching at this time will serve as a getter for unbalanced IP defects injected into the bulk and diffusing from the surface opposite to that under etching. Upon slicing, the first side to be etched is the side chosen as a front side. A by-side etching in the vicinity of a front side in slices helps to drastically diminish the metastable PD aggregates, i.e. sources for oxidation–induced stacking faults in silicon, and reduce the amplitude of the residual deformation in the lattice upon chemicomechanical polishing of crystals.

The low–temperature gettering by structurally–damaged layers can be made significantly more efficient, if abrasive–chemical treatment is combined with other external actions able to create unbalanced defects in semiconductors. Into this number of actions falls a processing of crystals and structures by high hydrostatic pressure in liquid or gaseous ambients.

4.2
Structural changes in hydrostatically–compressed semiconductors

In structurally perfect crystals, external hydrostatic pressure does not induce shifting stresses (strains), and as a result, all barric effects turn out to be reversible. In hydrostatically–compressed crystals, where compression induced no phase convertions, the residual effects usually experimentally recorded through changes in different physicochemical properties are conditioned by transformations in the structural defects of the material.

In structurally compressed crystals, the spectrum, the spatial distribution profiles, the concentrations of original defects and new structural distortions are usually thought [Redcliff (1973); Pilyankevitch (1987)] to be regulated by the below aspects:

• the pressure–induced changes in the elastic constants of the material and in the intrinsic elastic stresses of defects;

• a birth of shift stresses that determine a conservative recombination of defects and relate to anisotropy of elastic moduli and structural and phase unhomogeneity of the material;

• the changes in the kinetic parameters of birth, interaction, multiplication (breeding) and annealing of defects.

It should be also added here that in structurally compressed crystals the structural distortions are greatly affected by a real surface of solids whose macro– and microrelief, the layers of natural oxides and adsorbed impurities affect the distribution picture of internal stresses and behaviour of defects in the material. The transformation process in defects is also stimulated by elastic waves born in crystals both by the changes in the external pressure (a sharp rise or fall, or fluctuations) and also as a result of recombinations in the defects themselves leading to locally–fast changes in the fields of internal mechanical stresses [Indenbom (1979); Skoupov and Tetel'baum (1987)].

This section considers possible applications of this high hydrostatic compression to microelectronics, first of all, for regulating the extrinsic defect composition of device components. This section does not touch the design problems for the units to treat the materials by hydrostatic pressure in gaseous or liquid working ambients, because these problems were discussed in details in the monographs [Pilyankevitch (1987); Beresnev and Martynov, et al. (1970)] and periodic literature as well.

4.2.1
Effects of hydrostatic compression on structural defects in crystals

An increase of external hydrostatic pressure is known [Gegouzin (1962)] to diminish the balanced concentration of IP defects in crystal. As for vacancies in structurally–compressed crystals their concentration is described as

$$C_p = C_0 \exp\left(-\frac{P\Omega}{kT}\right) \qquad (4.6)$$

or for $p < \dfrac{kT}{\Omega}$ we have $C_p = C_0\left(1 - \dfrac{p\,\Omega}{kT}\right)$.

Here, C_0 is a concentration of vacancies for the pressure $P = 0$; and Ω is an atomic volume. If a rise of the pressure rate is sufficiently high and exceeds the diffusing rate of IP defects, then crystal obtains oversaturations in vacancies and in intrinsic interstitial atoms. A balanced concentration is restored at the expense of their drift to the surface and to the internal drains, i.e. the extended defects like dislocations, clusters, cracks and the like. As a result, spatial distribution, density and sizes of these defects may change significantly, thus inducing residual phenomena when the external pressure ceases.

It should be emphasized here that the rate of nonconservative recombination of defects in hydrostatically–compressed (HC) crystals depends on the magnitude and the distribution picture of pressure–induced mechanical stresses. Diffusional interactions of edge dislocations with vacancies in hydrostatically–compressed crystals were studied in [Paniotov and Toky (1977)]. The authors jumped from the supposition that vacancies are generated or absorbed by a dislocation within the area of its nucleus (this area being $r_0 \approx b$ in size; b is the Bürgers vector modulus) and the overcrawling rate of dislocations is less than the rate at which a balanced concentration of vacancies is restored near the nucleus. The dislocation overcrawling rate, not considering the diffusion of the vacancy along its nucleus, is determined as

$$V = ND_v(C - C_d) , \qquad (4.7)$$

where
N is a coefficient determined by the geometry of the problem;

D_v is a vacancy diffusion coefficient;

C and C_d is a concentration of vacancies, respectively, far from and near the dislocations.

In equilibrium, when the chemical potentials of vacancies near the dislocations and on the surface are equal, no overcrawling occurs. However, if the dislocation is acted upon by some elastic force f, then a local concentration of vacancies near the dislocation nucleus will change, and the overcrawling rate will be

$$V = N \cdot D_v \cdot C \left[1 - \exp\left(\pm \frac{fb}{kT} \right) \right], \qquad (4.8)$$

the signs in the exponent stand for the travell of dislocations with emitting or absorbing the vacancies. From (4.6 – 4.8) we get

$$V = N \cdot D \cdot C_0 \left[1 - \exp\left(\pm \frac{fb}{kT} \right) \right] \exp\left(-P \cdot \frac{\omega + \omega_m}{kT} \right), \qquad (4.9)$$

where ω and ω_m are the activation volumes of vacancy generation and migration, respectively.

If a rise of pressure does not change the force f, then, as it follows from (4.9), the dislocation overcrawling rate under balanced conditions will diminish. If an unbalanced concentration of vacancies is maintained during hydrostatic compression process, for example, at the expense of plastic deformation of the hydrostatically–compressed crystal, then the crystal deformation at constant rate, i.e. a shift stress is increased with an increase of external stress, will induce in the material such a vacancy oversaturation that, despite a drop of the vacancy diffusion coefficient [Shinyaev (1973)], the vacancy fluxes onto the dislocations (drains) will grow.

In other words, the dislocation overcrawling rate rises, if a rise of hydrostatic pressure in crystal simultaneously causes a rise in the amplitude of shift stresses.

A dislocation sliding and the associated interactions, multiplication and vanishing constitute one of the basic mechanisms of structural transformations "remembering" an external action upon the crystal. Naturally, the energetic and kinetic properties of dislocations are dependent on the type of the action. To a full extent, it also pertains to a hydrostatically–compressed crystal when nonlinear elastic properties of materials start to reveal.

As noted in [Zaitzev (1983)], with the dislocation sources getting intensive (i.e. for $\delta\rho / \delta P = 0$, where ρ is a density of dislocations) the cell structures start to be actively formed. This process is initiated by a thermodynamical instability in the hydrostatically–compressed crystal where the dislocations are distributed chaotically and with a rise of compression their internal stresses will grow. A dislocation ordering and a wall formation will diminish elastic fields and reduce the crystal internal energy. The transversely sliding processes or polygonization by sliding is supposed to be crucial for such structures, accordingly at high or average values of energy in stacking faults in the crystal. In the course of hydrostatic compression, the effect of polygonization of the dislocation structure was observed in the materials with face– and spatially–centred lattices and was qualitatively explained in some papers (for example, see [Zaitzev (1983)]). Though, this does not mean at all that as of to–day the mechanism of ordered dislocation ensembles in hydrostatically–compressed crystals is clear. Many aspects of this problem are left debated, though the practical significance of the hydrostatic compression in establishing an ordered dislocation infrastructure in crystals casts no doubt.

Practically significant are also the interrelations between the dislocations and the external surface of hydrostatically–compressed crystals. In [Zaitzev and Toky (1987)], this problem was considered for the case of a linear spiral dislocation located in the cylinder of radius R at the distance ξ from its axis. It is shown here that within a presurface region $\xi > \sqrt{1 - \dfrac{R\sqrt{2}}{2}}$ the dislocation is acted upon by a pushing force increasing with an increase of pressure. A similar result was obtained for an edge dislocation as well.

The elastic interaction of dislocations, increasing with pressure, increases the amplitude of the long–range fields of stresses hampering the dislocation travell in a hydrostatically–compressed crystal. As was found theoretically and experimentally in [Zaitzev and Toky (1987)], a rise of hydrostatical pressure increases the deformation hardening rate in mono– and polycrystals of metals and in a number of alkalihalogen compounds.

How hydrostatic compression affects the mobility of dislocations was the subject of numerous studies (for example, see [Zaitzev and Toky (1987)]). Here, the dislocation mobility under hydrostatical pressure is shown to fall, due to a growing interaction of dislocations with stoppers. Pressure shifts the function $\ln v(\tau)$ into the side of high shifting stresses, the stronger it shifts the higher is the crystal doping level. With a rise of temperature, the effects of hydrostatic pressure weaken. On alkali–halogen crystals, the material fluidity limit was found to linearly increase with pressure and this dependence is the stronger, the lower the values of elastic constants and the less impurities in crystal. The question touching a specific mechanism of dislocation resistance in hydrostatically–compressed crystals still retains open. The mechanism regulating the travell of dislocations is supposed to change as the pressure grows – at low pressure the impurity atoms will function as stoppers, and at high pressure this role is played by steps on dislocations.

A rise of density of mobile dislocations with a rise of hydrostatical pressure is related both to a transfer of partial dislocations to full dislocations and to a generation of new dislocations arising, for example, on elastic inhomogeneities held in the hydrostatically–compressed crystal.

Hence, if there is a spherical inclusion in a solid body (say, second–phase particles or pores) of the radius R_b with the spatial compression modulus k_b, then the pressure around such inclusion will induce shift stresses determined by various compressions of this inclusion and the matrix material. The dependence of the amplitude of these compressions on the distance to the inclusion may be expressed as

$$\sigma = \left[\frac{3\mu(k - k_b)}{k(4\mu + 3k_b)} \right] P \left(\frac{R_b}{r} \right)^3 , \qquad (4.10)$$

where

P is an external pressure;

μ and k are a shift modulus and a matrix spatial compression modulus, respectively.

Near the inclusion, there emerge maximal shift stresses, i.e. for $r \approx R_b$ and depending on the ratio of elastic moduli the following magnitudes may be obtained:

for a pore ($k_b = 0$), $\sigma_{max} = 3P/4$;

for a hard inclusion ($k_b \gg k$), $\sigma_{max} = \dfrac{MP}{k}$;

for an elastic inclusion ($k_b \approx k$), $\sigma_m = \dfrac{3M(k - k_b)P}{k(4\mu + 3k_b)}$.

In a hydrostatically–compressed crystal, the stresses in the "inclusion–matrix" interface will induce dislocation loops capable of conservatively travelling from the generation centre.

Here, it is expedient now to describe such aspect of the problem as generating the dislocations in porous materials, for this aspect immediately touches the importance of hydrostatical compression for increasing the density and adhesion of the dielectric and metallic films in semiconductor structures.

4.2.2
Effects of pressure on pores in solids

Macro– and micropores in film elements of semiconducting structures are one of basic reasons of failures and instability in device functioning. A use of hydrostatic compression to decrease porousness in films was theoretically and experimentally proved in simulations on "curing" the pores in compressed massive specimens. In general, the curing processes in massive and thin–film materials under compression are different slightly, except the fact that in the films there arises some complementary shift stress of the value tP/h (here, t and h is a wafer and film thickness, respectively). Therefore, the experimental results obtained to-day may be analysed and explained on the basis of the results obtained for massive poly– and monocrystals.

In a series of studies by Y.E. Gerouzin and his co-authors (for example, see [Gegouzin (1962)]) it was investigated how high temperatures and hydrostatic pressures simultaneously affect the porousness of the NCl salt crystals and polycrystalline metals. A drop in porousness observed by the researchers was explained by them through the external pressure that increases a vacancy oversaturation in crystal and speeds up a pore coalescene process. In the hydrostatically–compressed crystal, the vacancy concentration near a pore of the radius R will be:

$$C_R = C_0 \exp\frac{2\gamma}{R}\cdot\frac{\Omega}{kT} \approx C_0 \left(1+\frac{2\gamma}{R}\cdot\frac{\Omega}{kT}\right),\qquad (4.11)$$

where γ is a surface energy, and far from the pore it will be equal to

$$C_p \approx C_0 \left(1-\frac{P\Omega}{kT}\right).\qquad (4.12)$$

For the oversaturation case we have from (4.11) and (4.12):

$$\frac{\Delta C_p}{C_0} = \frac{\Omega}{kT}\left(P+\frac{2\gamma}{R}\right).\qquad (4.13)$$

In the vacancy subsystem of the hydrostatically–compressed crystal, equilibrium is restored at the expense of the exchange by vacancies between the pore and the volume being performed by diffusive fluxes in the way proportional to the concentration gradient:

$$j_v \approx -D_v\nabla C_p \approx \frac{C_0}{l}\left(P+\frac{2\gamma}{R}\right)\frac{\Omega}{kT},$$

where D_v is a vacancy diffusion coefficient; l is a characteristic linear size of the region where a balanced concentration occurs. Here, as shown in [Gegouzin (1962)], in the absence of pressure (R_0^*) and in the compressed material (R_p^*) there exists a below relation between the critical sizes of pores:

$$R_p^* = \frac{R_0^*}{1-\beta P},$$

where $\beta = \dfrac{R_0^*}{2\gamma}$. Under the pressure $P = \dfrac{2\gamma}{R_p^*}$ all the pores held in a hydrostatically–compressed crystal must get "cured" through a vacancy–by–vacancy dissolution and by a drift of vacancies toward the external surface. As was experimentally shown in [Gegouzin (1962)], a hydrostatic compression and a rise of annealing temperature actually clean the crystals from pores; and, besides, at high hydrostatic pressure there occurs an annealing. The external pressure also suppresses a diffusive porousness in the cases of mutual diffusion of metals, of both two–layered and three–layered compositions like $Cu - \alpha -$ brass, $Cu - Ni$ and $Ni - Cu - Ni$.

In [Gegouzin (1962)], the gases available in pores were also noted to play some definite role in a pore curing process. The pressure of gases rising as the pore sizes are diminished must stabilize the pore sizes at the level

$$r_e = \left(\frac{P_0 \cdot r_0^{\,3}}{2\gamma} \right)^{1/2} ;$$

upon reaching this level, the pores cease to participate in the coalescene process. Such situation must encounter, say, when the material is saturated with fast–diffusing hydrogen.

The state of the art in cleaning the crystals from pores with a use of the diffusion–dislocation mechanism of their dissolution was described in [Gegouzin and Kononenko (1982)]. It was noted by them that in hydrostatically–compressed alkali–halogen monocrystals near the pores there are being deformed mainly not prismatic but shifting dislocation loops. Interactions between the dislocations born on pores can induce cracking. Alongside with this, in the hydrostatic treatment the cracking probability may be substantially reduced depending on the type of the material's crystalline structure.

When speaking about the effects exerted by pressure on the pores in thin films deposited onto a massive wafer, it is necessary to emphasize that in this case the shift stresses induced by the difference in the elastic constants of the conjugate materials serve as some additional factor for increasing the pore curing efficiency. Processing such structures by pressure brings positive results even in a low–temperature ambient transferring the pressure.

4.2.3
Compression–induced structural changes experimentally observed in semiconductors

In extensive literature on hydrostatic compression, the results concerning the irreversible structural changes in all–sided compressed semiconductors are much poorer than those obtained on the materials with ionic and metallic bonds.

The majority of the results was usually analysed and generalized from the point of view of the behavioural regularities in structural defects in hydrostatically–compressed metals and alkali–halogen crystals. Such approach is to some extent limited, for it neglects semiconductor's specific properties induced by distortions in the crystalline lattice. First, among such properties must be mentioned a sufficiently high structural perfection of the materials themselves having a low density of growth dislocations, a special sensitivity of physicomechanical properties of semiconductors to the ID composition, an increased probability of aggregation with a use of IP defects, a dependence of energy parameters of kinetic processes on the charge states of unbalanced vacancies and IP defects. These specificities when taken into consideration can not only help, as will be shown later, to correctly interpret the experimental results but also help to develop particular technologies for controlling structurally–sensitive properties in hydrostatically–compressed semiconductors.

4.2.3.1
Hydrostatically–compressed monocrystals and epitaxial structures

Developing a more sophisticated technique for growing ingots and manufacturing dislocation–free materials, firstly, silicon, has made urgent a problem of cleaning crystals from growth microdefects – aggregates of IP defects and impurity atoms [Kock (1977)]. How hydrostatic compression affects the microdefects in silicon was studied in [Jung and Saynovsky, et al. (1979)]. Using the X-ray topography and electronic microscopy, the authors studied the dislocation–free silicon being hydrostatically compressed by the pressure $P \leq 3\ GPa$ and simultaneously heated up to $1500\ K$. Structural changes in compressed crystals were revealed to be induced only by a total action, compressive and thermal. In X-ray topograms made upon the hydrostatic compression, there have been spotted the defects identified by the authors as dislocation loops and "spherical defects". It was found that the temperatures equal to $1000\ K$ and $P = (0.6 - 0.7)\ GPa$ create dislocation loops mostly, and with the pressure growing the "spherical defects" start to dominate. At higher temperatures ($T \geq 1300\ K$) and under the pressure $0.3\ GPa$ these two types of defects start to increase their concentration. The nature of these defects is related by the authors to the dislocation aggregates arising in the hydrostatically–compressed crystals near the second–phase particles (SiO or SiO_2). A specific mechanism was not considered by them. Dislocation aggregates may be supposed to be generated both during coagulation of small clusters (microdefects like $B -$ type) to form the Frank loop [Zeeger and Fell, et al. (1979)] and at the expense of the loops emitted by growth extrinsic defect (ED) clusters.

In the silicon treating regimes used in [Jung and Saynovsky, et al. (1979)], it is also possible, according to the results in [Milevscky and Vysotzkaya, et al.

(1980)], to consider the $B-$clusters together with the excessive point defects migrating to the external surface of the crystal. The absence of information about this in [Jung and Saynovsky, et al. (1979)] is possibly explained by the insufficiency of the techniques which should meet severe demands rather in recognizing the B–clusters than in investigating large–size $A-$type defects [Zeeger and Fell, et al. (1979)].

We have studied the effects of hydrostatic compression on the extrinsic defect composition and mechanical properties of presurface layers in silicon monocrystals.

Subjected to study were the slices of dislocation–free $p-Si(001)CZ$ ($\rho=12\ Ohm\cdot cm$) 450 mcm thick where for establishing a gettering layer one side of slices was polished by the Cab–O–Sil suspension and another by the microporous synthetic corundum powder ($7-mcm$ grains). In order to exclude the possible effects of the growth (ingot) inhomogeneity of distributions of impurities and microdefects in original slices, each of them was splitted into some specimens, one of which was used as a controlled one and others were hydrostatically compressed. Six specimens from different slices were simultaneously treated in one and the same regime. Hydrostatic compression was performed at room temperature for 1h in the 50–per cent mixture of isopropyl alcohol and deionized water. There have been studied a effect of static compression by the pressure $0.8\,GPa$ and by impulsive pressure under which the pressure in the chamber rose to $0.7\,GPa$ and then with the frequency $0.02\,Hz$ it was oscillating near the average with the amplitude $\pm 50\,MPa$. Upon hydrostatic compression, the specimens (together with controlled ones) were kept in the 49–per cent hydrofluoric acid, to have the silicon dioxide film removed; and then were washed in the flowing deionized water and dried at the centrifuge.

The impurity composition in the presurface layers of specimens from the polished side was analysed, prior to and after hydrostatic compression, by the SIMS spectroscopy unit where oxygen $6-keV$ ions constituted an original beam of the spraying rate equal to 0.5 Å/s. On each specimen, impurity distributions were recorded by the minor ion spectrums read in three points of the surface with the relative error ≤ 1 per cent. The effects of hydrostatic compression on the mechanical properties of crystals were recorded through changes in microhardness measured by the microhardness meter with the error ± 7 per cent and fiducial probability 0.98.

Fig. 4.7 [Aglaumov and Skoupov (1993)] shows typical dependences of the logarithm of the minor ion intensity as a function of the spraying depth.

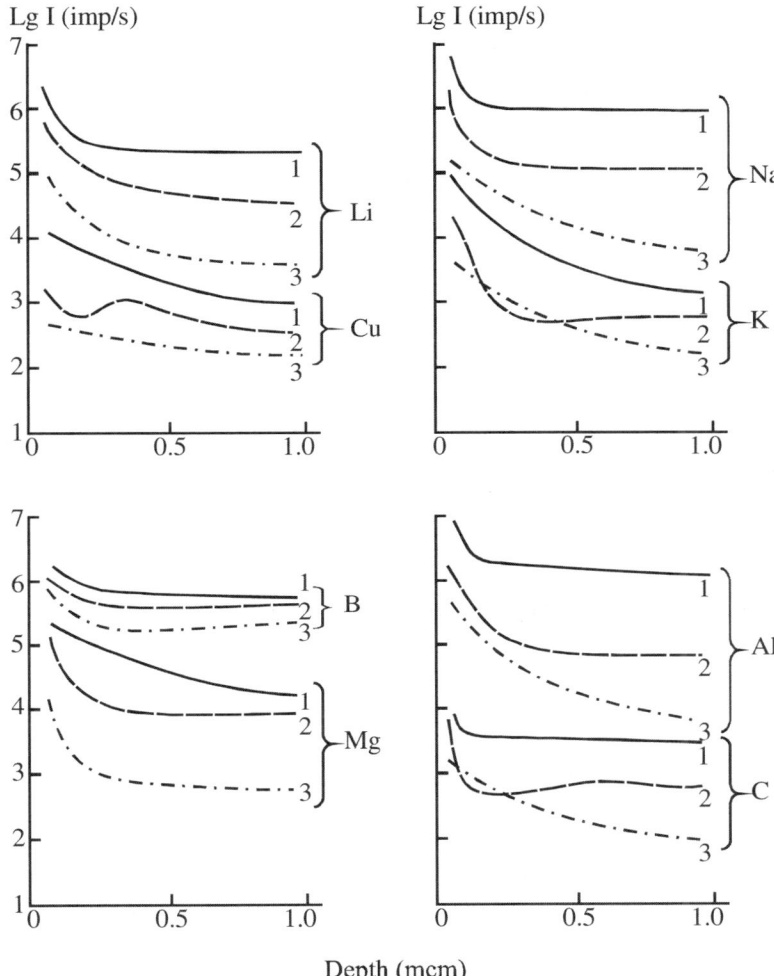

Fig. 4.7. Impurity distribution profiles for monocrystalline silicon prior to (**1**) and after hydrostatic compression (**2**) and under impulsive pressure (**3**).

Here, it is seen first that the hydrostatic compression in the presurface layer diminishes the concentrations of impurities in silicon both of high (lithium, copper) and low (boron, carbon, aluminium) coefficients of diffusion, and, second, the greatest effect is observed in the impulsively–compressed specimens. Besides, all the profiles exhibit a rise of concentration as they come nearer to the surface – this may be explained by the gettering action performed by the silicon dioxide film that is a local source of statically elastic stresses. Anomalous distributions of copper, potassium and carbon are seemingly accounted for the specificity in the aggregating of these impurities with original structural disturbances in silicon, and also with IP defects emerging during hydrostatic compression. The curves for relative changes of microhardness induced by indenter loads (Fig. 4.8) [Aglaumov and Skoupov (1993)] and describing the profiles of in-depth changes in elastic–plastic properties of silicon qualitatively correlate with the impurity distributions.

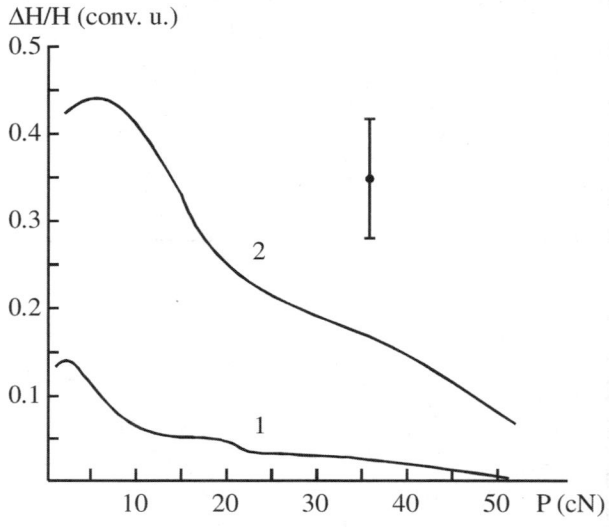

Fig. 4.8. Dependence of a relative complement of microhardness on the indenter load in silicon crystals hydrostatically (**1**) and impulsively (**2**) compressed.

The silicon hardening observed upon hydrostatic compression may be, on the one hand, caused by a drop in the mobility of the incorporated dislocations due to a fall in the concentrations of electrically–active impurities, and, on the other hand, it may be explained by a fall in the density of internal concentrators of elastic stresses, i.e. growth microdefects and clusters induced by polishing and promoting a generation of dislocations under the indenter. As it is seen here, these both factors act most efficiently in impulsively–compressed crystals.

The results obtained may be explained by elastic waves being excited and playing an active role in structural transformations under pressure. First, in the course of hydrostatic compression these elastic waves are induced by the crystalline defect recombinations occurring near the grinded side and in the vicinity of the clusters arising in the bulk during an ingot growth and later technological stages of slice preparation. A field of elastic waves arising in the hydrostatically–compressed crystal, alongside with stimulating a drift of impurity atoms, also sustains an unbalanced concentration of IP defects, chiefly, the vacancies which control an athermal heat-free surface–bound diffusion of impurities and their capture by the gettering layer and the SiO_2 film. In particular, this is proved by the gettering process most efficiently running during an impulsive compression. The very fact that a both–sided compression causes changes in the extrinsic defect composition agrees with the model of the locally–damaged solids. According to this model, the asymmetrically growing amplitudes of atomic thermal fluctuations in the lattice can give rise to such microprocesses as heating, recrystallization, cluster dissolution and elastic waves.

How the hydrostatic compression affects the perfection of the epitaxial silicon was studied in [Bourago and Skoupov, et al. (1985)]. Such studies are, on the one hand, interesting, since epitaxially–grown structures constitute a basis in the contemporary semiconductor manufacture, and, on the other hand, they unveil new effects, not specific for monocrystals, induced by hydrostatic compression, for example, such as misfit dislocations within the interface of mated layers.

In [Bourago and Skoupov, et al. (1985)], there have been studied the $n - n^+ -$ and $p - p^+ -$ type structures grown by the chloride technique. Prior to and after the hydrostatic compression by $0.5\,GPa$, the residual stresses in the compositions were measured by the X-ray three–crystal spectrometer and their distributions along the film–wafer interface were recorded. The compression was found to change the residual stresses in the structures making their values close to the stress values in zero–defect compositions, while the elastic stresses are determined only by a deference in the thickness and in the doping level of the film and the wafer. In the event of $p -$ type structures, similar changes must cause additional misfit dislocations, as was proved by the X-ray topography. In explaining these effects, the authors [Bourago and Skoupov, et al. (1985)] also emphasize such aspect as a dependence of semiconductor elastic constants on a doping level.

The results from [Barch and Chang (1967)], where under the compression approximately equal to $0.5\,GPa$ a rise of concentration in a silicon dopant changes the elastic constant C_{11} as much as three orders, were used in [Bourago and Skoupov, et al. (1985)] to derive a compression complement:

$$\Delta\chi = \frac{\Delta C_{11}}{3k^2} \,,$$

where $k = \dfrac{C_{11} + 2C_{12}}{3}$ is for silicon. It is shown that during a hydrostatic compression a misfit in the lattice periods for the film and the wafer

$$\delta = \frac{a_{wafer} - a_{film}}{a_{wafer}} \,,$$

for the structures under study, will be, respectively:

$$\delta (n - n^{+}) = \beta_{Sb} \cdot N_{Sb} - \Delta \chi P$$

and

$$\delta (p - p^{+}) = \beta_{B} N_{B} - \Delta \chi P \ ,$$

where

β_{Sb} and β_{B} are the compression coefficients for the silicon lattice by the antinomy and boron atoms ($\beta_{Sb} = 0.16$, $\beta_{B} = -0.2$);

N_{Sb} and N_{B} are the concentrations of the dopants;

P is an external pressure.

For $N_{B} = 3 \times 10^{18} \ cm^{-3}$, $N_{B} = 2 \times 10^{18} \ cm^{-3}$ and $\Delta \chi = 1.74 \times 10^{-14} \ Pa^{-1}$ for   the   pressure $P = 0.5 \, GPa$ we   get $\delta (n - n^{+}) = 9 \times 10^{-7}$ and $\delta (p - p^{+}) = 1.87 \times 10^{-5}$.

Through the value of δ there was measured in [Bourago and Skoupov, et al. (1985)] a linear density of misfit dislocations $N = \delta / a$, which turned out to be equal to $16.3 \ cm^{-1}$ for $n -$ type and $3.4 \cdot 10^{2} \ cm^{-2}$ for $p -$ type structures. These estimates satisfactorily agree with the values of dislocation densities calculated through changes in microscopic curvature of the structure.

A variance of internal stresses within the film–wafer interface was estimated [Bourago and Skoupov, et al. (1985)] through measuring the microscopic curvature of the structures by the X-ray technique, with the basic distances being different between the beams. The measurements were conducted in two mutually perpendicular directions, k_{x} and k_{y}. Through k_{x} and k_{y}, they calculated the difference $\sigma_{x} - \sigma_{y}$ where the stresses were calculated as

$$\sigma_{x(y)} = \frac{E \cdot 4 \cdot h^{3} \cdot k_{x(y)}}{(1 - v) \cdot 3 \cdot t \cdot (2h - t)} \ ,$$

where $2h$ and t is the thickness of the entire structure and epitaxial layer, respectively. The results exhibiting a fall in the inhomogeneity of the mechanical stress distributions in epitaxial $n - n^{+} -$ type compositions upon their 1.5h compression by $P = 0.5 \, GPa$ are given in Table 4.1.

Table 4.1

The effects of hydrostatic compression on the homogeneity of mechanical stress distributions in epitaxial silicon

Basic distance (mm)	$(\sigma_x - \sigma_y)\cdot 10^{-5}(Pa)$ (prior to hydrostatic compression)	$(\sigma_x - \sigma_y)\cdot 10^{-5}(Pa)$ (after hydrostatic compression)
5	5.8	2.5
10	4.7	2.9
20	2.5	2.0

That a variance of stresses in hydrostatically–compressed structures drops is additionally proved by the changes in the halfwidths of diffractional curves. Upon the hydrostatic compression, the halfwidths, recorded in [Bourago and Skoupov, et al. (1985)], fall in average by 30 – 40 per cent.

The unstable residual stresses in the hydrostatically–compressed 12-mcm thickness $n-Si(001)CZ$ ($\rho=1.3\ Ohm\cdot cm$, phosphorus–doped) – $250-mcm$ thickness $n-Si(001)CZ$ ($\rho=0.01\ Ohm\cdot cm$, $Sb-$ doped) (the compression was in the range of $0.5-1.1\ GPa$) and in $GaAs-$ based heterostructures $5\,Al_{0.19}Ga_{0.81}As/500\,GaAs$ were studied in [Skoupov and Sherban' (1986)]. The specimens underwent compression, and later the X-ray measurements of the relative complement in the epilayer lattice period ($\Delta d/d$) and the curvature (K) of the structures were performed at the room temperature.

Upon compression, the structural parameters were found to change not monotonically and for some days the relaxational curves of the specimens compressed by different regimes correlated between themselves. The complement $\Delta d/d$ and curvature K as a result of relaxation turn out to be 2.5 – 3 times higher than immediately after compression. Both types of structures exhibit a fall in the misfit of the lattice periods for the film and wafer and also a fall in the inhomogeneous distributions of residual stresses located near the interface of the conjugative layers.

It has been also revealed that the characteristic time of relaxation τ and the maximal stresses in the structures σ are interrelated and this relation is described as $\tau = \tau_0 \exp(-\sigma/\sigma_0)$. The constants τ_0 and σ_0 for the $Si-$ and $GaAs-$ based structures are equal, respectively, to $\tau_0(Si)=1.2\times10^7\,s$, $\sigma_0(Si)=3.43\ MPa$ and $\tau_0(GaAs)=9.8\times10^6\,s$, $\sigma_0(GaAs)=83.02\ MPa$. With the internal stress amplitude being similar, the heterostructures showed the least relaxation rate upon hydrostatic compression.

This agrees with the idea that elementary defects in semiconducting compounds are more stable than in the elementary semiconductors.

After compression, the internal stresses in structures are stabilized via 1h annealing at $520-570\ K$. Combining a hydrostatic compression with a thermal treatment makes it possible to vary the gradient of in–thickness stresses in epitaxial compositions.

The low–temperature relaxation is thought [Skoupov and Sherban' (1986)] to relate to a recombination of metastable defects induced both at the stage of growing an ingot for a wafer and at the stage of building up an epitaxial layer. Without an external load, the relaxation rate is not significant. This process becomes activated under hydrostatic compression. According to authors, the activation volume per a single relaxing defect in silicon makes up 500 nm, i.e. it is close to the volume of the $B-$type clusters. In $GaAs$, such defect looks like an accumulation of 5–6 vacancies or interstitial atoms. Evidently, in a hydrostatically–compressed crystal there exists some spectrum of instable defects each of which having its own relaxation time. It is a simultaneous recombination of such defects that determines a nonmonotonous motion of relaxation curves.

Under habitual conditions, the natural relaxation, i.e. the transformation of genetic defects providing some relative minimum of free energy in crystals, is either retarded and revealed only upon lengthy periods of time (the aging effect) or is fully suppressed, as, for example, in the growth dislocations blocked by impurity atmospheres. Though, under any sufficient external action (say, a hydrostatic compression) that provides a new metastable equilibrium the relaxation rate will grow. External pressures seemingly affect the defects through the two basic channels: through a field channel (at the expense of changes in the topology of the crystal's total elastic field) and through the unbalanced IP defects that interact with the original structural distortions and with each other, before new unstable aggregates arise (upon a cease of excitation). In a hydrostatically–compressed crystal, one of the sources for unbalanced IP defects is a surface near which (even when the external pressure fluctuates slightly /up to 5 percent from the average/) there arises an excess concentration of vacancies and interstitial atoms [Perevostchikov and Skoupov (1992)]. The direct annihilation of vacancies and interstitial atoms in their region is hampered by a difference in the diffusive mobility [Vinetski and Kholodar' (1979)] and, seemingly, by the Frenkel pairs arising uncorrelated in space and time.

As was noted in [Bourago and Perevostchikov, et al. (1987)], the zone to be mostly affected by the hydrostatic compression is the presuface layers, i.e. the zone where the active regions for the semiconducting devices are being formed. Therefore, a compression process must precede the basic technological processes, and this refers not only to the heterostructures but also to the original, comparatively homogeneous, monocrystalline wafers; or the compression is arranged at the intermediate stages of mechanical treatment of wafer surfaces. Here, it is more efficient to have the hydrostatic compression followed with a short–time annealing, to speed up the relaxational processes (this helps to correctly classify the wafers according to their defectiveness).

4.2.3.2

Effects of hydrostatic pressure on the MOS-structure elements

The semiconductor–dielectric film structure is one of the basic topological elements in the work pieces of the contemporary micro– and nanoelectronics; this element determines to a large extent the properties and the reliability of semiconductor devices. As practice reveals, one of failures in such structures is a structural instability and a presence of defects in the dielectric and semiconducting substrate. In this connection, there arises an urgent problem to find the ways of controlling the structural perfection in both the components themselves and the entire composition. For this purpose, one may use a high hydrostatic compression.

It is shown in [Bourago and Skoupov, et al. (1985)] that in the $Si - SiO_2$ structures on the wafers of $n - Si$ (001) CZ ($\rho = 0.3 \, Ohm \cdot cm$, phosphorus–doped) and $n - Si$ (111) CZ ($\rho = 0.005 \, Ohm \cdot cm$, phosphorus–doped), hydrostatically compressed by $P = 0.4 - 0.5 \, GPa$ for 60 – 90 s, the level of residual compressing stresses in SiO_2 films drops significantly and a hardening occurs (Table 4.2); also the porousness in the dielectric diminishes and a distribution of structures in porousness becomes clearly nonmonotonous (Fig. 4.9).

Table 4.2

Changes in elastic stresses and microhardness in the hydrostatically–compressed $Si - SiO_2$ (pirolytic) structures

Measuring stage	Stress in SiO_2 (MPa)	Microhardness (GPa) when the indenter is under load (cN)		
		1.0	2.0	3.0
Prior to hydrostatic compression	-314 ± 26	4.6 ± 0.6	5.3 ± 0.4	4.9 ± 0.3
Upon hydrostatic compression	-52 ± 20	3.8 ± 0.6	6.9 ± 0.7	6.7 ± 0.7

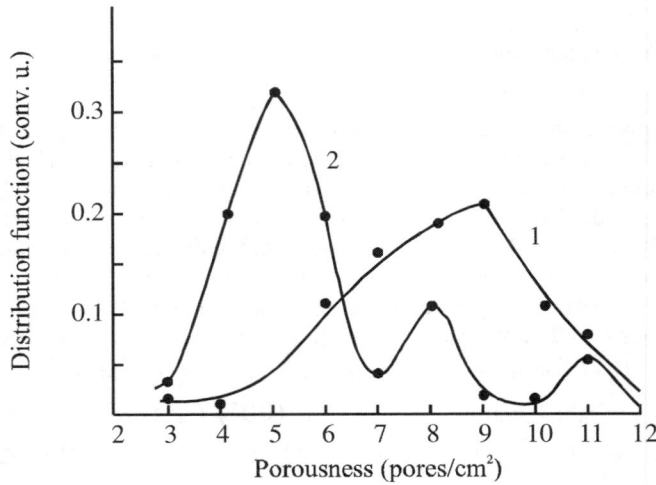

Fig. 4.9. Distribution of the $Si - SiO_2$ structures (thermal) in dielectric porousness: **1** prior to hydrostatic compression, **2** after hydrostatic compression.

In the $[111]$ − oriented hydrostatically–compressed specimens, the anisotropy of the elastic stress distribution within the plane of the $Si - SiO_2$ interface, $<110>$ − and $<112>$ −oriented, was noticed to drop. Prior to compression, the difference in stresses $\sigma_{<110>} - \sigma_{<112>}$, seemingly regulated by anisotropic and inhomogeneous distribution of structural defects in the silicon wafer, in average made up $107 \pm 25 \ MPa$. Upon the compression, it did not exceed $44 \, MPa$.

In all the structures under study, the elastic stresses after compression were found nonstable and they proceed diminishing with a characteristic relaxation time from 37 to 80 days. Here, no changes in porousness and microhardness was observed. The external hydrostatic pressure, except that it compresses the dielectric film and diminishes the defectiveness in the semiconducting wafer, especially near the interface with silicon dioxide, evidently serves as a starter in activating a recombination of unstable structural distortions in silicon [Bourago and Perevostchikov, et al. (1987)]. In particular, these processes diminish the growth and technological clusters in the substrate responsible rather for the anisotropy of elastic stresses in the $Si - SiO_2$ structures.

In studying the hydrostatically–compressed polycrystalline aluminium films up to 1.0 −1.1 mcm thick deposited onto the silicon wafers thermally and through electronic and magnetic spraying it was found that under the compression of 0.2 − 0.6 GPa the films obtain visible structural changes at the pressure more than 0.3 GPa and with the compression time more than 60s [Milevscky and Vysotzkaya, et al. (1980); Bourago and Skoupov, et al. (1985); Barch and Chang (1967); Skoupov and Sherban' (1986)].

Such changes are betrayed by diminished sizes of grains (if the films were obtained thermally or through electronic deposition) or by increased sizes of grains (if magnetronic spraying was used) and by a reduced texture. High hydrostatic compression increases the film microhardness; the in-depth changes in such films are not monotonous and most strongly expressed near the interface with the wafer. Simultaneously, this hydrostatic compression reduces mechanical stresses in films and diminishes the anisotropy of their distribution within the interface plane. The stresses also proceed to diminish upon a cease of compression. The hydrostatic compression was observed to diminish the oxidation rate on the surfaces subjected to aluminium metallization. High hydrostatic compression is mostly efficient on the films undergoing, prior to compression, a 30–min anneal at 780 – 800 K.

In these studies it was also observed that during an all–sided compression the components of the MOS-structures interrelate with the processes occurring in them. There has been revealed a positive effect of impulsive compression on structures. During the impulsive compression, the pressure was set at a given level and then it was raised periodically with the frequency $\approx 1\,Hz$ and dropped at $\pm 0.1\,GPa$ from the average.

On silicon slices with a single–sided chemicomechanical polishing, a thickness of the structurally–damaged layer and its residual elastic stresses were estimated through relative complements $\Delta d / d$ of the lattice period measured by the X-ray three–crystal spectrometer. The results obtained are given in Table 4.3.

Table 4.3

Mechanical stresses and the thickness of defective layer
in hydrostatically–compressed silicon
(0.5 GPa) upon a 10 min polishing

Type of treatment	Stress (MPa)	Thickness of a structurally– damaged layer (mcm)
Prior to hydrostatic compression	– 2.2 ± 0.5	16 ± 6
Continuous hydrostatic compression	–1.7 ± 0.3	11 ± 3
Impulsive hydrostatic compression	– 0.9 ± 0.3	6 ± 2

As seen in Table 4.3, upon compression the amplitude of residual compressing stresses in the layer structurally damaged by polishing and its thickness will diminish, and besides, most visible changes encounter in the impulsively–compressed specimens. It is worthy of mentioning here that the thicknesses obtained by X–ray measurements in a structurally–damaged layer exceed, almost by an order, the values habitually published in the literature concerning the penetration depth of structural defects in silicon in the course of a chemicomechanical polishing. This is evidently explained by the X–ray technique being able to detect both the lattice deformations by the dislocation–type distortions located in the layer less than 2 mcm in thickness and the deformations induced by the IPD clusters generated by a polishing process.

Since the high hydrostatic compression was performed at the room temperature and the stresses in the damaged layer are largely less than the limit for the silicon fluidity, it is supposed that in the hydrostatically–compressed crystals the majority of structural changes are mainly induced by unconservative recombination of defects interacting with unbalanced IP defects. This process embraces not only deeply–lying clusters (whose reduction in sizes or full dissolution decreases the residual stresses and diminishes the thickness in the damaged layer) but the dislocation structure in the presurface layer as well. The latter is supported by the measurements of microhardness in the specimens of silicon and gallium arsenide. In particular, in the impulsively–compressed specimens the microhardness maximum, corresponding to a region of the greatest density of structural defects, was found to be shifting to the crystal surface.

Compression–induced structural changes in the wafer is evidently one of the reasons for diminishing the residual stresses in the $Si - SiO_2$ heterosystems. For such structures, the impulsive compression also turns out to be more efficient than continuous. The results in Table 4.4 show this.

Table 4.4

Effects of 10 min hydrostatic compression (0.6 GPa) on the residual stresses in films and the deformation in $Si - SiO_2$ wafers

Type of treatment	Stresses in the SiO_2 films (MPa)	Deformation of the wafer 10^{-6} (conv. u.)
Prior to hydrostatic compression	– 107.6	2.8
Continuous hydrostatic compression	– 71.7	1.9
Impulsive hydrostatic compression	– 36.9	–0.6

One of possible reasons for diminishing the stresses in hydrostatically–compressed films is a drop in porousness, in particular, a fall of sizes in pores, at the expense of vacancies emitted by them. As the pores of lesser sizes have a large coefficient of material compression, a dissolution of pores will partially compensate the structural film – wafer misfit that determines residual stresses in the system. This process is initiated and supported by extending stresses in the film induced by the difference in special compressibility moduli in hydrostatically–compressed silicon and silicon dioxide. In general, this agrees with [Gegouzin (1962)] if some supplementary condition is taken into account, due to inhomogeneously distributed stresses in the film.

The data suggested in Table 4.4 indicates one more interesting regularity: a positive sign for the wafer deformation and its drop after compression; and in the event of impulsive compression the complement in the period of silicon lattice in the presurface layer will change the sign.

The specificity of this result lies in the fact that according to the theory of elasticity for two–layer $Si - SiO_2$ structures the wafer deformation in the direction normal to the interface is to be negative and its estimated value must be 1.3×10^{-8} *conv. u.* That real values differ from theoretical is, first, explained by the residual defects not fully removed after polishing, and, second, by structural changes induced by chemical etching. The positive sign of the deformation expresses an accumulation of mostly interstitial defects in the presurface layer in the course of polishing and etching operations. During a hydrostatic compression, especially impulsive, the deformational "power" of such accumulations will be reduced because their sizes and concentrations will drop.

When discussing a nature of structural transformations under low–temperature compression, it may be noted that the experimental results distinguish in hydrostatically–compressed crystals a dominating role of unbalanced vacancies whose generation from the energy point of view is more profitable than intrinsic interstitial atoms. During a continuous compression, as was told above, the vacancies restore their balanced concentration at the expense of their travelling to their external (surface) and internal (defects) drains.

During an impulsive compression, the compression–expansion cycles are accompanied with periodic excessive vacancies and their "pumping" to the drains – this process intensifies a structural defect recombination as compared against a continuous hydrostatic compression. For $\Omega = 2 \times 10^{-23} \, cm^3$, $P = 0.5 \, GPa$ and $T = 300 \, K$, the oversaturation with silicon crystal vacancies is found by (4.6) to be equal to $C \approx 11.3 \cdot C_0$. In the case of impulsive compression, a single compression–expansion cycle ($\Delta P = 0.1 \, GPa$) raises the excessive concentration to $C \approx 11.3 \cdot C_0$. If there are $n = \tau v$ cycles in the compressing operation (v is a frequency of coming cycles), then at $\tau = 10 \, min$ and $v = 0.05 \, Hz$ the equivalent concentration of vacancies will make up $C = 48.7 \cdot C_0$. The impulsive hydrostatic compression will most efficiently affect those defects near which, in the draining period $\Delta \tau < \tau$, there will be restored a balanced concentration of vacancies. We, supposing the surface to be a basic source of vacancies and exploiting the results from [Perevostchikov and Skoupov (1992)], shall find the depth at which the defect recombination runs most efficiently: $\Delta x = \sqrt{D_v \Delta \tau} = 1.2 \, mcm$, where $D_v = 2.5 \times 10^{-10} \, cm^2 / s$ and

$\Delta \tau = 15 \ s$. This value is comparable with the thickness of the layer structurally damaged by polishing.

To explain the effects occurring deeper than 1 mcm, the sources of unbalanced vacancies are supposed to be available also in the bulk of the crystal. In the capacity of such sources there may be the growth microdefects and defects, induced by abrasive–chemical treatment of crystals at the stages of its brittle–elastic damage (cutting, grinding, polishing), and also during chemicomechanical and chemicodynamic polishing. The number of vacancies in the bulk of a hydrostatically–compressed crystal needed for complete dissolution of the clusters of interstitial atoms induced by polishing is equal to $(0.2 - 2.0) \times 10^{18} \ cm^{-3}$. Such concentration is not possible to be established only at the expense of annihilation of vacancy aggregates [Perevostchikov and Skoupov (1992)]. Hence, it may be supposed that "annealing" of the clusters of interstitial atoms is strengthened at the expense of the collective rearrangement of the defects within the cluster [Skoupov and Tetel'baum (1987)]. According to [Skoupov and Tetel'baum (1987)], each act of the vacancy and interstitial atom annihilation creates an elastic wave of the amplitude $\sigma = B\varepsilon'$, where B is an elasticity modulus, ε' is a deformation in the lattice near the vacancy ($\varepsilon' \approx 0.1 - 0.2$). The defects forming the interstitial clusters relate to an elastic interaction whose energy is $E_c \leq 0.1 - 0.2 \ eV$. When a wave determined by a first defect annihilation approaches a second defect, it will drop its detachment (separation) barrier by the magnitude $\Delta E \approx \Omega_{at} \dfrac{r_0 \sigma}{r}$, where r_0 is a radius at which the static field of elastic stresses is acting in the vicinity of the defect; r is an average distance between the defects in the interstitial cluster. If to accept $r_0 \approx 2 \times 10^{-10} \ m$, $r \approx 10^{-9} m$, $B = 10^{11} Pa$, then we have $E = 0.3 - 0.6 eV$; and this exceeds the value E_c. The detachment of the second defect changes the local elastic field and induces an elastic wave, which stimulate a detachment of the third defect, etc. Therefore, to have an interstitial cluster disassociated, a single act of annihilation per a cluster will be enough, i.e. the number of vacancies needed to annihilate an interstitial cluster will be reduced n times, where n is an average number of defects per a cluster. An external pressure makes this process favourable from energy point of view. The oscillating pressures will increase the dislocation intensity, i.e. periodically establish in the hydrostatically–compressed crystal an oversaturation and undersaturation in vacancies thus stimulating a travell of vacancies from the sources onto their drains.

This thinking is qualitatively supported by the results the authors have obtained for the hydrostatically–compressed $Si - SiO_2 - Al$ structures.

These structures were built on monocrystalline and epitaxial silicon, via depositing a pirolytic or high–temperature SiO_2 and then vacuum spraying of aluminium. Structural changes in the compositions were recorded by X-ray three–crystal spectrometer, by the electronograph unit, through measuring the microhardness and slice bending. Porousness in SiO_2 was estimated electrographically.

Residual mechanical stresses in SiO_2 and Al films were calculated through the microscopic bending of the structures, when they undergo a by–layer etching off from the wafer side as deep as Δx. They were calculated as

$$K_i = 3(1 - v_{Si}) \frac{\left| \sigma_{SiO_2} \, t_{SiO_2} \, (2h_i - 2t_{Al} - t_{SiO_2}) + \sigma_{Al} \, t_{Al} (2h_i - t_{Al}) \right|}{4h_i^3 \, E_{Si}},$$

where

K_i is a curvature after the i–th etching step;

t_{SiO_2} and t_{Al} is, respectively, a thickness of the silicon dioxide film and aluminium film;

$2h_i$ is a thickness of the structure after the i–th etching step.

As was revealed, in hydrostatically–compressed films the stresses are diminished and the microhardness of conjugate layers increases (Table 4.5). The effects of compression of SiO_2 are more efficient if no metallization is done (Table 4.6). The hydrostatic compression reduces both an average bending in structures and the variance of its values, within a set of one–type specimens (Table 4.7). The compression also drops the amplitude and the variance of residual deformations in the silicon wafer lattice near the interface with the silicon dioxide film. The largest effect is obtained in the case of impulsive compression (Table 4.8). In the SiO_2 films, porousness diminishes irreversibly (Table 4.9) and a level of its texture falls as well.

Table 4.5

Effects of the 1 min hydrostatic compression (0.5 GPa) on the residual stresses in films

Measuring stage	Elastic stresses (MPa)	
	In the SiO_2 films	In the Al films
Prior to hydrostatic compression	-467 ± 15	214 ± 13
Upon hydrostatic compression	-450 ± 11	184 ± 10

Table 4.6

Changes in the microhardness in the elements of the hydrostatically–compressed
MDS–structures (0.5 GPa, 1 min)

Measuring stage	Microhardness (MPa)					
	Load upon the intender (cN)					
	0.5	1	2	3	4	5
Prior to hydrostatic compression	85 ± 5	134 ± 4	247 ± 9	340 ± 14	351 ± 20	410 ± 8
Upon hydrostatic compression	87 ± 4	147 ± 5	273 ± 10	362 ± 10	435 ± 22	493 ± 15

Table 4.7

Changes in the stresses and microhardness of hydrostatically–compressed
structures
(0.5 GPa, 1 min)

Measuring stage	Stresses in the film (MPa)	Microhardness (GPa)		
		Load upon the intenter (cN)		
		1	2	3
Prior to hydrostatic compression	-314 ± 26	4.6 ± 0.6	5.3 ± 0.4	4.9 ± 0.3
Upon hydrostatic compression	-52 ± 20	3.8 ± 0.6	6.9 ± 0.6	6.7 ± 0.6

Table 4.8

The bending and the variance of its values on the surface upon various
technological operations for the structures on the wafers of
$n - Si\,(001)\,CZ\,(\rho = 0.3\;Ohm \cdot cm$) (76 mm in diameter,
400 mcm in thickness, $t_{SiO_2} = 0.1\;mcm$, $t_{Al} = 1.0\;mcm$)

Operation	Average bending (mcm) ($\pm 2.5\,mcm$)	Selective variance in the surface (mcm)
a) Chemical etching	2.4	1.8
Thermal oxidation	2.4	1.9
Aluminium deposition	23.7	12.4
Burning–in	18.8	10.7
Hydrostatic compression (0.4 GPa, 1 min)	12.5	8.8

b) Chemical etching	3.5	2.0
Thermal oxidation	3.8	2.3
Aluminium deposition	22.6	12.0
Burning–in	20.6	10.8
Hydrostatic compression (0.5 GPa, 1 min)	13.0	7.0

Table 4.9

Curvature changes in the $12 - mcm$ thickness $n - Si$ CZ ($\rho = 1.3$ $Ohm \cdot cm$, $P -$ doped) / $250 - mcm$ thickness $n - Si$ CZ ($\rho = 0.01$ $Ohm \cdot cm$, $Sb -$ doped) structures upon a 10 min hydrostatic compression (0.5 GPa)

Measuring stage	Curvature (10^{-3} m^{-1})	
	Average value	Variance of values
Prior to hydrostatic compression	1.7	11.8
Continuous hydrostatic compression	2.6	7.2
Impulsive hydrostatic compression (load frequency is $(4-5) \cdot 10^{-2}$ Hz; amplitude $50-60$ MPa)	6.6	3.7

Table 4.10

Effects of hydrostatic compression on the porousness in SiO_2 $0.3 - mcm$ films produced by thermal oxidation in wet oxygen

Pressure (GPa)	Length of hydrostatic compression	Concentration of pores, micropores per sq. cm. (± 5 per cent)
Prior to compression		11.3
0.4	60	10.7
0.4	120	10.3
0.4	1200	8.6
0.5	60	9.0
0.5	120	7.3
0.5	1200	4.1

4.2.3.3
Strength of microwelded hydrostatically–compressed joints

Residual mechanical stresses in metallic film contact pads deposited onto an oxidized structure substantially affect the reliability of metallization and interelement joints in semiconducting workpieces [Oikawa (1977); Sinha and Sheng (1978)]. For example, an aluminium film placed on a stepped relief wafer turns out to be of different thicknesses and elastically stressed. This causes a local distortion in the continuousness of metallization, strengthens an inhomogeneity in its structure, and in general this becomes one of the reasons in rejecting a given group of elements or an entire workpiece as a whole. Stresses in the film stimulate silicon diffusion along the metallization and induce silicon discharges on the aluminium grain boundaries. This process results in some reduction of residual stresses in the films, and simultaneously it reduces a mechanical strength of metallization and its electrical conductivity [Jowett (1979)]. Film resistance against destruction may be increased by the following two ways of hydrostatic compression: through lessening the internal stresses and increasing the density at the expense of reducing a film defectiveness. Besides, hydrostatic compression may be expected to facilitate a partial smoothing of the surface relief subjected to metallization and, accordingly, it will diminish the probability of local breaks.

Microunevennesses in the relief (bulges and hollows of curvature minimal radius) are smoothed via relaxation of internal stresses concentrated in these regions of the surface; this relaxation is determined by microplastic deformation and diffusive redistribution of the material within the presurface zone.

As was experimentally shown by authors, the aluminium metallization of ICs hydrostatically compressed prior to ultrasonic welding increases the strength of joints by 4 – 22 per cent and simultaneously significantly reduces a variance of strength values and the number of peelings in welded points of the crystal. The data concerning the strength of welded $Au - Al$ joints in ICs upon 1.5 min hydrostatic compression ($0.5\ GPa$) is given in Table 4.11.

Table 4.11

Effects of hydrostatic compression on strength characteristics
in the $Au - Al$ joints

Type of treatment	Number of destructions of joints under continuous compression (pc.)				Number of peelings of the welded point (pc.)
	Compression (cN)				
	3	4	4.5	5	
Without hydrostatic compression	8	14	28	140	20
Upon hydrostatic compression	0	8	22	120	11
Without hydrostatic compression	22	22	42	116	3
Upon hydrostatic compression	0	28	34	108	0

Upon the hydrostatically–compressed metallization, prior to welding, a drop in the transient resistance of $Al-Al$ contact pads made up 30 per cent, in average in the batch.

The thermocompressive $Au-Al$ joints, upon compressed aluminium metallization, increase their strength by 3 –22 per cent, with the variance and the number of peelings in the welded points on the crystal being simultaneously diminished. The changes in the strength of welded $Au-Al$ joints that underwent a 1–min compression by $0.48\ GPa$, prior to welding, are suggested in Table 4.12. Specificity of local destructions is also shown here.

Table 4.12

Changes in the strength of welded $Au-Al$ joints hydrostatically compressed

Batch number		Specific cross–sections in the destructions of microwelded joints			
Con–trolled batch	Hydro–static compres–sion	In the $Au-Al$ interface	Along the average cross–section	Along the output (near the edge of the joint)	On the cross-arm
		Number of destructions (pc.)			
1-A		21	235	16	8
	1-B	4	174	52	50
2-A		15	193	15	57
	2-B	0	180	56	44

Table 4.12. shows that in hydrostatically–compressed joints the number of destructions within the interface has dropped in average as compared against the controlled batch not compressed. Due to hydrostatic compression, the sample variance of destructive forces in $Au-Al$ joints has dropped from $3.2\ cN$ to $1.1\ cN$ (batches 1-A and 1-B) and from $7.9\ cN$ to $2.2\ cN$ (batches 2-A and 2-B). It was also observed that in the thermal annealing ($470\ K$ and $770\ K$) the transient resistance of joints, hydtrostatically compressed prior to welding, is in average $30-60$ per cent less (depending on the batch) than the uncompressed have. This betrays a slow-down of diffusion and a suppression of intermetallic $Au-Al$ compounds at the expense of film compression.

4.2.3.4
Changes in electrical properties of hydrostatically–compressed semiconductors

There are many publications devoted to theoretical and experimental studies of the effects exerted by hydrostatic compression on the electrophysical properties of semiconductors. Such studies were first systematized, actually in a chronologic way, in [Pall and Varshower (1966)]. In later reviewing papers [Baransky and Klotchkov, et al. (1975)] the changes in parameters in all-sided hydrostatically–compressed semiconductors were described; these papers also discussed a use of these effects in the manufacture of converters and pressure meters.

The effects of mechanical stresses on the electrical properties of semiconductors may be manifested, first, as a piezoeffect in crystals having no symmetry centre and, second, through changes in free current carriers caused by structural transformations in the zone. The latter effect specific to all semiconductors independently of the symmetry in their lattice is most often revealed through changes in specific electrical resistance in a mechanically–treated crystal; this effect is called thermoresistive.

Hydrostatic pressure does not change a crystal symmetry but diminishes a lattice period, and, as a result, increases a level of wave function overlapping and shifts energy levels, firstly, in a valency zone and a conductivity zone. Accordingly, this leads to a change in the width of the prohibited zone and to a concentration of charge carriers in semiconductors [Bir and Pikus (1972)]. A tensoresistive effect is reversive if there are slight deformations in the sufficiently structurally perfect crystals – this is used in measuring various mechanical magnitudes by meters with semiconducting elements [Polyakova A.L. (1979)]. However, practically needed are the irreversible changes in the electrophysical characteristics of the deformed semiconducting crystals and structures; studying them could, first, help to deeper understand the interrelations between the properties of real crystals and, second, allow to appropriately exploit these interrelations in the technologies.

As noted in [Polyakova (1979)], the semiconductors' electrical characteristics are changed irreversibly under heavy external loads when these loads are close to a material elasticity limit and the volt–ampere characteristics upon deformation differ from the original. Irreversible changes are commonly measured by locally loading the semiconductor with some concentrated force. This case creates in the semiconductor high mechanical stresses inducing structural distortions, say, dislocations causing additional energy levels in the prohibited zone. This increases a recombination rate and drops the lifetime of minor current carriers. How the direct and reverse currents in the $p-n-$junctions are changed by the concentrated load was studied in [Polyakova and Shklovskaya–Kordi (1969)]. It was shown here that there exist some critical values of load at which there comes a sharp rise of the reverse and a slight fall of the direct current. The ratio suggested in [Polyakova (1979)] describes a relation between a critical load onto the indenter and a radius of its curvature, a shift modulus of the material under study and a depth of the $p-n-$junction. This ratio well agrees with the experimental results for the junctions lying deeper than $0.5\ mcm$. It was emphasized that the loads above a critical value will change not only a current of the $p-n-$junction but also the value F_{cr} itself, i.e. under all further loads the current will change reversibly, provided the external force in the first test is equal to F_{cr}.

How an axial pressure affects the surface properties in silicon was studied in [Kantchyukovsky and Presnov, et al. (1978)]; it was observed here that such load increases a density of surface states and diminishes a hight of the potential barrier in the silicon–nickel contact pad. In studying the volt–ampere characteristics in the Schottky diodes under the axial load [Kantchyukovsky and Moroz, et al. (1980)] showed that prior to mechanical stresses being equal to $1.2\,GPa$ the reverse current will be changing reversibly. Under loads exceeding $1.2\,GPa$, the volt–ampere characteristics were observed to be not restored, and the reverse current also was growing; and in the vicinity of the breakthrough at shift voltages $20-60\,V$ a microplasma was observed. At this, the sensitivity to pressure was increasing.

To unvail the reasons of the observed effects, the authors [Kantchyukovsky and Moroz, et al. (1980)]] made use of silicon etching selectively. In the specimens compressed by more than $1.2\,GPa$, a density of dislocations was found to be increasing.

How a hydrostatical compression of up to $1\,GPa$ affects the volt–ampere characteristics in silicon diffusive junctions was investigated in [Koutchyukov and Shapovalov, et al. (1979)]. These characteristics were measured at the room temperature under pressure in the chamber immediately (a cylinder–piston in the ambient of transformer oil) and upon a cease of external pressure. It was revealed that a rise in the hydrostatic compression diminishes the reverse current through the $p-n-$junction and the direct current grows slightly. As was shown by the authors, the hydrostatic compression makes volt–ampere characteristics irreversible and stable in time. The results gained were explained by a fall in the concentration of acceptors within in the region of the volume charge in the $p-n-$junction (boron diffusion with $N_b \approx 5\times 10^{20}\,cm^{-3}$ in $n-Si$ with phosphorus concentration $\approx 3\times 10^{-16}\,cm^{-3}$) that is, probably, similar to a fall of misfit dislocations. Incorporating boron into silicon gives birth to layer extending stresses of about $0.54\,GPa$, with the gradient being directed from the boundary of the $p-n-$junction to its surface. The misfit dislocations induced by stresses in the diffusive layer will partially penetrate into the $n-$region, if the junction is not deep ($\approx 5\,mcm$). When applying a reverse shift to the junction, a region of the volume charge will be widened to become the $n-$region (due to the difference in the dopant concentrations) and the reverse current will be determined by holes, i.e. by minor carriers for $n-Si$.

Under hydrostatic compression, an extension of the lattice in the diffusive layer is partially or fully compensated by the compression of the compressing gradient from the surface of the $p-n-$junction to its boundary with the $n-$region.

This creates a surface–bound travell by dislocations and determines a drop in the reverse current running through the junction. It was observed that the reverse current is heavily dependent on the pressure in the original section of its growth and this dependence becomes saturated when the external hydrostatic pressure approaches the values obtained in the boron-doped case. A slight increase in the direct current upon compression [Koutchyukov and Shapovalov, et al. (1979)] is explained by a reduction of recombination centres in the region adjacent to the $p - n -$ junction and also by a drop in the density of misfit dislocations. Upon hydrostatic compression, a fall in the reverse current is evidently conditioned by a growth in the width of the silicon prohibited zone and its additional relaxation because of present dopants.

In [Okoulitch and Panteleev, et al. (1982)], it was shown that hydrostatic compression decreases a depth of penetration of ionically–implanted phosphorus into silicon and simultaneously a low-concentration portion of the profile becomes smoothed. Simultaneously, a phosphorus concentration in the implanted layer will rise if there are annealed the hydrostatically–compressed crystals with the surface concentration more than $10^{20} \ cm^{-3}$. The results obtained are explained by the external pressure compensating the phosphorus–induced internal extending stresses. This also diminishes a portion of impurities in the interstitials. Through hydrostatic compression it becomes possible to regulate the distribution profile of dopants in silicon and also increase a difference of $p - n -$ junctions.

The analysis of kinetics in the recrystallization of silicon layers amorphized by ionic bombardment during a compressed annealing made it possible for the authors to suggest an interesting hypothesis that the crystal structure reduction is determined rather by a recombination of sufficiently large fragments of the material but not by changes on the atomic level. A reverse process – an amorphization – also seemingly relates to a recombination of sufficiently large atomic blocks.

The significance of this model lies in its ability to explain many effects initiated in semiconductors by comparatively slight elastic fields and by the all-sided compression of up to $1 \ GPa$, when changes in the process activation energy, in the order of magnitude, become less than or close to kT. As an example of how the "classically" light pressures can induce structural changes may serve the volt–ampere characteristics in the hydrostatically–compressed metal–semiconductor contact pads.

Subjected to the study were the $6 - mcm$ thickness $n - Si \ CZ$ ($\rho = 1.3 \ Ohm \cdot cm$, $P -$doped) / $250 - mcm$ thickness $n - Si \ CZ$ ($\rho = 0.01 \ Ohm \cdot cm$, $Sb -$doped) structures on which there had been formed the ohmic $Ti - Au$ contact pads, deposited in vacuum at $520 \ K$, ($d_{Ti} = 0.05 \ mcm$ and $d_{Au} = 0.1 - 0.2 \ mcm$), and on the epitaxial surface there was established the Schottky barrier – by spraying an Au film $0.15 \ mcm$

thick. The diameter of the barrier surface was $500\ mcm$. Volt–ampere characteristics were measured by TR-4801 curve tracer. The structures were hydrostatically compressed by $0.5\ GPa$ in the continuous and impulsive regimes (amplitude $\pm\ 50\ MPa$, frequency $0.1\ Hz$).

The volt–ampere characteristics measured demonstrate (Fig. 4.10) that in the hydrostatically–compressed diodes their parameters (and, first, the reverse branches of volt–ampere characteristics) have become substantially better: the reverse currents have diminished from $20\ mcA$ to less than $2\ mcA$, and the breakthrough voltages have grown to $32-35V$; and besides, the "entrance" of volt–ampere characteristics into the breakthrough has become abrupt. The most efficient positive changes were brought by the impulsive compression.

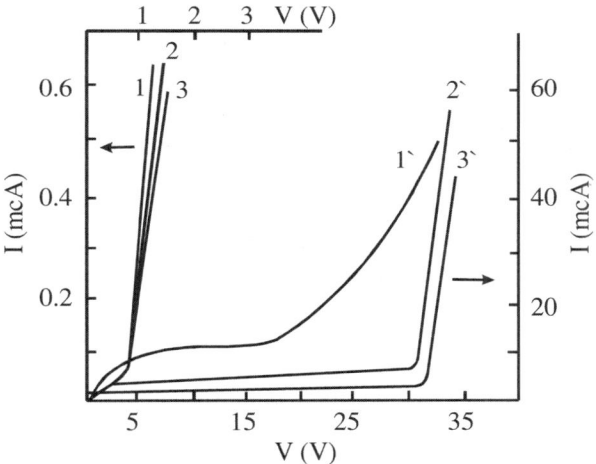

Fig. 4.10. Direct (1–3) and reverse (1' – 3') volt–ampere characteristics of Schottky barriers in the $Au - Si$ structures prior to (1, 1') and upon continuous (2, 2') and impulsive (3, 3') hydrostatic compression.

Hydrostatically–compressed diodes become better in their characteristics due to a fall in the density of crystalline defects in the epitaxial layer in the region of the spatial charge and due to a reduced amount of the related levels in the prohibited zone. Under impulsive compression, this process is additionally stimulated by elastic waves induced by elementary acts of defect recombination; these waves are able to grow when passing through the crystal's regions with damaged lattices [Gerasimenko and Mordkovitch (1987)].

It was told earlier that the high hydrostatic compression reduces mechanical stresses in the components of MOS-structures and porousness in dielectrics, increases a strength of microwelded joins and performs an ordering in the metallization structure. For example, upon a 20–min compression ($0.5\,GPa$) the Al films $0.8-1.2\ mcm$ thick on oxidized silicon showed a fall of resistance by $17-25$ per cent. This length of compression is enough to make resistance saturated and retain unchanged up to $1.1\ GPa$.

Since a reliability of an integrated circuit is dependent on the functioning by its each element, it becomes urgent to study the effects exerted by hydrostatic compression on specific combinations of homo– and heterostructures in the IC topology and to estimate the importance of compression in the structural–kinetic processes of the heavily loaded ICs. According to the accelerated tests at 570 K, it was revealed that upon hydrostatically–compressed metallization, prior to welding, the intermettaloids in the $Au-Al$ joints of ICs drop their growth rate 1.6 times. Storing at the same temperature the hydrostatically–compressed specimens and controlled specimens of the $Au-Al$ joints has shown that the number of compressed $Au-Al$ joints, whose resistance upon 10h storage grew two times and more, are two times less than the controlled specimens. The 4–pad aluminium menders hydrostatically compressed for 1.5 min by $0.5\ GPa$ to study an electrodiffusion rate [Morozov and Skoupov, et al. (1985)] showed a fall in resistance by $4-6$ per cent and an increase of reject time by 30 per cent.

An experimental examination of the total effects produced by hydrostatic compression on ICs has revealed that a $1.5-\min$ compression by $0.5\ GPa$ increased the number of ICs fit in their dynamic and static parameters, first of all due to a fall in the variance of their values within the control batch. As it has been strongly demonstrated by the tests in strengthened regime of thermocycling within the range of $470-210\ K$, the positive postcompression changes in ICs are irreversible.

In our opinion, high hydrostatic compression suggests wide and unique capabilities to purposefully affect a structure of the materials, to modify not only their elastic–plastic characteristics but also their electrical, magnetical and optical properties. As of to-day, far not the entire such possibilities have been already employed. For example, there are only few studies touching the problems of barric annealing of crystalline defects, an increase of adhesion and strength of joints in the heterosystems, a change of phase content in separate components in semiconductors, etc. Practically remains unstudied the effects of the pressure transferring ambients on the processes available under hydrostatic compression, the importance of additional specially deposited films and coatings having other elastic constants and serving as sources of shift stresses and – in the event of gettering layers – as sources of intrinsic point defects. It is a pity but there is no information on the consequences of exciting (simultaneously with the hydrostatic compression) the structural defects and/or electronic–hole subsystem in the material by acoustic–vibrational actions, by corpuscular and ionic irradiation.

The experimental results of the last years proving the efficiency of impulsive low–frequency compression of crystals and multilayer structures, in contrast to a continuous all–sided compression, have not so far received a needed theoretical thinking. Hence, the conditions and processing regimes are still chosen on a purely empirical basis; they need lengthy observations on large arrays of one–type objects. Alongside with this, the above given data prove that hydrostatic compression is possible to become an efficient technological tool for gettering point and extended defects in the components of semiconducting structures.

4.3
Structural changes in microdefects and properties of silicon and silicon–based ultrasonically–irradiated structures

In the contemporary manufacture of semiconducting devices, a liquid processing within an ultrasonic (US) field is mainly exploited for cleaning a surface of monocrystalline wafers from organic and nonorganic contaminations induced by some abrasive–chemical operations at the initial stages of slice preparation. The ultrasonic irradiation efficiently mixes up a washing solution and accelerates desorption and dissolution of contaminants, mainly, at the expense of cavitation and acoustic flows in the working liquid. By now, many physical problems of cavitation phenomena, in particular, the mechanism of formation, the kinetics of growth and collapse of gaseous bubbles, the effects of ambients, ultrasonic fields and external thermodynamical parameters on these processes have been studied experimentally and theoretically [Agranat and Bashkirov, et al. (1970)]. Numerous studies were also dedicated to different physicomechanical and chemical aspects of microdamages in foreign films and erosion of the solids' surface induced by the ultrasonic cleaning or held in occasional fluxes of a cavitating liquid. This made it possible to develop various cleaning technologies, being efficient in regimes, working conditions and composition of washing solutions, to eliminate pollutions of various adhesive properties from the surfaces of different materials.

An active action of high-power ultrasonic irradiation on the extrinsic defect composition in crystals is usually studied when they immediate contact with a sound emitter or concentrator, without any cavitating liquid interlayers [Koulemin (1978); Tyapounina and Naimi, et al. (1999)]. The changes observable in structures and in properties of the materials under irradiation are usually related to a high amplitude of ultrasonic alternate–sign elastic stresses activating diffusive processes and unconservative recombination of defects, at the expense of unbalanced IP defects and considerable heating of the irradiated specimens. The situation with an ultrasonic irradiation differs from a direct ultrasonic action, first, by a less amplitude in the initial sound field in the liquid created by ultrasonic sources in the commercial units used by semiconducting technologies, and, second, by practically room temperature needed for the crystals to be cleaned.

Certainly, in the ultrasonic irradiation it is impossible to fully ignore the role of original ultrasonic waves in structural transformations, since the travelling coefficient for elastic waves in pressure turns out to be higher but not more than by an order of magnitude, because of wave resistance in solids being higher than in liquids. Therefore, the sound pressure when used in liquids of the density close to water density does not exceed 1 MPa in commercial ultrasonic units, with the ambient shifting amplitude in the centre of the sound emitters being equal to 3 – 5 mcm [Agranat and Bashkirov, et al. (1970)]. Though the waves of such amplitude are able to structurally damage the crystals, but for this one will need a more lengthy irradiation and highly-concentrated original defects; with such concentration the defects are strongly unbalanced and undergo a natural relaxational recombination [Pavlov and Syemin, et al. (1986); Skoupov and Tetel'baum (1987)].

According to the above said, the ideas about an active role of original elastic waves, if taken alone, turn out to be insufficient in interpreting the results of some studies where the defectiveness in untrasonically–irradiated semiconductors was measured experimentally [Alekhin (1983); Perevostchikov and Skoupov (1992); Perevostchikov and Skoupov (1999)]. It should be said here that the results themselves, in spite of the identity of the objects studied, are contradicting and do not allow to uniquely predict the consequences of ultrasonic cleaning of the surface; these consequences can look like transformations in the intrinsic defect composition in crystals and, through this, can noncontrollably change the electrophysical properties of semiconductors. Since the ultrasonic irradiation, at the first technological stages, is able to immediately affect the structures (semifinished items) and, in the final end, the parameters of the ready devices, there arises a need in deeper studying the processes in the ultrasonically–irradiated semiconducting structures. The knowledge of regularities in the activation process and in the kinetics of such processes makes it possible to purposefully strengthen the positive and suppress the negative consequences in ultrasonically–irradiated semiconductors.

This subsection discusses one of the most probable mechanisms for transforming the intrinsic defect composition of ultrasonically–irradiated crystals; this mechanism is determined by minor elastic waves arising during a collapse of cavitation bubbles in the working liquid, mostly, on the surfaces of specimens. Active structural changes are evidently governed to a great extent by a compression phase in the wave impulses of mechanical stresses arising during microexplosions of bubbles on the surface of the solid. Here, there appears an analogy with irrradiation actions at which, as shown in [Pavlov and Syemin (1986)] with use of impulses of mechanical stresses, in the crystals irradiated by high–energy particles the first phase of the impulse is the strongest and in its amplitude exceeds by some orders of magnitude a further phase of expansion; this happens due to peaks in atomic shifts, i.e. during a compression of the material. The further phase, though to a lesser extent, also stimulates changes in the crystal's intrinsic defect composition and regulates in crystals the ratio of concentrations of balanced and unbalanced IP defects [Alekhin (1983)].

The amplitudes in a wave pressure field in ultrasonically–irradiated crystals will be estimated via the basic considerations in the theory of liquid cavitation. According to [Agranat and Bashkirov, et al. (1970)], the pressure in a vapour–gaseous mixture in a bubble prior to collapsing, i.e. when its radius becomes R_{min}, is calculated as

$$P_m = \frac{(P_0 + \alpha' P_A) \cdot R^3_{max}}{3R^3_{min}} ,$$ (4.14)

where

P_0 stands for an external static pressure acting onto the liquid (during ultrasonic irradiation being usually equal to $0.1\,MPa$);

P_A is a level of ultrasonic pressure calculated as $P_A = \rho \omega s A$ (ρ is a liquid density);

s is a sound velocity;

ω is a frequency; and

A is an amplitude of oscillations;

$$\alpha' = -\frac{(\cos \omega (t_{max} + \Delta t) + \cos \omega \cdot t_{max})}{\omega \Delta t}$$ (4.15)

t_{max} is the time for the bubble cavities to expand up to R_{max};

$$\Delta t = 0.9 \cdot R_{max} \sqrt{\frac{\rho}{\alpha P_A + P}} .$$ (4.16)

If to suppose that the entire work spent to grow a bubble of up to R_{max} and its compression of up to R_{min} is then spent for exciting an elastic (striking) wave and if to neglect the ultrasonic energy reflection losses, sonoluminescence and other minor cavitation–dependent phenomena, then the time τ_0 during which the bubble–accumulated energy will be converted into the energy of elastic wave will be:

$$\tau_0 = \frac{1.7 R_{min}}{\sqrt{\dfrac{P_m}{\rho}}}$$ (4.17)

or it may be estimated through Δt as:

$$\tau_0 = k\,\Delta t \left(\frac{R_{min}}{R_{max}}\right)^3 , \qquad (4.18)$$

where $k = 3.3$ [Agranat and Bashkirov, et al. (1970)].

In order to find an amplitude of the compressing impulse arising during a collapse of a bubble in a solid and a quantity of such impulses (per a unit of square) in the surface appearing per a unit of time (i.e. a density of the elastic wave flux), one needs to know a square of the contact between a bubble surface and a solid in the time when a bubble is being destructed. Jumping from the supposition that the contact is round in form and is of the radius a_0 and using the solution of the Hertz problem [Perevostchikov and Skoupov (1992)], it is not difficult to obtain:

$$a_0 = R_{min} \sqrt{3\pi\, P_A \left(\frac{1 - v_M{}^2}{E_M} + \chi\right)} , \qquad (4.19)$$

where

E_M, v_M is, respectively, an elasticity modulus and the Poisson coefficient for the material under irradiation;

χ is an isothermic liquid compressibility.

Suppose that at some fixed instant of time only the bubbles encountered within the zone of maximally–compressed waves of the length λ are collapsed on the surface of the solid. Then the density of elastic wave generating centres (a flux of impulses) will be

$$j = \frac{4\omega}{\lambda^2} = \frac{4\,\omega^3}{s^2} . \qquad (4.20)$$

With the elastic wave generating centres being distributed on the surface stochastically, the average resulting pressure for the wave field at the distance Z from the excitation place may be calculated [Pavlov and Syemin (1986)] as

$$P_M(Z) = \frac{P_m\, a\,[1 + 2\pi\, j\tau_0\, Z^2]^{0.5}}{Z} . \qquad (4.21)$$

Here, $a = a_0\left[1 - \exp\left(-\dfrac{a_0}{S_M \cdot \tau_0}\right)\right]$ is an efficient contact square; when introduced into (4.21) it allows to take into account, in the first approximation, a spreading of the elastic wave front during compression because of a high sound velocity S_m in the solid, as compared against the sound expansion velocity in a working liquid. Such a spreading drops the pressure amplitude during the collapse time τ_0 and this amplitude turns out to be somewhat less than P_m.

The expressions (4.14 – 4.21) have been employed to estimate the characteristics of wave fields in Si crystals being ultrasonically cleaned in standard conditions: a working liquid is water (other liquid ambients may be easily calculated with a use of their densities) and the ultrasonic field frequencies are 22 and $44\,kHz$, the ultrasonic pressure being equal to $P_A = 0.2$ and $0.4\,MPa$, respectively. In these calculations there have been used the experimental results from [Agranat and Bashkirov, et al. (1970)] and table data from [Mishenko and Ravdel' (1974); Kikoin (1976); Baransky and Klotchkov, et al. (1975)]. The results obtained are given in Table 4.13.

Table 4.13

Parameters of ultrasonically–irradiated silicon crystals

Ultra–sonic field frequ–ency (kHz)	$R_{max}\,/\,R_{min}$ (mcm)	P_m (GPa)	τ_0 (ns)	a_0 (mcm)	a (mcm)	j ($10^3\,cm^{-2}s^{-1}$)
22	582 / 9.66	29 7	3.1	2.0 8	0.1 85	2.74
44	292 / 4.95	29. 0	1.6	1.0 7	0.0 71	21.92

As seen in Table 4.13, in ultrasonically–irradiated crystals the density of elastic wave exciting centres on their surface is not great and the average distance $(j)^{-0.5}$ between them makes up hundreds of micrometers. This makes it possible to consider that the intrinsic defect composition in crystals is recombined under the action of the elastic wave isolated sources, and in estimating the amplitude P_M in (4.18) one should consider a first term only. In this case, the formula (4.21) will assume the form:

$$P_M(Z) = P_{ohm}\,/\,Z,$$

where $P_{ohm} = 5.5 \times 10^3$ and $2.1 \times 10^3\,Pa \cdot m$, with the ultrasonic field being irradiated by frequencies 22 and $44\,kHz$, respectively.

When analysing the experimental results, the expression (4.21) seems to have some physical sense for the crystals' layers at the depths $Z \geq a$. All the experimental data given above and in [Koulemin (1978); Perevostchikov and

Skoupov (1992); Perevostchikov and Skoupov (1999)] actually relate to large depths and so they may be interpreted through a recombined microdefect structure of silicon within the elastic wave field and through the action of unbalanced IP defects induced by the working liquid cavitation. At the depths $Z < a$, in particular, comparable with the thickness of SiO_2 films, the basic role is seemingly played by the nonlinear effects of interactions of striking waves with the components of the intrinsic defect composition that can not only decrease but also increase the residual defectiveness [Mayers and Mur (1984)]. It is most probable that it is a nonlinearity in the structural recombinations in SiO_2 films that determines the local phase transfers within the elastic wave field and there arise the inclusions of new phases (with the density exceeding that of amorphous SiO_2); these phases determine high values of the refraction coefficient n_k upon ultrasonic irradiation.

As for the microdefects, they may be transformed by the fluxes of unbalanced IP defects and elastic waves (mainly they change their sizes) and these processes may be controlled by the two basic phenomena – the condensation of point defects onto clusters and the emission of atoms (vacancies) from their surface [Pavlov and Syemin, et al. (1986); Skoupov and Tetel'baum (1987)]. The condensation may both diminish the sizes of microdefects (they will be dissolved) and lead to their growth. The phonon–induced emission of point defects from a cluster surface is always accompanied with their dissolution. It is the competition of these two processes that determines a crystals' reaction to any external action which can cause an oversaturation with IP defects and provide a diffusive–drifting mobility in vacancies, intrinsic interstitial atoms and impurities in the static or dynamical elastic fields.

The point defects are possible to be emitted from the surface of the microdefect if $P_m (Z) \, \Omega \geq E_c$ is valid, where E_c is a bonding energy of point defects in a cluster; Ω is an atomic volume. The thickness of the layer in the crystal where this mechanism runs efficiently may be calculated through (4.21). For example, if, according to [Italyantsev and Mordkovitch (1983)], the bonding energy in silicon atoms in the $A-$ type growth microdefects is taken equal to $2.2 - 2.8 \, eV$, then the $22 \, kHz$ ultrasonic irradiation will dissolve such clusters (due to emission of interstitial atoms) at the depths less than 0.31 and 0.25 mcm, respectively, i.e. within the presurface region of about a in thickness, where the elastic waves mainly interact with the microdefects nonlinearly. Though, there is another channel through which emission occurs with a larger than $Z = a$ region; this channel relates to the elastic waves born by the annihilation of closely placed Frenkel pairs. Such situation is most probable for the IP defects held in the Maxwell ambients encircling stable microdefects. Indeed, according to [Skoupov and Tetel'baum (1987)], a height of the barrier for annihilating vacancies and intrinsic interstitial atoms in silicon is $\Delta E \cong 0.25 \, eV$, and if the energy of the elastic wave reaching the microdefect turns out to be sufficient for overcoming this barrier, then the annihilation process gives birth to minor elastic waves of the amplitude $P_B \approx B\varepsilon_0$, where B is a spatial modulus of elasticity; ε_0 is a local deformation of the lattice in the neighbourhood of the defect ($\varepsilon_0 \approx 0.1-0.2$ [Skoupov and Tetel'baum (1987)]). The depth until which this mechanism is efficient will be $Z_B = \dfrac{P_{ohm}\,\Omega}{\Delta E} \cong 2.75 \; mcm$ for $\omega = 22 \; kHz$. The energy being transferred by the minor elastic waves to the atoms on the cluster surface will be equal to $1.25 - 2.5 \; eV$ – that is close to the bonding energy E_c.

Alongside with this, the estimates made show that the emission of point defects during ultrasonic irradiation may induce significant structural changes only in comparatively thin presurface layers of crystals, and through this emission mechanism it is difficult to explain experimentally observed microdefects dissolved at large depths. The transformations of microdefects within the bulk of the crystal most probably occur due to their interactions with the unbalanced IP defects induced by elastic waves within the presurface layer of the thickness $+C_M\tau_0 \approx C_M\tau_0$ and also in the ambients encircling the clusters [Perevostchikov and Skoupov (1992); Pavlov and Syemin (1986); Skoupov and Tetel'baum (1987)].

The concentration of unbalanced IP defects injected from the surface may be estimated as

$$ N_0 = \int_0^{C_M\tau_0} \frac{I}{D}\, dZ \, , \tag{4.22} $$

where

D stands for the diffusion coefficient of IP defects; and

I is a flux of IP defects described by the ratio

$$ I = jn_v = j\,\Delta V N_{Si} \exp\left(\frac{S}{K}\right)\exp\left[-\frac{U_f - P_M\Omega}{kT}\right] , \tag{4.23} $$

where

j is a density of elastic wave generating centres on the surface; this density is calculated with a use of (4.40);

n_v is a quantity of unbalanced IP defects (further, only vacancies will be considered, for certainty) generated by a single centre of the volume

$$ \Delta V = \frac{2\pi}{3}(C_M\tau_0)^3 ; $$

N_{Si} is a concentration of Si atoms in a unit of volume;

S and U_f is, respectively, the entropy and the generating energy of IP defects.

The idea that ultrasonic irradiation mostly gives birth to vacancies is explained by the fact that in an elastic wave impulse a compression phase exceeds in amplitude the expansion phase, and, accordingly, a local deformation (in the shape of a semispheric valley appearing in the place of microexplosion of the bubble) will be larger than a further extending deformation creating a semispheric hillock or leaving the surface practically flat.

The surface of microvalley is restored at the expense of the surface atoms recombination that results in a drop of local complement of the surface energy – this is possible when vacancies are drifting into the bulk of the crystal [Alekhin (1983)]. However, during ultrasonic irradiation a vacancy flux is seemingly dominating; it is determined by cavitation effects because at room temperature a spacial oversaturation with vacancies (being earlier thermodynamically balanced) within an elastic wave field (a compression phase) will be close to unit and will be restored fast.

With a due account of the above told, we, taking $S = 12\,K$, $U_f = 1.12\,eV$ (following the conclusions made in [Alekhin (1983)], suppose that the vacancy generating energy near the surface of the crystal is twice as less than in the volume), and taking $D_V\,(300\,K) = 2.5{\times}10^{-10}\ cm^2\,/\,s$, we shall obtain

$$I = 6.14{\times}10^{10}\ cm^{-2}\ s^{-1} \quad \text{and} \quad N_0 = 6.63{\times}10^{17}\ cm^{-3}.$$

The vacancy concentration at the depth Z from the surface to be cleaned will be:

$$N(Z) = \frac{2\,N_0\sqrt{Dt}}{Z\sqrt{\pi}}\exp\left[-\frac{Z^2}{4\,Dt}\right] \qquad (4.24)$$

Ignoring the aggregation process, in the first approximation the thickness of the layer, where the microdefect structure of the crystal is changed due to an interaction of the interstitial atom clusters with the unbalanced vacancies, may be calculated from the equality of concentrations of vacancies in the in-depth flux (4.24) and vacancies localized in the Maxwell ambients N_a. The magnitude N_a we find supposing that the vacancy atmospheres arise due to the dimensional first–order interaction existing between the vacancies and microdefects representing themselves small-size R – radius dislocation Frank loops of the introduced type with the Bŭrgers vector modulus b. According to [Runge and Wörl (1978)], the energy of such defect–vacancy interaction is determined as

$$E = \frac{2\pi}{3}\cdot\left[\frac{1+v}{1-v}\right]\cdot\mu b\,\delta r_0^{\,3}\,R^2\,\frac{1-3\cos^2\psi}{r^3}\,, \qquad (4.25)$$

where

μ is a shift modulus;

δ is a dilatation power of the vacancy ($\delta \approx \varepsilon_0$);

r_0 is a vacancy radius;

ψ is an angle between the direction of the Bűrgers vector and the radius–vector connecting the centres of the microdefect and the vacancy;

r is a current distance between a vacancy and a microdefect.

The profile of vacancy concentration in the ambient will be:

$$N_a(r) = N_\infty \exp\left[-\frac{A}{r^3 kT}\right], \qquad (4.26)$$

where

N_∞ is a crystal vacancy concentration far from the dislocation loop (a balanced concentration);

A is a constant which, according to (4.25) leaving the angle dependence ($\psi = \pm\frac{\pi}{2}$) beyond our consideration, may be calculated as

$$A = \frac{(1+v)\mu b \delta R^2 \Omega}{2(1-v)}.$$

Then, a total concentration of vacancies within the ambient encircling the microdefect will be

$$N_a = \int_V N_a(r)\, dV = 4\pi N_\infty \int_r^{r_2} r^2 \exp\left[-\frac{A}{r^3 kT}\right] \cdot dr . \qquad (4.27)$$

An upper integration limit r_2 is found from the equality condition that the interaction energy E from (4.25) is equal to kT; and a low limit r_1 is found from E coinciding with the energy of vacancy migration to the dislocation loop E_m, i.e. from the condition that vacancies are beyond the zone of their possible disactivating drift to the defect surface, $A/r_1^3 = E_m$. Substituting numerical values into silicon parameters at room temperature for the dislocation loop of the radius $R = 0.5\,mcm$ we shall obtain $r_1 = 70$ and $r_2 = 160\,nm$ for $E_m = 0.33\,eV$. With a due account of these limits, taking the integral (3.13) will give:

$$N_a \cong \frac{4\pi r_1^{3}}{3} N_\infty \exp\frac{A}{r_1^{3} kT} \ . \tag{4.28}$$

Choosing the values of IP defect concentration within the bulk of the crystal is not so simple, and, as shown in [Alekhin (1983)], the estimates N_∞ may vary widely depending on the defect formation potential used when extrapolating into a low temperature region. However, if the ambients are considered to arise during high–temperature microdefect formation (and a further drop in temperature only favours the concentration of IP defects in the ambients (4.28)), then the equality $N_\infty = N(T_m)$ may be considered quite true (here T_m stands for the melting temperature of the material). Then, according to [Alekhin (1983)], for the vacancies in silicon we have $N_\infty = 5 \times 10^{16} \ cm^{-3}$, and the total number of the defects within the ambient encircling a single microdefect will be $N_a \approx 2.2 \times 10^{7}$ vacancies. The average number of atoms forming a cluster of

the radius $0.5 \ mcm$ will be equal to $n_1 = \dfrac{4\pi R^3}{3\Omega} \approx 2.6 \times 10^{10} \ at$.

The depth Z_{depth} until which the microdefects are dissolved because of the interaction with unbalanced vacancies migrating from the crystal's surface may be estimated as

$$N_\mu N_a = \frac{2N_0 \sqrt{Dt}}{Z_{depth} \sqrt{\pi}} \exp\left[-\frac{Z_{depth}^{2}}{4Dt} \right], \tag{4.29}$$

where N_μ is a spatial concentration of microdefects in the crystal ($N_\mu \cong 10^{9} \ cm^{-3}$).

Upon transformations, we find from (4.28) and (4.29) that

$$Z_{depth} = \frac{2N_0 \sqrt{Dt}}{\sqrt{N_0^{2} + \pi N_\mu^{2} N_a^{2}}} \ .$$

If to neglect a complement to the coefficient of IP defect diffusion within the ultrasonic field (though it is known to grow [Alekhin (1983); Gorshkov and Perevostchikov, et al. (1989)]), then a 1h ultrasonic irradiation, according to this mechanism, will efficiently dissolve the microdefects in silicon as deep as $Z_{depth} \approx 19 \ mcm$.

At greater depths another dissolution mechanism is possible to run – at the expense of direct exchange by IP defects between clusters and their Maxwell ambients; this exchange is conditioned by the transient stress field created by elastic waves of the original ultrasonic field (P_A) and born by the cavitation in the working liquid. The amplitude of the first-type waves in the crystal (P_C) will be approximately twice as higher than in the liquid (P_A) – due to the difference in their wave resistances, η_C and η_L , since these parameters are interrelated by the ratio

$$P_C = \frac{2\eta_C}{\eta_c + \eta_L} P_A ,$$

where $\eta_L = \eta_1 = \rho_1 c_1$.

For example, for silicon we have $\eta_C \approx 2.1 \times 10^6 \, g/cm^2 s$, and hence when processing it in water ($\eta_L = 1.5 \times 10^5 \, g/cm^2 s$) the amplitude of the original ultrasonic field in the crystal will be $P_C \leq 0.8 \, MPa$. For microdefects to be effectively dissolved this will be insufficient. Therefore, it is true to suppose that this process is controlled by the second-type cavitation–induced waves. Therefore, if to follow the conclusions in [Pavlov and Syemin (1986); Skoupov and Tetel'baum (1987)] and consider that a cluster–ambient exchange by IP defects starts from the magnitude $P_M \geq 1 \, MPa$, then the extreme depths at which the consequences of such exchange can be seen will be 2.1 and $5.5 \, mm$, for the case of ultrasonic irradiation by frequencies 44 and $22 \, kHz$, respectively. These estimates naturally represent the upper depths, without accounting the balanced IP defect concentration that is being restored in the ambients after each drift of vacancies or interstitial atoms onto the clusters. The depths, until which the efficiency of this mechanism is possible to be seen (with an ultrasonic irradiation length being chosen reasonably), are constrained by the values $Z \leq Z_{depth}$ if a free surface of the crystal serves as a source of unbalanced IP defects. If there are available such sources of internal mechanical or electrical stresses as the interface boundaries and the extended dislocation–type defects, the defects may be simultaneously dissolved by the minor elastic wave field across the entire bulk of the crystal. In more details, this microdefect dissolution mechanism and its (numerically described) efficiency in silicon are described in [Pavlov and Syemin, et al. (1986); Skoupov and Tetel'baum (1987)].

Thus, the above given estimates, though approximate to some extent, demonstrate that in ultrasonically–irradiated crystals the defect dissolution process and associated changes in the physical properties of crystals, say, in microhardness, may be fully explained by the known theories which describe the interactions between the properties and the thermodynamically unbalanced IP defects performing an in-diffusion and/or arising in the Maxwell ambients. On the basis of these considerations it is also possible to understand the structural changes in the ultrasonically–irradiated silicon [Alekhin (1983); Perevostchikov and Skoupov (1992); Perevostchikov and Skoupov (1999)]. In [Alekhin (1983)] it was detected that vacancy–type clusters increase their concentration in the presurface region in the crystals subjected to a powerful ultrasonic irradiation without a use of a cavitation interlayer. This proves the idea that this kind of irradiation gives birth to excessive vacancies.

A growth of microdefect concentration, and more exactly, a growth of selective etching wells in the liquid ambient ultrasonically irradiated (this was reported in [Perevostchikov and Skoupov (1990); Damask and Dienes (1963)]), may be naturally related to an increase in the cluster recognizing capabilities of the selective etching technique, at the expense of partial dissolution of Cottrell ambients encircling the microdefects within the field of elastic waves. Such ambients compensate local mechanical stresses arising in the vicinity of microdefects and this reduces the difference in the rates of etching the defective and defectless regions in the crystal. The ultrasonic irradiation leads to a diffusion–drifting dispersal (clean–out) of ambients and makes it easier for unbalanced vacancies to approach a cluster's surface. The facts given in [Perevostchikov and Skoupov (1992); Perevostchikov and Skoupov (1999)] relate rather to the irradiation lengths at which the ambients are being dispersed efficiently but the microdefect dissolution process has not yet become activated properly. In other words, for the "aged" crystals there exists some latent period of ultrasonic irradiation during which the microdefects get rid of impurity ambients and then intensively interact with the unbalanced IP defects. This latent period may be significantly shortened via a previous high–temperature annealing of crystals with a fast cooling. In its essence, the oxidation of silicon in the above described experiments served as such annealing process. Certainly, alongside with a presurface microdefect dissolution (due to highly oversaturated IP defects, mostly, the vacancies) there should arise the simplest aggregates and accumulations of defects; the latters, if sufficiently large in size, can be detected via the selective etching, against the background of the falling concentration of original clusters. During ultrasonic irradiation a defect formation scenario seems qualitatively similar to a generation of vacancy aggregates under the implanted layer in the ionically–irradiated crystals [Morozov and Skoupov, et al. (1985)].

In ultrasonically–irradiated crystals, foreign films on the surface seemingly play an important role in transforming the extrinsic defect composition of crystals; these films create inhomogeneous static mechanical stresses which, depending on the sign (compression, extension) and amplitude, are able to suppress or activate structural transformations in the irradiated materials. Solutions of this problem have a very important application, for a use of special topologically–distributed (multilayered, in a general case) heterosystems on the surface will make it possible to exploit the ultrasonic irradiation for locally controlling the extrinsic defect composition and the properties of semiconducting structures at various stages of the device manufacturing line.

As an illustration of "long–range" effects of ultrasonic irradiation on the microdefectiveness and the properties of silicon and $Si - SiO_2$ structures, let us discuss our two experiments.

Experiment 1

Under study were the slices of the dislocation–free $p - Si(001)CZ$ ($\rho = 12\ ohm \cdot cm$) 460 mcm thick. The front side in the slices underwent chemical–mechanical polishing by Cab-O-Sil (aluminium oxide) suspension and the back side was polished by the diamond synthetic paste. On both sides of a slice there was built a SiO_2 layer of the thickness $d_{ox} = 0.17 - 0.24\ mcm$, first by thermal oxidation at $T_{ox} = 1223\ K$ in the flux of wet (adding 5 per cent of HCl) and then dry oxygen. The oxidized structures were then ultrasonically irradiated in the commercial unit УМ-4 (power 160 W, frequency of the ultrasonic field $25 \pm 2\ kHz$); into the water–filled bath of this unit there was placed a fluoroplastic working camera with specimens submerged into isopropyl alcohol. During irradiation, the temperature of the specimens did not exceed 337 K. In experiments, the ultrasonic irradiation duration varied from 30 to 150 min. Here, the specimens cut by scribing from already irradiated slices served as controlled specimens.

The density of microdefects and their in-depth distribution profiles prior to and upon the ultrasonic irradiation were recorded through a selective chemical etching in the chromium solution of the composition $CrO_3\ (50\%) : HF\ (49\%) = 1 : 1$ (vol. portions). The density of microdefects was measured in 12–15 fields of the Neophot-32 microscope and the values were averaged with the probability 0.95. Besides, sample variance of density values across the entire surface was estimated. Alongside with metallographic measurements, the by-layer microhardness of crystals was also measured, by the microhardness measuring unit at various loads on the indenter. Average values of microhardness of the specimen surface was calculated along the diagonals 20 – 25 of prints, with reliability being equal to 0.95. The effects of ultrasonic irradiation on silicon dioxide were recorded through changes in film optical characteristics being measured by the automatic digital ellipsometer at the wave length $\lambda = 0.633\ mcm$ with the accuracy not less than $\pm 0.1\ \%$.

The experimental results are the following.

The average values and sample variances of the microdefect densities measured at various depths in the specimens, prior to and upon the ultrasonic irradiation, are given in Table 4.14. This table shows the following regularities.

The microdefect densities in original crystals have an in–depth inhomogeneous distribution and this tendency retains upon the irradiation, though the defect concentration gradient, especially close to the surface, drops significantly.

b) The ultrasonic irradiation diminishes both the average concentration of microdefects and the variance of its values across the surface the more significantly, the more the ultrasonic irradiation length is.

c) Upon the ultrasonic irradiation in the course of the period Δt ,

The microdefect dissolution rate calculated as $V_{\mu} = \dfrac{N_0 - N_1}{\Delta t}$ (where N_0 and

N_1 are, respectively, the values of defect densities in the original specimens and in ultrasonically–irradiated specimens during Δt) has in the presurface region the least value $4.97 \times 10^2 \ cm^{-2} \ min^{-1}$ and becomes $9.27 \times 10^2 \ cm^{-2} \ min^{-1}$ at the depth $10 \, mcm^1$. As the ultrasonic irradiation length increases, the maximum rate is detected at large depths. At $\Delta t = 120 \, min$ the rates of microdefect   dissolution   for   the   observed   depths   are

$V(x = 5 \, mcm) = 1.84 \times 10^2 \ cm^{-2} \ min^{-1}$;

$V(x = 10 \, mcm) = 3.84 \times 10^2 \ cm^{-2} \ min^{-1}$ and

$V(x = 20 \, mcm) = 4.34 \times 10^2 \ cm^{-2} \ min^{-1}$.

Table 4.14

Densities of microdefects in silicon crystals exposed to ultrasonic irradiation of various lengths

USI length (min)	Distance from the crystal surface (mcm)					
	5		10		20	
	Density of microdefects ($10^4 \ cm^{-2}$)					
	Average value	Vari– ance	Average value	Vari– ance	Average value	Vari– ance
0	2.53 ± 1.05	1.32	5.00 ± 1.32	1.72	6.33 ± 1.73	2.32
30	1.04 ± 0.42	0.50	2.22 ± 0.35	0.44	3.64 ± 1.03	1.07
60	0.83 ± 0.13	0.23	1.28 ± 0.32	0.41	3.22 ± 0.71	0.88
90	0.42 ± 0.009	0.18	0.51 ± 0.09	0.11	2.47 ± 0.65	0.76
120	0.32 ± 0.12	0.19	0.39 ± 0.09	0.12	1.12 ± 0.17	0.21

The data describing the changes in the microdefect structure of the specimens upon the ultrasonic irradiation correlates with the measurements of microhardness. Microhardness in original specimens is of minimal value in the presurface region and increases with an increase of depth and load onto the indenter. Upon ultrasonic irradiation, the in-depth microhardness distribution profile will retain and the microhardness values will increase proportionally to an ultrasonic irradiation length (Figs. 4.11 – 4.14) [Perevostchikov and Skoupov (1999); Kiselyev and Perevostchikov, et al. (2002)].

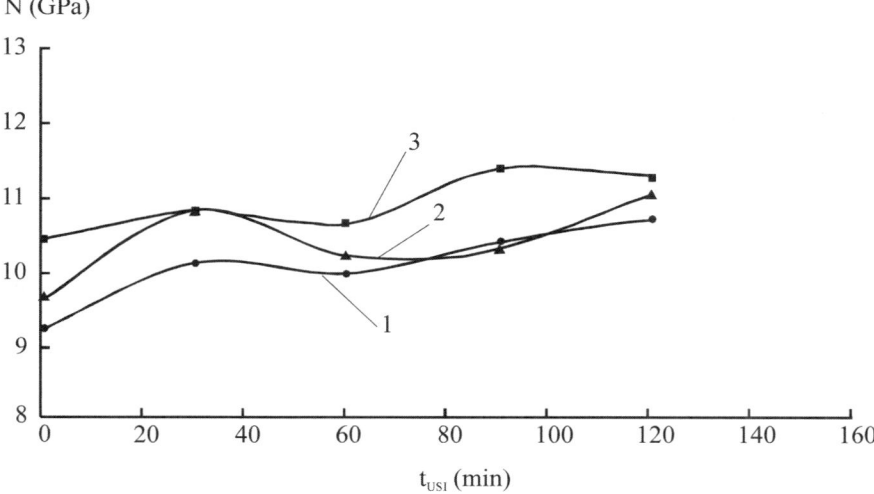

Fig. 4.11. Dependence of microhardness in ultrasonically–irradiated Si crystals on the irradiation length at the depth $x = 0$. The load upon the indenter (cN) is **1** – 10, **2** – 20 and **3** – 30.

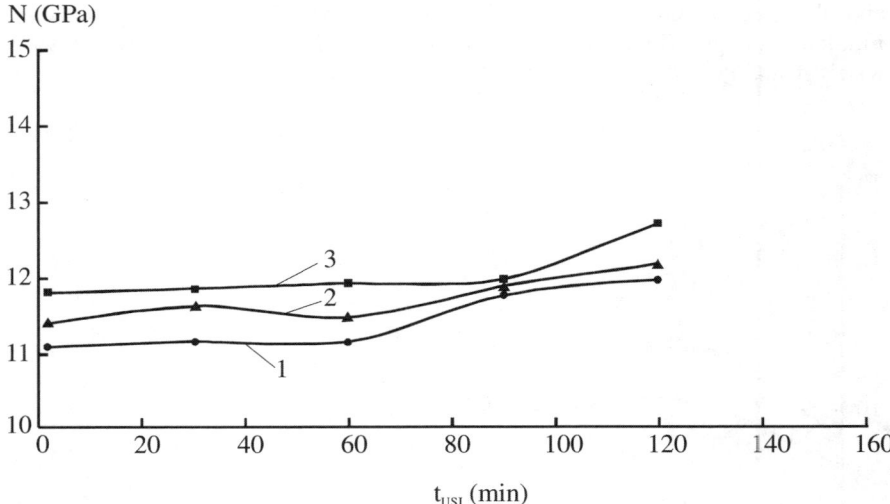

Fig. 4.12. Changes in microhardness in ultrasonically–irradiated Si crystals at the depth $x = 5\,mcm$ according to the irradiation length. The load upon the indenter (cN) is **1** – 10, **2** – 20 and **3** – 30.

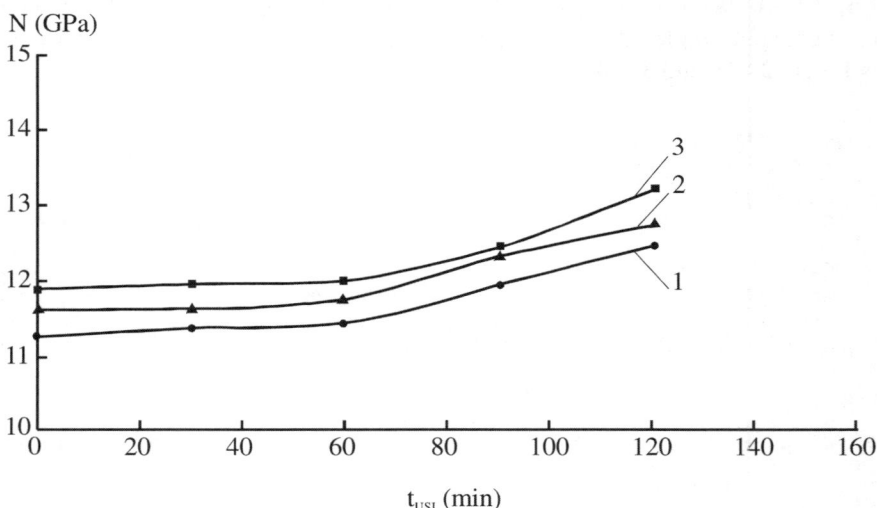

Fig. 4.13. Changes in microhardness in ultrasonically–irradiated Si crystals at the depth $x = 10\,mcm$ according to the irradiation length. The load upon the indenter (cN) is **1** – 10, **2** – 20 and **3** – 30.

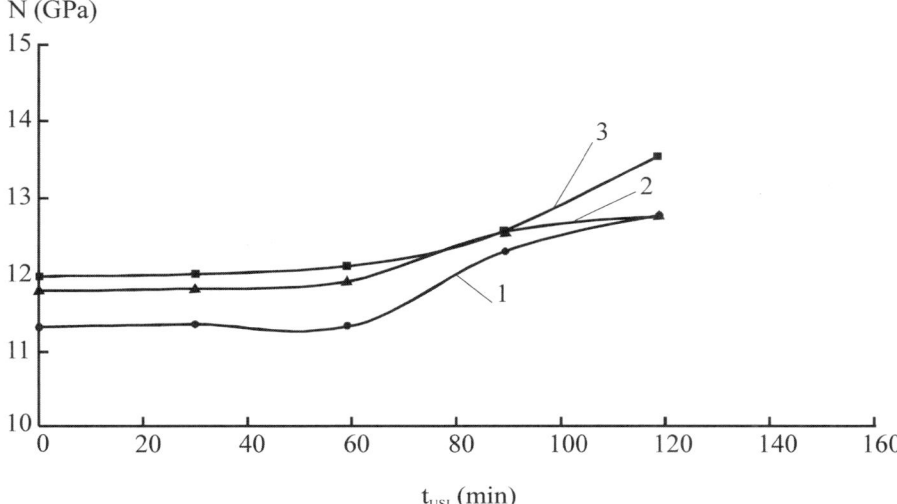

Fig. 4.14. Changes in microhardness in ultrasonically–irradiated Si crystals at the depth $x = 20\,mcm$ according to the irradiation length. The load upon the indenter (cN) is **1** – 10, **2** – 20 and **3** – 30.

Simultaneously, a drop in the variance of surface microhardness is also observed (Figs. 4.15-4.18) [Perevostchikov and Skoupov (1999); Kiselyev and Perevostchikov, et al. (2002)].

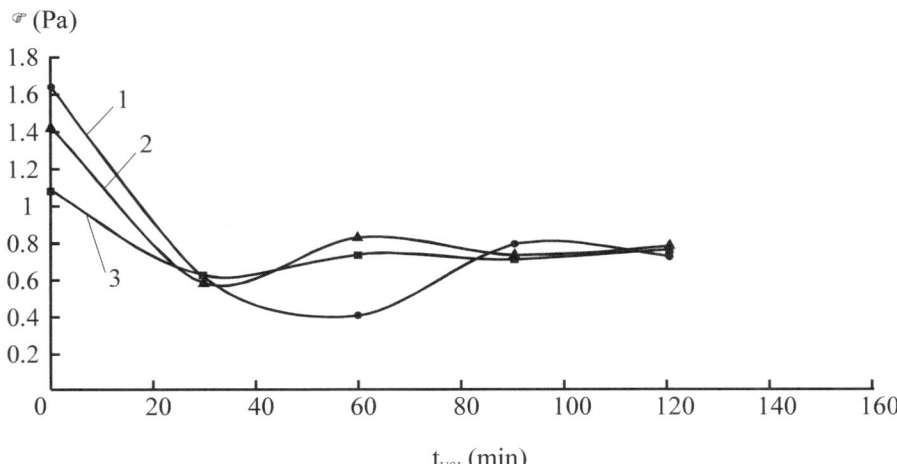

Fig. 4.15. Changes in microhardness variance of ultrasonically–irradiated Si crystals at the depth $x = 0\,mcm$ according to the irradiation length. The load upon the indenter (cN) is **1** – 10, **2** – 20 and **3** – 30.

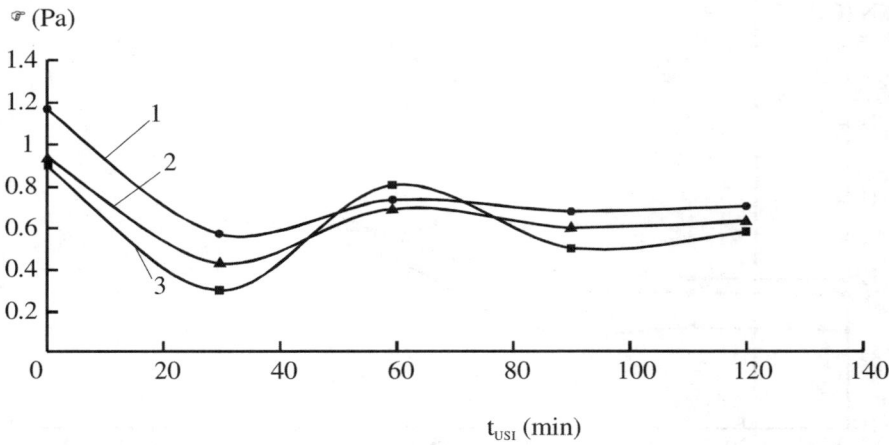

Fig. 4.16. Changes in microhardness variance in ultrasonically–irradiated Si crystals at the depth $x = 5\,mcm$ according to the irradiation length. The load upon the indenter (cN) is **1** – 10, **2** – 20 and **3** – 30.

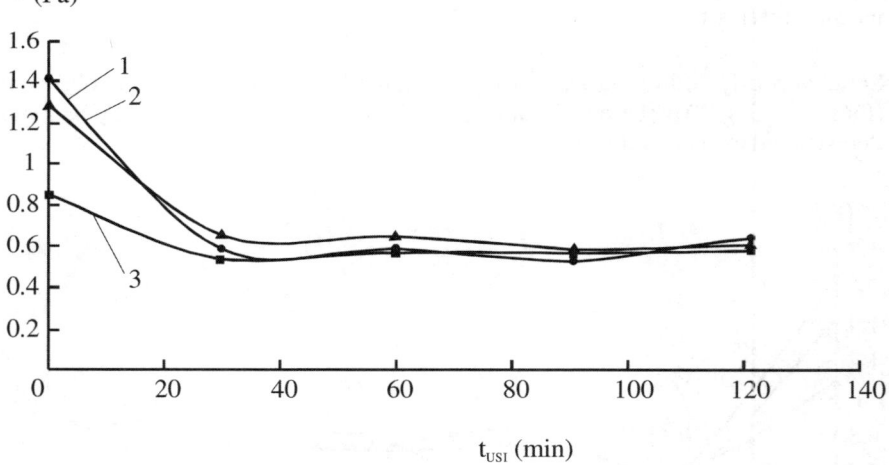

Fig. 4.17. Changes in microhardness variance in ultrasonically–irradiated Si crystals at the depth $x = 10\,mcm$ according to the irradiation length. The load upon the indenter (cN) is **1** – 10, **2** – 20 and **3** – 30.

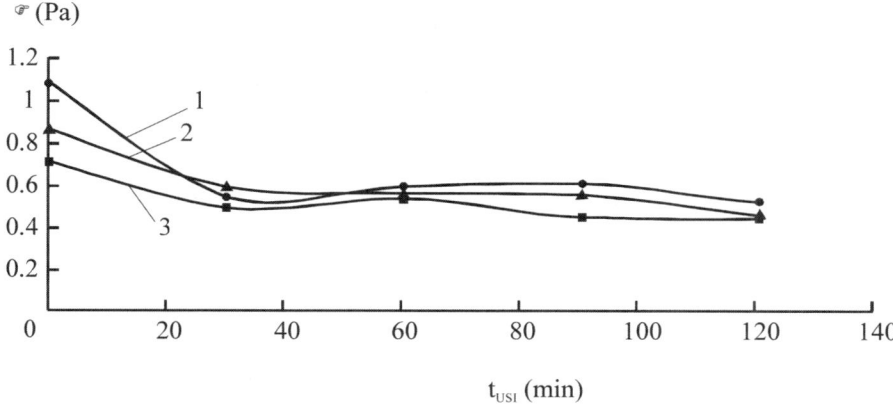

Fig. 4.18. Changes in microhardness variance in ultrasonically–irradiated Si
crystals at the depth $x = 20 \, mcm$ according to the irradiation length. The load
upon the indenter (cN) is **1** – 10, **2** – 20 and **3** – 30.

A nonmonotonous change in the elastic–plastic properties across the entire surface
and in depth relates rather to a inhomogeneous spatial distribution of growth and
technological microdefects in the specimens.

An active role of microdefects in forming the prints is proved by measurements
of microhardness relaxation depending upon the length of indenter's constant load
in the crystals prior to and upon the ultrasonic irradiation; these measurements
were done with a use of the technique from [Kiselyev and Perevostchikov, et al.
(2002)].

The microhardness relaxation was measured in the specimens with the indenter
being loaded by $20 \, cN$ at the depths $x = 0$ and $x = 10 \, mcm$ prior to and upon
the ultrasonic irradiation for 60 min.

The experimental data describing a change of microhardness according to the
irradiation length is approximated with the accuracy $\pm 11\%$ by the function:

$$ H(t) = H_{\infty} + (H_0 - H_{\infty}) \exp\left(-\frac{t}{\tau} \right), $$

where
$H(t)$ stands for the value of microhardness measured upon the load for the time
   t ;
H_0 and H_{∞} , respectively, the original ($t \rightarrow 0$) and asymptotic ($t \rightarrow \infty$)
   values of microhardness; and
τ is a characteristic time for microhardness relaxation.

The microhardness $H(t)$ was measured at discrete times of loads,
$t = 10, 30, 60, 90$ and $120 \, s$. In calculating τ and H_{∞}, as H_0 there was

accepted the microhardness value measured at $t_0 = 10\,s$. Supposing that $t_0 < \tau$, the characteristic time of relaxation was found as

$$\tau = \frac{t_{i+1} - t_i}{\ln\left(\dfrac{d\,H}{d\,t}\right)_i - \ln\left(\dfrac{d\,H}{d\,t}\right)_{i+1}}\,,$$

where $\left(\dfrac{dH}{dt}\right)_i$ and $\left(\dfrac{dH}{dt}\right)_{i+1}$ are the microhardness changing rates,

respectively, at the load lengths t_i and t_{i+1} ($t_i < t_{i+1}$).

The numerical values of τ make it possible to estimate the efficient energy of the activation of the microhardness relaxation process from the ratio:

$$\tau = \frac{h}{k\,\theta}\exp\left(\frac{U}{kT}\right),$$

where

h and k are the Planck and Boltzmann constants;

θ is the Debaev temperature (for calculations we used $\theta = 588\ K$ [Alekhin (1983)]);

T is the temperature for measuring the microhardness; in our case it was equal to $300\ K$.

The parameters describing the kinetics of crystal microhardness relaxation prior to and upon ultrasonic irradiation are given in Table 4.15.

Table 4.15

Effects of ultrasonic irradiation on parameters
of semiconducting crystals

Parameter	Prior to ultrasonic irradiation		Upon ultrasonic irradiation	
	Distance from the surface of the crystal (mcm)			
	0	10	0	10
H_0 (GPa)	9.63 ± 1.13	11.63 ± 1.04	10.2 ± 0.64	11.83 ± 0.41
H_∞ (GPa)	9.51 ± 1.05	11.56 ± 1.27	10.14 ± 1.12	11.78 ± 1.30
$\tau\ (s)$	36 ± 11	87 ± 23	74 ± 9	148 ± 16
$U\ (eV)$	0.87 ± 0.13	0.89 ± 0.12	0.89 ± 0.12	0.91 ± 0.12

The ultrasonic irradiation, as shown in Table 4.15, increases the microhardness relaxation time almost 2 times. This result, together with an increase in microhardness and a decrease in microdefect densities of ultrasonically–irradiated crystals, manifests an increase of elasticity in the silicon presurface layers, with the impurity-defect clusters being available in them. This agrees with the conclusions made in [Alekhin (1983)] concerning abnormal mechanical properties in silicon crystals encountered within the presurface layers of the thickness $2 - 5\,mcm$ where the Peierls – Nabarro barriers are lowered and the activation energy for dislocation motions is equal to $0.84 - 1.4\,eV$, and microdefects serve as centres for heterogeneous nucleation of dislocations. Besides this, according to [Skoupov and Tetel'baum (1987); Kiselyev and Perevostchikov, et al. (2002)], microdefects may serve as sources of unbalanced IP defects within the field of locally high mechanical stresses under the indenter; these stresses will stimulate not only a sliding but also an unconservative drift of the print frame dislocations causing a microhardness relaxation. It is evident that most contribution to this process will be made by small clusters (of $B-$, $C-$ and other types) which are not practically detected via the selective etching, though they are always present in the zero-dislocation Si crystals and have a lesser energy of dissociation onto the free interstitial atoms (vacancies), as compared against the $A-$ type microdefects whose majority are dislocation loops.

As a result of microdefects dissolved by ultrasonic irradiation, the concentration of active sources of IP defects under the indenter will diminish and this will cause a slow down of microhardness relaxation. In this case, the relaxation will be generally determined by the concentration of vacancies which become unbalanced within the field of compressing stresses under the indenter and are absorbed by the dislocations of the print frame thus determining their overcrawl under load [Alekhin (1983)]. As shown in Table 4.15, the greatest relaxation occurs in the presurface layers of crystals that is sooner determined by the inhomogeneous in-depth distribution of metastable microdefects able, when locally loaded, to dissociate onto free IP defects.

As was shown by ellipsometric measurements, in the original $Si - SiO_2$ structures there takes place an oscillating change in their average values and in the variances of refraction and optical thickness in SiO_2 across the entire surface in specimens. In ultrasonically–irradiated films, the oscillation amplitudes of the ellipsometric parameters diminish drastically (Tables 4.16 and 4.17). The information concerning a rate of changes in the refraction indicator during ultrasonic irradiation (this rate was estimated as $\left| V_{n_i} \right| = \dfrac{n_i - n_{0i}}{\Delta t}$, where n_{0_1} and n_i are, respectively, the refraction indicator values prior to and upon the ultrasonic irradiation for Δt) helped to derive – through a linear approximation by the averaging technique – the below empiric function:

$$V_n = 1.52 \times 10^{-2} \cdot n_0 - 2.29 \times 10^{-2}.$$

This dependence shows that there is some definite value of the refraction indicator n_k which retains constant during the irradiation regimes $V_n = 0$ and $n_c \approx 1.507$. This function may imply that in the SiO_2 films with a definite extrinsic defect composition there is some structural equilibrium a transfer to which (in the films of various extrinsic defect compositions) is stimulated by ultrasonic irradiation in the given regimes. It may be expected that such state, determined, in particular, by some definite value of the refraction indicator n_c, is sooner individual for the SiO_2 layers and dependent on the film manufacturing technology, the length and conditions of preirradiation storing of structures, the parameters (including the extrinsic defect composition) of the wafers and the like. It is quite possible that a change in the irradiation regimes will cause a jump to another equilibrium (or quasiequilibrium) state with another value of n_c.

Table 4.16

Effects of ultrasonic irradiation on the SiO_2 film refraction indicator

USI length (min)	Refraction indicator (n_c)			
	Prior to ultrasonic irradiation		After ultrasonic irradiation	
	Average value	Variance	Average value	Variance
30	1.472	0.293	1.498	0.077
60	1.614	0.139	1.517	0.054
90	1.488	0.148	1.494	0.129
120	1.531	0.151	1.523	0.085
150	1.517	0.229	1.503	0.093

Table 4.17

The effects of ultrasonic irradiation on the optical thickness of the SiO_2 films

USI length (min)	Refraction indicator (n_c)			
	Prior to ultrasonic irradiation		After ultrasonic irradiation	
	Average value	Variance	Average value	Variance
30	0.172	0.087	0.240	0.033
60	0.247	0.059	0.219	0.008
90	0.211	0.064	0.183	0.048
120	0.219	0.048	0.225	0.075
150	0.206	0.083	0.216	0.011

It is clear that a detailed structure and extrinsic defect composition in SiO_2 films in equilibrium may be detected only via a wider choice of experimental techniques, firstly, by an immediate analysis of the extrinsic defect composition. The ellipsometric parameters, chiefly, a refraction indicator (the optical thickness was experimentally found to be less sensitive to ultrasonic irradiation), though being integral in nature, make it, nevertheless, possible to suggest some conclusion about the quality of the films under study. First, since the original layers are of the dominating value $n > 1.458 - 1.485$ (a range from glass to crystabolite [Gerasimenko and Mordkovitch (1987)]), the films may be supposed to hold microinclusions of the density more than the SiO_2 density, for example, the density of impurity aggregates of metallic atoms inhomogeneously distributed across the entire bulk of the layers. Second, that the ultrasonic irradiation causes changes in the refraction indicator and optical thickness betrays an active role played by the transient regions in the vicinity of the $Si - SiO_2$ interface, both in the film and the wafer. The latter is proved, in particular, by the relation $V_n d_0 = A$ existing between the optical thickness of the films (d_0) prior to ultrasonic irradiation and the changing rate of the refraction indicator during irradiation; here A is a constant whose value for the structures in question is close to 1.4×10^{-3}. Such dependence implies that in thin films with properties largely determined by the extrinsic defect composition in the transient layers the ultrasonic irradiation makes the processes of transition to a new quasiequilibrium state more intensive.

In general, how the refraction indicator depends on the ultrasonic irradiation length may be approximated as

$$n(t) = n_C + (n_0 - n_C) \exp\left(-\frac{t}{\tau}\right),$$

where n_0 is a refraction indicator prior to the irradiation; τ is a characteristic time of the refraction indicator relaxation depending on the original value n_0. For the structures under study, $\tau = \tau(n_0)$ in its first approximation is described as $\tau(n_0 - n_c) \approx 7.6 \times 10^4$. In both formulas, the difference ($n_0 - n_c$) is taken in its absolute value, since, as was told above, the refraction indicator in the original state may be both more and less than the value of n_c, but ultrasonic irradiation brings it close to n_c. If the relaxation time is supposed to obey the exponential law

$$\tau_1 = \tau_0 \exp\left(\frac{U_1}{kT}\right),$$

then it becomes possible to estimate the effective activation energy responsible for changing the refraction indicator during ultrasonic irradiation. Taking as τ_0 the period of valency oscillations of the $Si - O$ bond in the silicon dioxide (the oscillation frequency is $v \approx 1100\, cm^{-1}$) we obtain $\tau_0 = 3.35 \times 10^{-14}\, s$, and the activation energy, calculated through the experimental values n_{0_1} and $n_i(t_i)$, will be $\overline{U} = (1.05 \pm 0.02)\, eV$. As the difference ($n_0 - n_c$) increases, the activation energy will diminish, though for the structures in question it remains constant, within the limits of the definition. The activation energy found coincides with diffusion activating energies of impurities in SiO_2 (Na, Li, B, P and others), which determine the low-temperature instability of parameters in the SiO_2 layers [Mayers and Mur (1984)].

It was detected experimentally that the ultrasonically-induced structural changes occur not only during irradiation but also proceed in the same kinetic direction, though at a lesser rate and after ultrasonic irradiation. For example, during ultrasonic irradiation the refraction indicator will change, depending on its original value n_0 (see above), within the interval $(0.1-5.0)\times10^{-5}\ s^{-1}$, and the relaxation rate after irradiation will change within the interval $(0.13-1.6)\times10^{-9}\ s^{-1}$. Upon ultrasonic irradiation, there has been detected some instability in microhardness values of specimens, though it will change much less than the refraction indicator. Investigating the effects of relaxation on structurally–sensitive parameters in ultrasonically–irradiated semiconducting structures constitutes itself a relatively independent problem to be attacked and this problem needs further thorough investigation. The first results gained manifest that the intensiveness of relaxation depends both on the ultrasonic irradiation regimes (a length, first of all) and on the original defectiveness in the components of the structure under study. However, to explain the empirical regularities in the relaxational changes one needs to understand their activation mechanisms and this is impossible without knowledge of nature and mechanisms of structural transformations occurring in the homo- and heterostructures immediately during a surface cleaning process.

Experiment 2

The purpose of this experiment is to determine how a protective coating on the surface affects the formation of microdefects in ultrasonically–irradiated silicon [Kiselyev and Perevostchikov, et al. (2002)].

Subjected to the study were the zero-dislocation slices of $p-Si\,(001)\,CZ$ ($\rho=12\ ohm\cdot cm$) 450 mcm thick. The front sides of the slices (i.e. the side to be studied) underwent chemicomechanical polishing by the Cab-O-Sil suspension (with particles less than 0.1 mcm in diameter), and the back sides underwent a loose abrasive treatment (with particles being about 8 mcm in size). Ultrasonic irradiation was performed in the UM-4 ultrasonic bath, with the ultrasonic frequency being equal to 25 kHz. A front side of the silicon slice was divided into four regions: the first was coated by the $150-200-mcm$ film, the second region was left unprotected, and the other two regions, during ultrasonic irradiation, had a contact with the sputnik-slices (one with the both-sided polished and the other with the both-sided grinded slices).

Ultrasonic irradiation was done for 100 min in the isopropyl alcohol. Structural changes induced by ultrasonic irradiation in various regions of the slice were recorded through the Sirtl–Adler pictures of selective chemical etching

HF $(49\%):CrO_3$ $(10\%) = 2:1$, in combination with the by-layer chemico-dynamical polishing.

The in-depth distribution profiles of microdefect concentrations for all specimens are shown in Figs. 4.19 and 4.20 [Kiselyev and Perevostchikov, et al. (2002)]. It is seen that in the presurface layer in the film–coated region there is the highest density of defects. This is probably explained by the impurity ambients being most efficiently dissolved around the microdefects. Further, there comes a sharp drop of density (practically to zero), because of elastic waves being suppressed by the film.

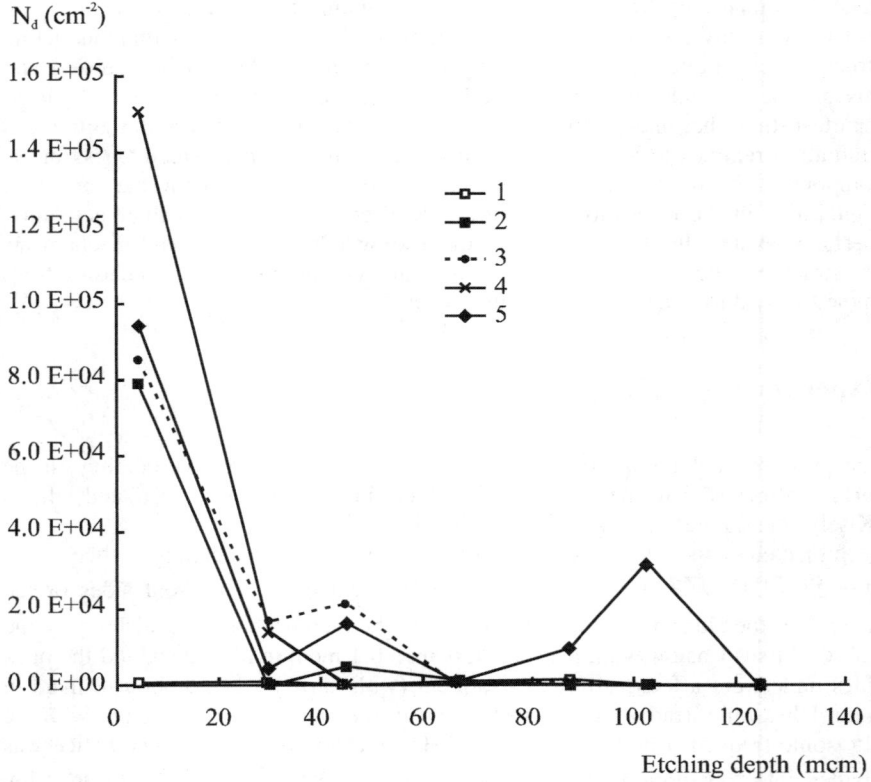

Fig. 4.19. The in-depth microdefect density profiles in Si crystals: **1** prior to ultrasonic irradiation; **2** upon irradiation under a grinded wafer; **3** upon irradiation under a polished wafer; **4** upon irradiation under a film; **5** an open surface upon irradiation.

$N_d \cdot 10^4$ (cm^2)

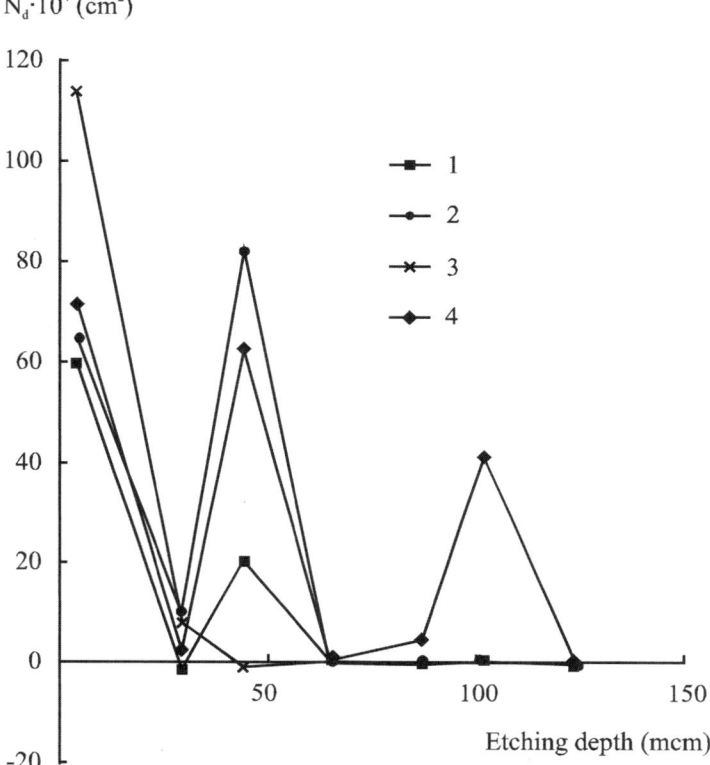

Fig. 4.20. Relative in-depth changes in microhardness density upon ultrasonic irradiation: **1** under a grinded wafer; **2** under a polished wafer; **3** under a film; **4** an open surface.

In the unprotected region, a higher density of defects is observed at larger depths, in contrast to other regions. In this case, the Cottrell ambients are primarily dissolved in the depth of the crystal.

To prove the supposition that the ultrasonic irradiation dissolves the impurity ambients, the specimens were annealed in all the regions of the controlled slice, and a later selective etching as deep as 5 mcm was performed. An one-hour annealing was done in the diffusion furnace at the pressure 5×10^{-5} mm mercury column (Figs. 4.21 – 4.24) [Kiselyev and Perevostchikov, et al. (2002)].

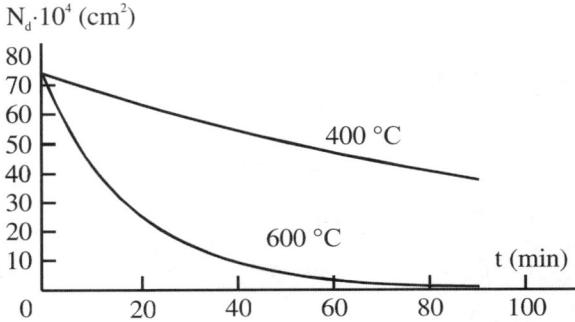

Fig. 4.21. Dependence of microdefect densities on the annealing length in the specimens with open surface.

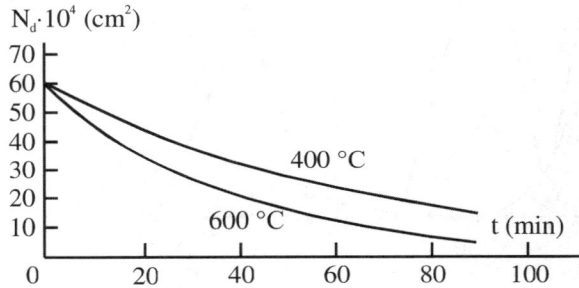

Fig. 4.22. Dependence of microdefect densities on the annealing length for the surface under a grinded wafer.

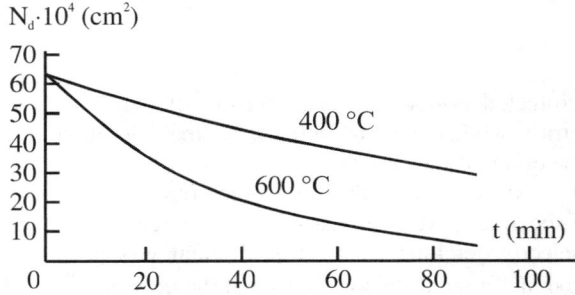

Fig. 4.23. Dependence of microdefects on the etching length for the surface under the polished wafer.

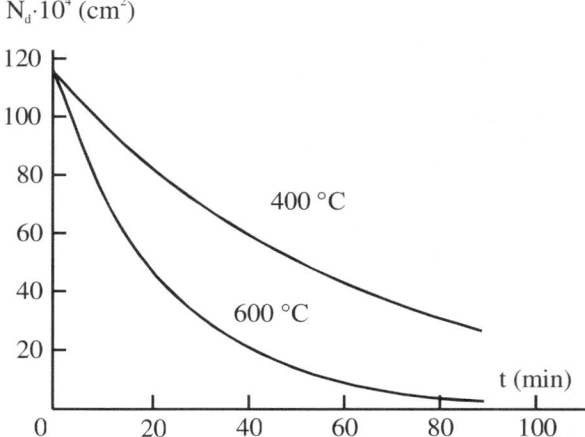

$N_d \cdot 10^4$ (cm^2)

Fig. 4.24. Dependence of microdefect density on the etching length for the surface under the polished wafer.

That upon annealing the defect concentration was reduced may be related to the mechanism of ambient restorations at the expense of the impurities diffusing from the surface.

In all the specimens, upon removing a 5 mcm front-sided layer through selective chemical etching, the topography of the surface was then studied with a use of the scanning probe TMX – 2100 "Accurex" microscope, in the non-contact regime.

As demonstrated by the atomic microscopy (Figs. 4.25-4.26), the ultrasonic irradiation diminishes microroughness as much as two times both in the unprotected and coated region.

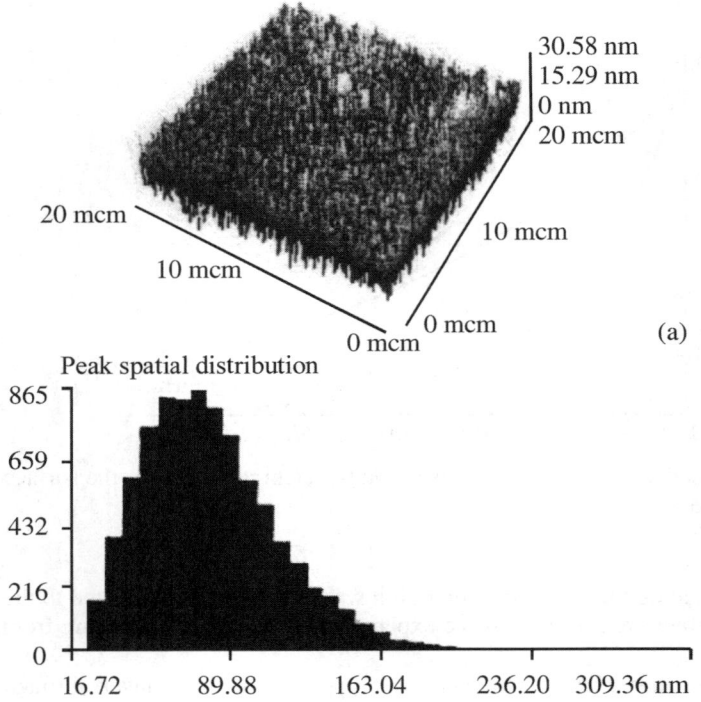

(a)

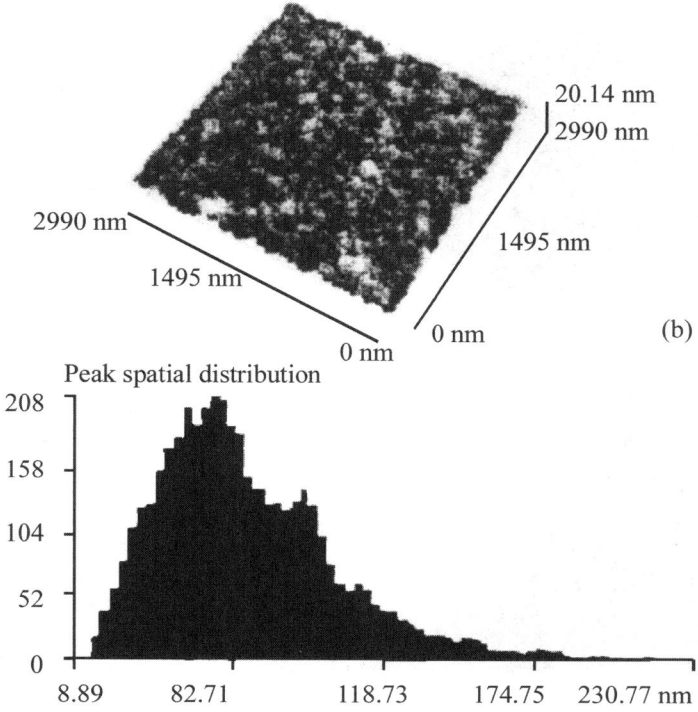

Peak spatial distribution

(b)

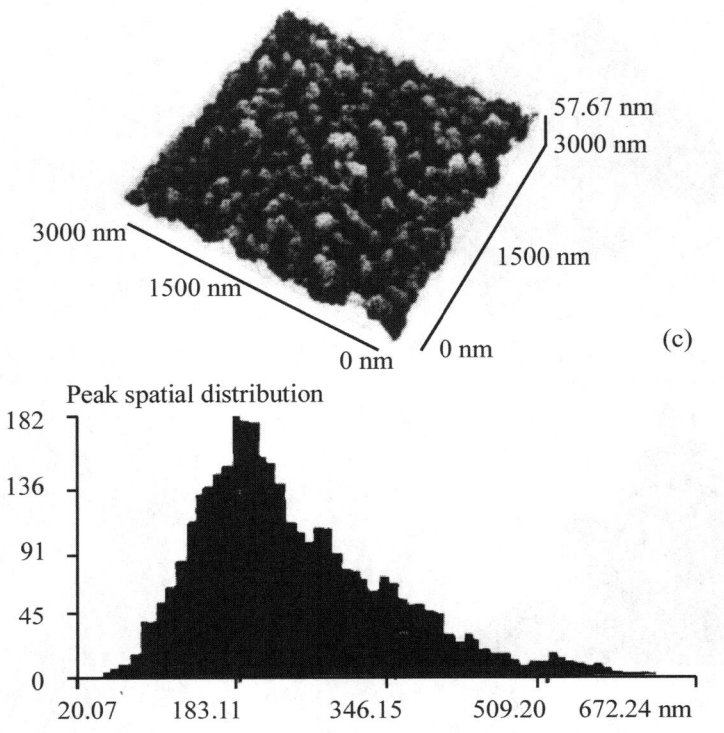

Fig. 4.25. Photopictures and histograms for variance of lateral sizes of surface microrelief in $p - Si\,(001)\,CZ$ ($\rho = 12$ $ohm \cdot cm$, boron – doped) selectively etched as deep as 5 mcm: **a** an original specimen (prior to ultrasonic irradiation); **b** a specimen with an open surface; **c** a specimen contacting with a polished wafer.

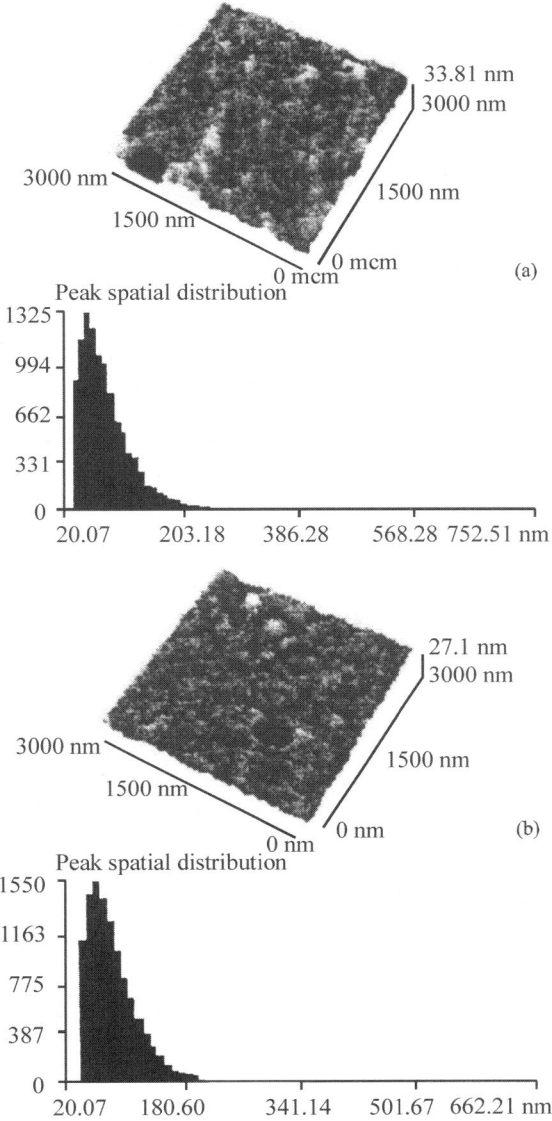

Fig. 4.26. Photopictures and histograms for variance of lateral sizes of surface microrelief in $p - Si\,(001)\,CZ$ ($\rho = 12\;ohm \cdot cm$, boron – doped) selectively etched as deep as 5 mcm: **a** a specimen contacting with a polished wafer; **b** a surface under a film.

Upon the ultrasonic irradiation, the surface contacting with the grinded slice shows the greatest increase in roughness and in lateral sizes. This region produces the greatest width of the histogram of lateral sizes. This seemingly relates to some supplementary action of minor elastic waves induced by the layer damaged by the grinding process and also to a cavitation of the working liquid in-between the slices; the microroughness of the grinded slice favours this process. Besides the changes in the states of microdefects (and this may be used in developing the ways of their gettering), ultrasonic irradiation, as shown experimentally, is able to make the selective chemical etching more sensitive to the structural defects in silicon crystals, since ultrasound favours the diffusion of impurity ambients encircling a defect. Here, the best effect is obtained near the crystal surface protected, during ultrasonic irradiation, by a chemically stable varnish. And the greatest penetration (with a lesser dissolution of Cottrell ambients) of this effect (100 – 200 mcm) is observed on the surface being open to ultrasonic irradiation.

During ultrasonic irradiation, the atomic microscopy has made it possible to also detect a fall in lateral sizes of the microrelief in the region under the polished slice – and this may be exploited for additionally increasing the quality of the semiconducting wafers during abrasive–chemical irradiation.

4.4
Irradiation-stimulated gettering

Among highly perspective technologies for gettering impurities and extended defects in semiconducting structures, a special place is occupied by the techniques in which a spectrum and a concentration of components in the extrinsic defect composition are purposefully changed by exposing the materials to corpuscular or photonic radiations of various nature. The basic advantage of these techniques lies in their ability to control a gettering process through varying the kind, the energy, the irradiation intensity and the irradiation dose itself as well. This subsection discusses some techniques exploited to modify by irradiation the extrinsic defect composition and properties in semiconducting structures.

4.4.1
Irradiating crystals by accelerated ions

Ionically doping the materials, and, mainly, semiconductors, represents itself one of basic operations in the technological line of manufacturing contemporary workpieces in micro- and nanoelectronics. Practically, this ionic implantation used as a doping technology began to be exploited for gettering the fast–diffusing metallic impurities by the radiationally–damaged layers during a high–temperature treatment of structures (for silicon, at $T > 1023 K$) [Sigson and Cspedi, et al (1976); Geipl and Tice (1977); Runge and Wörl (1978); Kolger and Peeva, et al. (2003)]. Low-temperature irradiating techniques of gettering relate to a so–called "long–range effect" born by ionic implantation, i.e. structural changes and changes in the properties of the irradiated materials at the depths substantially exceeding the runs of the implanted ions. This long–range effect was detected in materials of various physicochemical properties and crystallographic structures irradiated by ions of a wide spectrum of masses, energies and doses [Martynenko (1993)]. Below, a long–range effect will be described mainly in the ionically–irradiated semiconducting structures where controlling the extrinsic defect composition becomes very crucial, since the active and passive components in the workpieces to be constructed should be of submicronic and nanometric sizes.

Spatial structural changes in ionically–irradiated silicon were seemingly detected first by [Pavlov and Pashkov, et al. (1973)]. Here, the authors via the X-ray topography were observing the behaviour of dislocations incorporated by scribing into silicon crystals 600 mcm thick (having an electronic conductivity and $\rho = 15\ Ohm \cdot cm$); the crystals were irradiated by Ar^+ ions, of $E = 40\,keV$ and of the doses from 6×10^{16} to $6 \times 10^{17}\ cm^{-2}$. The transparent topograms of one and the same region in the specimen were shot prior to and upon irradiation, and also a month and a half later when the irradiated crystals were kept under normal conditions. It was experimentally detected that irradiation increases a contrast of separate dislocation lines which can drift from the centre of the scratch at the distance of up to 10 mcm. The scribe mark when being irradiated from the opposite side of the crystal was splitted into separate dislocations having efficient widths of zones equal to 150 – 200 mcm (Fig. 4.27).

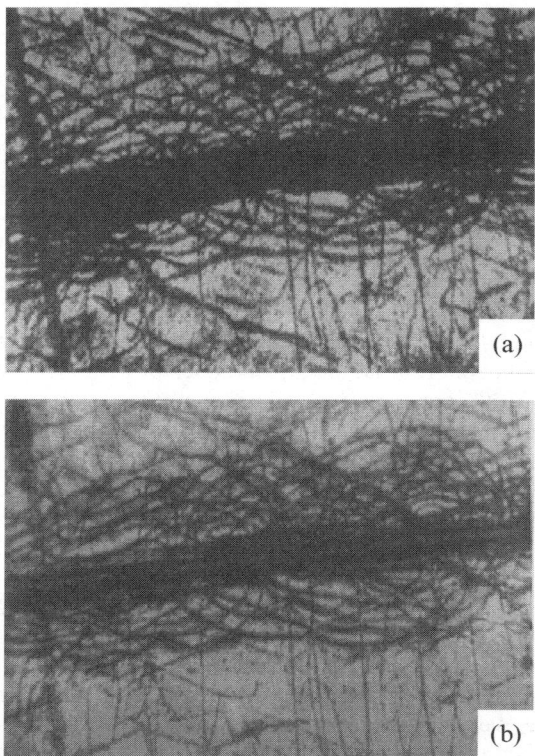

Fig. 4.27. Structure of dislocations in a silicon wafer (x70): **a** immediately after irradiating by argon, dose $6 << 10^{17} cm^{-2}$; **b** in a month and a half after argon irradiation.

As demonstrated by topograms, structural changes encounter not only in irradiated specimens but also during their storage. In the latter case, the dislocations have additionally drifted as far as $10 - 15$ mm.

This drift of dislocations is explained in [Pavlov and Pashkov, et al. (1973)] by their interactions with the unbalanced IP defects generated by the ionic beam, and a growth in contrast is explained by an annihilation of impurity ambients and by a dislocation detachment from these ambients. During a storage period, the dislocations in crystals are relaxed due to a decomposition of metastable irradiation–induced defects into free vacancies and intrinsic interstitial atoms that determine an unconservative recombination of dislocations. With the help of the X-ray two-crystal diffractometry the authors [Pavlov and Pashkov, et al. (1973)] managed to prove the changes in the ionically–irradiated dislocations in silicon to be determined by the dislocations absorbing chiefly the excessive vacancies. According to the authors, combining an ionic irradiation with a further annealing at temperatures like $773 K$ will create a possibility to efficiently control a spatial distribution of dislocations within the bulk of the crystal, and even to eliminate them completely. As an indirect proof of this serve the below internal mechanical stresses in the epitaxially–grown silicon structures, prior to and upon ionic irradiation, studied by the three–crystal diffractometry [Skoupov and Uspenskaya (1975)].

Subjected to study were the $n-$conductivity structures obtained by the gas transported chloridization technique, and namely, the films of $\rho=1\,Ohm\cdot cm$ and $6\,mcm$ thick on the $250-mcm$ wafers with $\rho=0.01\,Ohm\cdot cm$. The specimens were irradiated from the side of the wafer by the argon ions with $E=40\,keV$ and of the dose $4\times10^{16}\,cm^{-2}$. As a characteristic of the stress there has been taken a microscopic curvature (C) in the structures and a relative increment for the wafer lattice period $\left(\dfrac{\Delta d}{d}\right)$ in the direction normal to the $(111)-$surface. Through these parameters, the stresses in the film–wafer interface (σ) and the film deformation (ε) were estimated as

$$\sigma=\frac{E}{1-v}\left[\varepsilon-\frac{\Delta d}{d}\right];$$

$$\varepsilon_0=\frac{\Delta d}{d}+\frac{4a^3}{3t(2a-t)}C_0;$$

$$\varepsilon=\frac{\Delta d}{d}+\frac{4a^3}{3t(2a-t)}\left(C-8{,}2\cdot10^{-3}\right),$$

where

ε_0 and ε, respectively, a film deformation prior to and upon irradiation;

$2a$ and t is the thickness of the entire structure and that of the epitaxial layer, respectively;

E is the Young modulus , and

v is the Poisson coefficient.

The experimental results obtained on 70 structures were then processed through smoothing the static rows by the moment method. The densities of parameter variance in the observed array are presented in Figs. 4.28, 4.29 [Pavlov and Skoupov, et al. (1987); Zorina and Popov, et al. (1983)] and 4.30 [Pavlov and Pashkov, et al. (1973)].

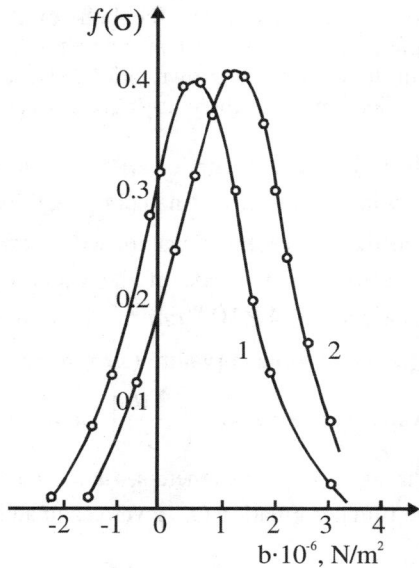

Fig. 4.28. Density functions of stress variances in the film–wafer interface for a batch of epitaxial structures: **1** prior to argon irradiation of the collecting side; **2** upon irradiation.

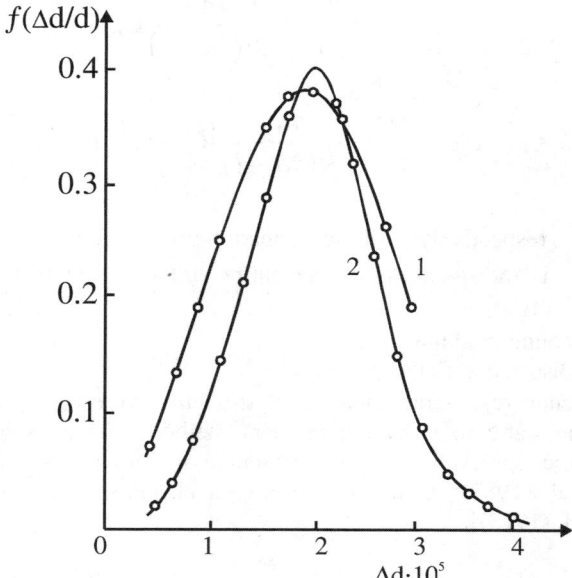

Fig. 4.29. Density functions of deformation variances for the lattice in the wafer: **1** prior to irradiation; **2** upon irradiation.

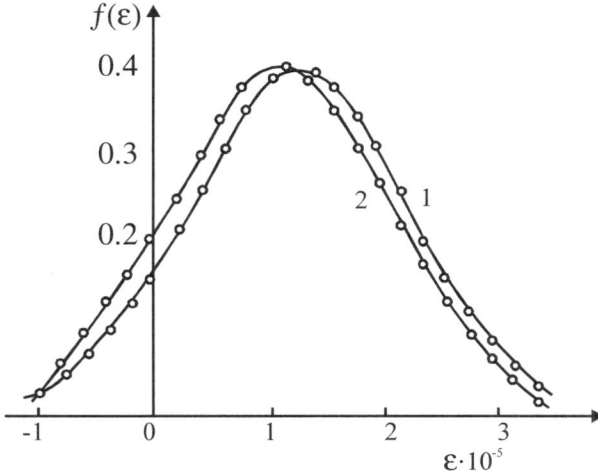

Fig. 4.30. Density functions of deformation variances for the lattice in the epitaxial layer: **1** prior to irradiation; **2** upon irradiation.

The stresses in the interface are seen here to tend, in average, to the theoretical estimate $1.7 \times 10^6 \, Pa$ obtained for an ideal case, when no defects are available in the structures and a curvature is determined only by the misfit periods in the phosphorus–doped film and the antimony–doped wafer. In this case, a misfit in periods makes up 9.6×10^{-4} per cent and a curvature in the structures is equal to $C_0 = 5.6 \times 10^{-3} \, m^{-1}$. The irradiation process shifts the wafer lattice deformation values to great values; this may be, in particular, as a consequence of "drawing" the antimony distribution profile towards the film–wafer interface. By this there is also explained a fall in the film deformation (Fig. 4.30) caused by recompensating the phosphorus elastic fields by antimony atoms. An actual extrinsic defect composition in a wafer and epitaxial layer is, of course, richer and, therefore, the irradiation–induced processes in them cannot be reduced to a diffusion–drift redistributions of impurities only.

In [Skoupov and Pashkov (1979)], the mechanical stresses were experimentally proved to affect the formation of residual defectiveness in crystals subjected to ionic irradiation and low–temperature annealing.

Under study were the slices of monocrystalline dislocation–free $p - Si\,(001)\,CZ\,(\rho = 15 \, Ohm \cdot cm)$ and the silicon epitaxial structures grown by the chloridization technique: the films of $n - Si\,(111)\,CZ\,(\rho = 1.3 \, Ohm \cdot cm)$ 12 mcm thick, on the wafers of silicon (of electronic conductivity, antimony-doped, $\rho = 0.01 \, Ohm \cdot cm$), 250 mcm thick. The crystalline surfaces were oriented to the (111)–plane.

During the irradiation and annealing process, the specimens were elastically deformed either by a four–point bending or a one-sided depositing (through reactive spraying) of silicon nitride films $20-22$ nm thick or by growing the anode silicon dioxide 70 nm thick.

Structural changes in crystals were recorded through the pictures of selective chemical etching in the Sirtl etcher, by the X-ray graphic "pass-through" technique (shooting the $Mo-$irradiation), and also through measuring the residual deformation in the specimen lattice by the X-ray three-crystal spectrometer [Skoupov and Uspenskaya (1975)]. The spectrometer was operating with a use of CuK_α – irradiation in the symmetric Bragg diffraction regime according to the scheme (3 – 3.3). To make the spectrometer more sensitive to slight deformations, the technique of scanning a working beam was used [Pavlov and Skoupov, et al. (1987)].

During irradiaton, the importance of static elastic fields in creating residual defectiveness in the implanted layer becomes mostly vivid in the crystal's regions where mechanical stresses are spatially inhomogeneous. If a specimen is ionically–irradiated not across the entire surface, then a zone of maximal amplitude of stresses lies within the presurface layer (its thickness is close to a run of incorporated ions), near the "irradiated region – nonirradiated region" interface. It is in this very zone that the difference in defectiveness is possible to be detected via a usual selective chemical etching.

In Fig. 4.31 [Demidov and Latysheva, et al. (2000)] there is a micropicture

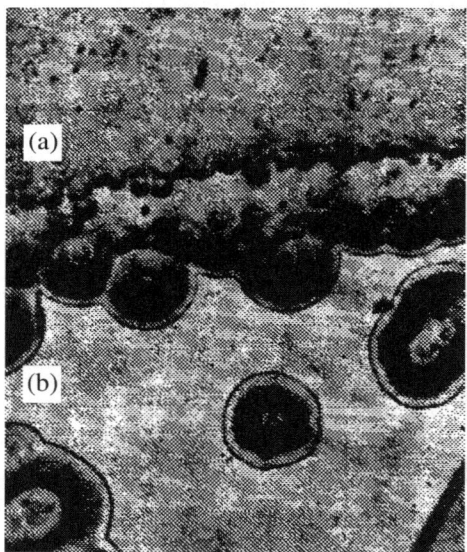

Fig. 4.31. The "nonirradiated **a** – irradiated **b**" interface on the silicon surface irradiated by argon ions of energy $E = 40\,keV$, $\Phi = 3 \cdot 10^{16}\,ions\,/\,cm^2$; x340.

of the etching process at 1 mcm depth within the "nonirradiated (**a**) – irradiated (**b**)" interface upon irradiating the crystal surface by argon ions of energy $E = 40\,keV$ and dose $\Phi = 3\times10^{16}\,ions\,/\,cm^2$. During irradiation the region **a**) was protected from the ionic beam with a copper foil.

Upon irradiation, the selective etching distinguishes the following specificities in the defectiveness of the crystal. First, the etching wells under the implanted layer get rid of the growth microdefects (the region **b**) – this effect was earlier detected by the high-temperature decoration of clusters [Zorina and Popov, et al. (1983)]. Second, as seen in Fig. 4.27, the most density of the etching wells, and, accordingly, the structural defects inducing them, is concentrated in the vicinity of the 'region **a** – region **b**" interface, where, according to [Gerward (1973)], the stresses may take the value $0.1 - 0.2\,GPa$. The third specificity implies the defects being inhomogeneous in their concentrations and sizes within the interface itself (in our case, this interface is $26\,mcm$ wide). This event may be regarded as a consequence of the complicated distribution of stresses near the boundary of the implanted zone.

In measuring the residual mechanical stresses in the film – wafer interface in the epitaxial structures, the elastic fields were found to affect the structural perfection in the crystal regions lying from the irradiated surface at the distances sufficiently exceeding the runs of ions. The specimens were deformed through depositing the Si_3O_4 or SiO_2 films onto an epitaxial layer, and the ionic irradiation was performed from the side of the wafer. The Si_3O_4 film compressed the epitaxial layer and created the extending stresses of the value $\sigma \approx 1.2\,MPa$ in the vicinity of the irradiated surface. The SiO_2 film deflected the slices in the opposite direction ($\sigma \approx 1.1\,MPa$). Upon irradiation, the specimens had a 1h anneal in vacuum at 800 K, and then the Si_3O_4 and SiO_2 films were removed in hydrofluoric acid (HF).

The residual stresses within the epifilm – wafer interface were calculated by X-ray measurements of the relative difference in the lattice periods of the conjugate layers and through a microscopic bend in structures. Table 4.18 suggests the measuring results for the specimens irradiated by argon ions ($E = 40\,keV$, $\Phi = 6\times10^{16}\,ions\,/\,cm^2$).

Table 4.18

Changes in the residual elastic stresses in silicon epitaxial structures in the film –
wafer interface upon irradiation and annealing

Kind of treatment	Stresses in the epilayer – wafer interface (MPa)	
	Original*	Upon treatment
Without deformation (controlled)	1.8 ± 0.3	2.1 ± 0.3
With deformation: by the SiO_2 film by the Si_3N_4 film	5.1 ± 0.2 1.7 ± 0.4	3.8 ± 0.2 2.7 ± 0.3

* The stresses in the coated specimens were measured prior to a film deposition.

According to the data suggested, the irradiation–thermal treatment of the specially deformed structures induces a stronger change in residual stresses, in contrast to controlled specimens. The latters betray a clear tendency to approach the value $\sigma_0 = \dfrac{\varepsilon E}{1-v} = 3.3\,MPa$ (here, $E = 169\,GPa$ and $v = 0.262$ is the Young modulus and the Poisson coefficient, respectively) which are thought to represent the defect–free structures, jumping from the supposition that stresses are only determined by a difference in the type and the concentration of dopants in the film ($N_f \approx 2.5 \times 10^{16}\ cm^{-3}$) and wafer ($N_{sub} \approx 4.6 \times 10^{18}\ cm^{-3}$). At this, the effects of Si_3O_4 and SiO_2 films on the original stresses turn out to be reversal – the nitride deformed by the film during irradiation and annealing will favour a growth of stresses in the interface, whereas the deformation by the SiO_2 film will diminish these stresses. This manifests the situation that during the irradiation–thermal action the compressing stresses generated by the Si_3N_4 film in the epitaxial layer stimulate a recombination (annealing) of defects, primarily, of the interstitial type, and the extending stresses in the SiO_2 film stimulate the vacancy–type defects mainly. In both cases, the concentration and power of the associated defects will drop; this is betrayed by the stresses approaching close to the level σ_0.

The recombination process covers not only the point defect aggregates but also a dislocation structure of the material. How the ionic bombardment affects the dislocation structure in the deep layers in silicon was observed earlier in [Pavlov and Pashkov, et al. (1973)]. We have detected a recombination of dislocations in the course of annealing the ionically–irradiated crystals. Fig. 4.32 shows a series of X-ray topograms of the specimens where, upon irradiating by boron ions

($E = 40\,keV$, $\Phi = 3.1 \times 10^{16}\,ions\,/\,cm^2$), a high density of dislocations was created via scribing along the $<112>$ − direction and vacuum annealing ($T = 800\,K$, $t = 1h$).

During annealing, one portion of the specimens was deformed by a four–point bending technique at which within the implanted layer there were created either compressing (Fig. 4.32b) or extending (Fig. 4.32d) stresses of the magnitude $14\,MPa$. When comparing the topograms one can see that the annealing results in a drop of dislocation densities, and here most crucial changes occur in the thermally–deformed crystals.

The results obtained may be explained by the Frenkel pairs deeply migrating and interacting with original structural damages; during irradiation and later annealing, these pairs are diffusing from the implanted layer [Morozov and Skoupov, et al. (1985)]. Another possible reason for structural changes may be the recombination of defects under the action of elastic waves arising during ionic retardation and also during defect annealing (for example, in annihilating the vacancies and interstitial atoms) near the irradiated surface [Pavlov and Syemin, et al. (1986); Pavlov and Tetel'baum, et al. (1986a)].

The role played by statically inhomogeneous elastic stresses in the defect transformations during ionic bombardment and annealing is qualitatively understandable: the vacancies must be attracted to the compressed regions in the lattice, and the interstitial atoms must repulse. However, as the simplest quantitative estimates reveal, both the known and above given experimental data are difficult to be explained by the commonly used thinking concerning the effects of elastic fields on the activation of decomposition, recombination or diffusion of defects. Indeed, if a recombination is thought to be related to atomic processes, then, with the elastic fields being overlapped, a change in the activating energy of these processes must be equal to $\Delta E = \sigma V_{at}$, where V_{at} stands for an atom volume. For $\sigma = 10\,MPa$ and $V_{at} = 2 \times 10^{-23}\,cm^3$ the activation energy will change as much as $1.2 \times 10^{-3}\,eV$ only, that is much lower than not only

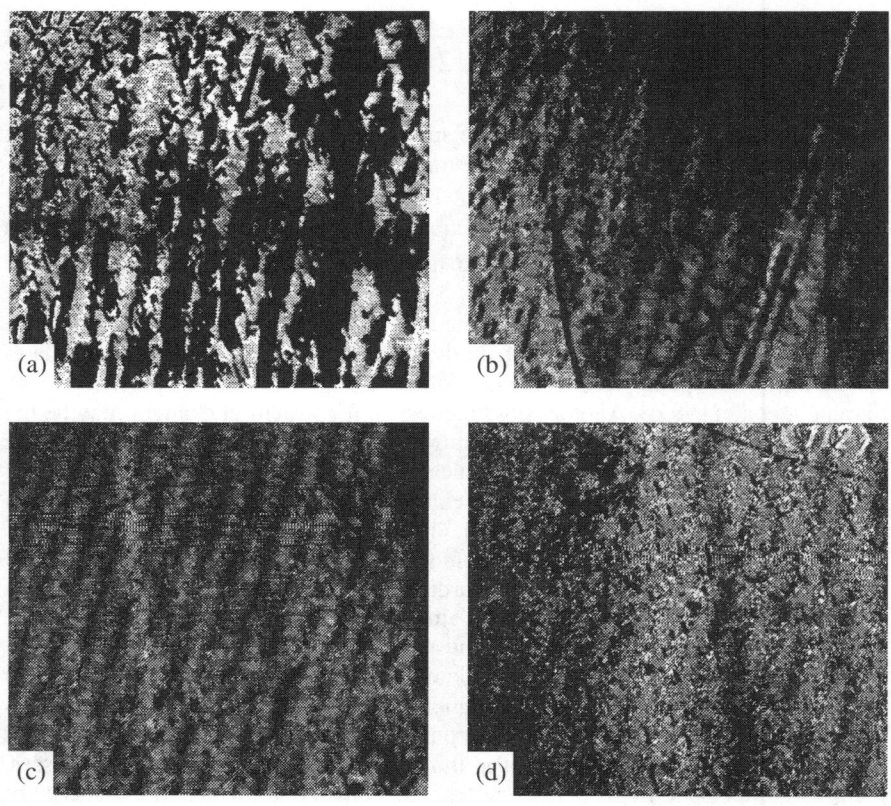

Fig 4.32. Dislocation structure of silicon upon irradiation (**a**) by boron ions ($E = 40\,keV$, $\Phi = 3.1 \times 10^{16}\ ions\,/\,cm^2$) and annealing (**b–d**) at 800 K during 1h; $\sigma\ (MPa)$: **b** – 0, **c** – -14, **d**– + 14; x20.

the typical values of the activation energy ($> 0.1\,eV$) [Sigson and Cspredi, et al. (1976)] but also kT . Even in the vicinity of the "irradiated – nonirradiated region" interface, where $\sigma \approx 0.1\,GPa$, we obtain $\Delta E = 10^{-2}\ eV$.

A strong effect of elastic stresses can be explained by the supposition that a defect recombination process is not atomic but rather collective in its nature, i.e. a sufficiently large amount of elementary defects and crystal atoms are simultaneously take part in this process, and this makes the activation volume of the process largely exceeding the magnitude V_{at}.

Indeed, if a length of the recombination (annealing) process of the cluster is less than the time during which there occurs a deformational response of the crystal to the recombination of each separate defect within the cluster, then a sum of activation volumes of defects held in the cluster will serve as an activation volume.

Annealing the defects inside the cluster may be accelerated by such agents as elastic waves induced by elementary recombinations of defects [Indenbom (1979)]. Let us suppose that a thermal fluctuation has resulted in a recombination (annihilation) of a single defect. This process has emitted, say, a wave of the amplitude $\rho_0 \approx \varepsilon_0 B$, where ε_0 stands for a deformation in the vicinity of the defect, and B is an elasticity modulus. Having approached the second defect being at the distance of R, the wave will diminish its recombination barrier by the value of $\Delta E_1 \approx \dfrac{\rho_0 r_0 V_{at}}{R}$, where r_0 stands for a radius of the stress region near the defect. If this drop is sufficiently large, then the second defect will be annihilated, and at this there will be emitted a wave that, summing up with the first, will suppress the barrier for recombining the third defect, etc. The estimations made with a use of calculated values ε_0 and r_0 show that for silicon at $R = 10^{-1}\, cm$ we obtain $\Delta E_1 \approx 0.2\, eV$, i.e. this magnitude is comparable with the typical values of barriers [Smirnov (1980a)]. A resulting action of the waves emitted by elementary recombinational acts on the defects held in the cluster of the size $l \le 10^{-8}\, cm$, is more crucial. The duration of the collective recombination in the cluster is equal to $t_0 = \dfrac{l}{u}$, where u is a sound speed. For example, in silicon at $l = 10^{-8}\, cm$ we have $t_0 \approx 10^{-12}\, s$, that is less than t_1 of the acoustic wave attenuation (in its order, t_1 is equal to the time of a free run of the phonon) forming a deformational "response" of the crystal to the recombination process.

According to the above said, for sufficiently small clusters the activation volume of the recombination may be accepted to be equal to $V_{act} \approx n V_{at}$, where n stands for a quantity of defects in a cluster. The ratio of the number V of the clusters being decomposed (or recombined) per a unit of time, in the presence of the elastic stress σ, to the associated number during the absence of stresses will be expressed as

$$m = \frac{V(\sigma)}{V(0)} = \exp\left(\pm \frac{\sigma n V_{at}}{kT} \right).$$

For example, for $\sigma = 10^6\,Pa$, $T = 900\,K$ we have $m^{\pm} \gg 1$, if $n \gg 100$. The sign of the power (± 1) is determined by a specific type of the defect and by the sign for σ.

These estimates indicate that in real conditions the elastic stresses can significantly affect the rates of defect recombination in silicon.

The thinking suggested above makes it possible to understand also the effects of local stresses near the boundary of the irradiated region: these stresses can stimulate a decomposition of small clusters followed by a coagulation of point defects into larger clusters recognized through the etching wells (Fig. 4.31). An accelerated annealing of defects within an implanted layer, with the deformation available, will increase a flux of the Frenkel pairs that will diminish the dislocation density.

In the last years, microdefects in dislocation–free crystals, primarily in silicon, are exploited as a "simulating environment" for studying the long–range effect. Microdefects being highly sensitive to various external actions, on the one hand, make it possible to specify the mechanisms of possible structural changes in materials, and, on the other hand, these microdefects need to be gettered, for they are potential sources of aging and degrading of parameters and properties in the device structures. How the ionic irradiation affects the concentration of growth and technological microdefects in the $n-$ and $p-$ silicon crystals was studied in [Demidov and Latysheva, et al. (2000)].

For experiments, there were used the slices of dislocation–free $n - Si(001)CZ$ ($\rho = 4.5\,Ohm \cdot cm$) 100 mm in diameter and 460 mcm in thickness. The slice surfaces to be irradiated was polished chemically and mechanically by the Cab-O-Sil suspension; the back side of $n - Si(001)CZ$ ($\rho = 4.5\,Ohm \cdot cm$) slices was polished by the diamond paste (1–mcm grains), and the surface in $p - Si(001)CZ$ ($\rho = 12\,Ohm \cdot cm$) specimens was polished by the loose abrasive, the suspension of the synthetic corundum micropowder (5-mcm grains). All the slices were ionically–irradiated at room temperature by the boron ions of energy $60\,keV$ and dose $\Phi = 6.25 \times 10^{15}\,ions\,/\,cm^{-2}$. Exposed to irradiation was only a half of the surface, the other half served as a controlled one.

Microdefects in the irradiated and controlled specimens were detected through a selective chemical etching in the solution $CrO_3\,(10\%) : HF(49\%) = 1:1$ at the dissolution rate $2\,mcm\,/\,min$. The in-depth defect distribution profiles in the slices were recorded by the by-layer chemicodynamical polishing of the observed surface of the specimens in the etcher of the composition (vol. portions):

$$HNO_3(70\%) : HF(49\%) : CH_3COOH\,(99.8\,\%) = 40:1:1$$

with the etching rate being not higher than $1 \, mcm / \min$. The pictures of selective chemical etching and densities of microdefects across the surface and at various depths were analysed by the micrographic microscope through averaging the number of etching wells measured in not less than 10 scanning fields. When recording the in-depth defect distribution profiles, combining a polishing operation and a selective chemical etching helped to suppress a "memory effect". In each step of the by-layer etching away, the microhardness of the crystals, in addition to metallographic investigations, was measured by the microhardness meter with a use of 20 indenter prints obtained for the load of $0.2 \, N$.

As shown by the analysis of the crystal surface prior to irradiation [Perevostchikov and Skoupov (1992)], the density of growth and technological defects (i.e. induced by abrasive–chemical preparation of surfaces in specimens [Perevostchikov and Skoupov (1992)]), the inhomogeneity of their in-depth and surface distributions and also their spread in sizes in the original slices of $p - Si\,(001)\,CZ\,(\rho = 12 \, Ohm \cdot cm)$ is many times more than in the specimens of $n - Si\,(001)\,CZ$ ($\rho = 4.5 \, Ohm \cdot cm$) (Figs. 4.33 – 4.35).

In the irradiated slices the microdefect density near the implanted layer drops significantly, the most changes occur in the boron–doped crystals (Fig. 4.36). Here, upon irradiation there retains an oscillating change (specific for original specimens) of the in-depth defect distribution profiles (Fig. 4.33). The thickness of layers, where a dissolution of ionically–irradiated microdefects is recorded metallographically, turns out in $p - Si\,(001)\,CZ\,(\rho = 12 \, Ohm \cdot cm$, boron– doped) to be more than in the specimens of $n - Si\,(001)\,CZ$ ($\rho = 4.5 \, Ohm \cdot cm$, phosphorus–doped).

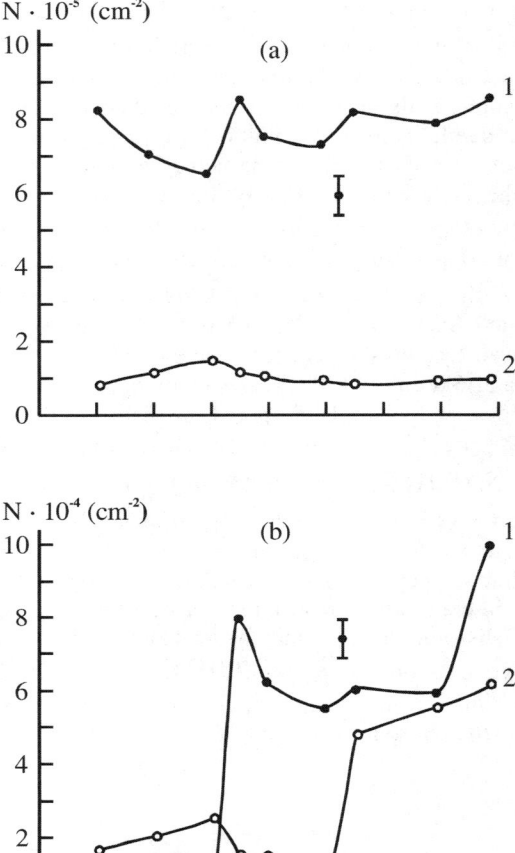

Fig 4.33. Microdefect depth profiles in **1** $p - Si\,(001)\,CZ$ ($\rho = 12\ Ohm \cdot cm$, boron–doped) and **2** $n - Si\,(001)\,CZ$ ($\rho = 4.5\ Ohm \cdot cm$, phosphorus–doped) wafers **a** prior to and **b** after boron implantation.

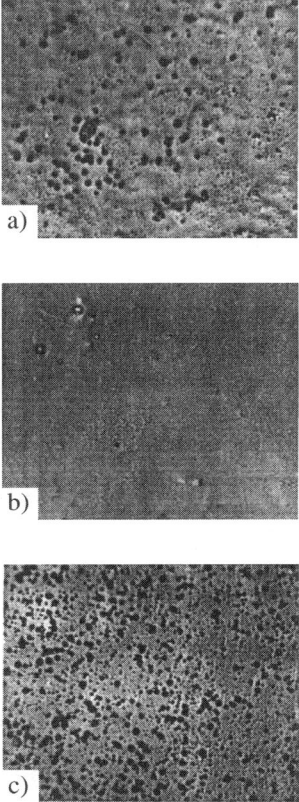

Fig 4.34. Selective etching patterns on $p - Si\,(001)\,CZ$ ($\rho = 12\ Ohm \cdot cm$, boron–doped) (x180) **a** prior to irradiation, **b,c** after irradiation and etching to 2- and 8-mcm depths, respectively.

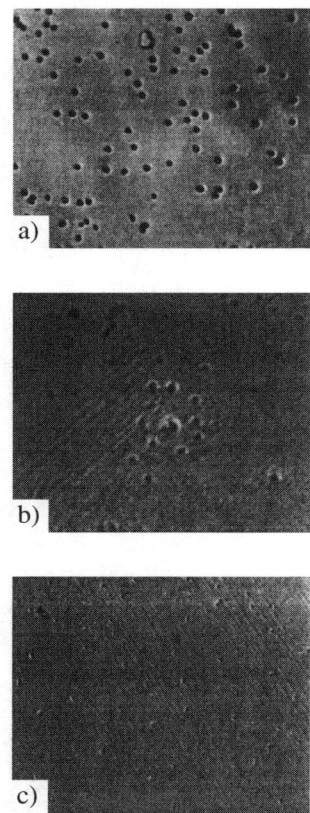

Fig 4.35. Selective etching patterns on $n - Si\,(001)\,CZ$ ($\rho = 4.5\ Ohm\cdot cm$, phosphorus–doped) (x180) **a** prior to irradiation, **b,c** after irradiation and etching to 2- and 8- mcm depths, respectively.

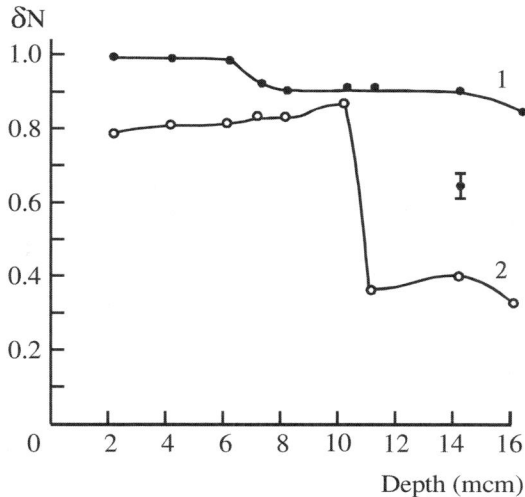

Fig 4.36. Depth profiles of the relative changes in microdefect density, $\delta N = (N_0 - N_i)/N_0$ where N_0 and N_i are the microdefect densities prior to and after irradiation, for **1** $p - Si(001)CZ$ ($\rho = 12$ $Ohm \cdot cm$, boron–doped) and **2** $n - Si(001)CZ$ ($\rho = 4.5$ $Ohm \cdot cm$, phosphorus–doped).

Indirectly, this agrees with the measurements of microhardness that, as seen from Fig. 4.37 [Demidov and Latysheva, et al. (2000)], rises upon irradiation, and, besides, with a larger amplitude of changes in the specimens of $p - Si(001)CZ$ ($\rho = 12$ $Ohm \cdot cm$, boron–doped), in contrast to the specimens of $n - Si(001)CZ$ ($\rho = 4.5$ $Ohm \cdot cm$, phosphorus–doped).

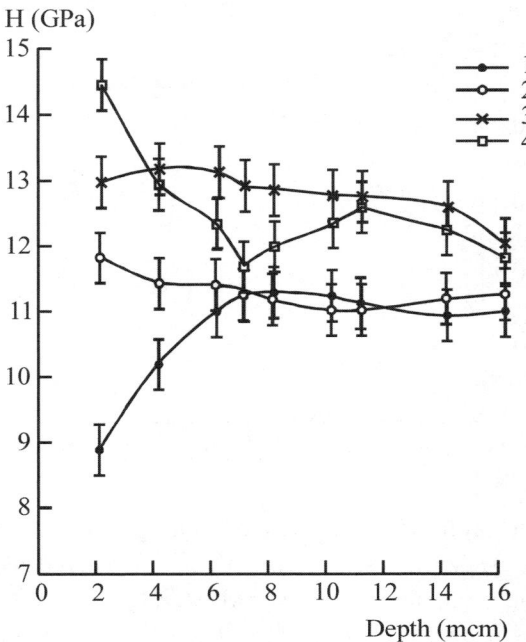

Fig 4.37. Microhardness (0.2-N load) in **1,3** $p - Si(001)CZ$ ($\rho = 12\ Ohm \cdot cm$, boron–doped) and **2,4** $n - Si(001)CZ$ ($\rho = 4.5\ Ohm \cdot cm$, phosphorus–doped) **1,2** prior to and **3,4** after irradiation by $60 - keV\ B^+$ ions, of the dose $6.25 \times 10^{15}\ cm^{-2}$.

According to [Pavlov and Syemin, et al. (1986); Morozov and Skoupov, et al. (1985)], the dissolution of the ionically–irradiated microdefects may be determined by the diffusing and/or acoustic mechanism. In the first case, the concentration of microdefects diminishes due to a saturation of unbalanced vacancies diffusing from the implanted layer during irradiation and in the course of the later low–temperature relaxational recombinations of radiation damages. During an impact–acoustic action of the ionic beams, the microdefects are dissolved due to a saturation of vacancies from the intrinsic Maxwell ambients initiated by a transient pressure field of elastic waves arising within the ion retardation zone. It is interesting to compare the efficiency of each of these mechanisms.

If an average radius of a microdefect is supposed to be equal to $0.5\ mcm$, then each such cluster will hold $n_c = \left(\dfrac{r_c}{r_0}\right)^3 = 7.9 \times 10^{11}$ silicon atoms

($r_0 = 0.117\, nm$ is a radius in a silicon atom). From Fig. 4.33 it follows that, for example, at the depth $\Delta X = 2\, mcm$ the irradiation–induced concentration of microdefects (N_M) for $p - Si\,(001)\, CZ$ ($\rho = 12\; Ohm \cdot cm$, boron–doped) and $n - Si\,(001)\, CZ$ ($\rho = 4.5\; Ohm \cdot cm$, phosphorus–doped) specimens will be $\Delta N_M = 4.18 \times 10^9$ and $3.72 \times 10^8\; cm^{-3}$, respectively. It means that during irradiation the concentrations of unbalanced vacancies $N_V = n_c \cdot \Delta N_M$ in a 2-mcm layer in $p - Si\,(001)\, CZ$ ($\rho = 12\; Ohm \cdot cm$, boron–doped) specimens should be sustained at the level of $3.3 \times 10^{20}\; cm^3$, and in $n - Si\,(001)\, CZ$ ($\rho = 4.5\; Ohm \cdot cm$, phosphorus–doped) at the level of $-N_V = 2.94 \times 10^{19}\; cm^{-3}$. On the other hand, the maximally possible concentration of vacancies injected from the implanted layer (i.e. without aggregations considered) will be equal to $N_V^{irr} = \dfrac{\Phi}{b}\left(\dfrac{E}{2Ed}\right)$, where Φ stands for an irradiation dose and b for a thickness of the implanted layer ($b = 0.3\; mcm$) and $E_d = 22\, eV$ for a threshold energy shift for silicon. For $E = 60\, keV$ and $\Phi = 6.25 \times 10^{15}\; ions\,/\,cm^{-2}$ we obtain $N_V^{irr} = 2.84 \times 10^{23}\; cm^{-3}$. Since $N_V^{irr} > N_V$, a dissolution of microdefects due to their saturation with vacancies, at least in the vicinity of the implanted layer, will be an absolutely possible process.

Taking into account the vacancy in-depth diffusion, i.e. supposing a drop of vacancy in-depth concentrations in accordance with the law

$$N_V^{irr}(X) = N_{V_0}^{irr} \cdot \exp\left(-\frac{X - b}{l_V}\right),$$

where l_V stands for a diffusion length of vacancies, for $l_V = 10 - 20\, mcm$ we shall have

$$N_V^{irr} = (0.38 - 0.62)\, N_{V_0}^{irr}.$$

The difference in the thicknesses of the regions cleaned from the microdefects in $p - Si\,(001)\, CZ$ ($\rho = 12\; Ohm \cdot cm$, boron–doped) and in $n - Si\,(001)\, CZ$ ($\rho = 4.5\; Ohm \cdot cm$, phosphorus–doped) is seemingly determined by the most efficient unbalanced vacancy trapping centres available in the bulk of the $n -$ type specimens. As such trapping centres, there may serve the phosphorous atoms which form more stable aggregates $P + V$ ($E -$ centres), being annealed off at $T > 400\,K$, than the aggregates $B + V$ ($T_{anneal} = 300\,K$). Besides, in the specimens of $n - Si\,(001)\,CZ$ ($\rho = 4.5\,Ohm \cdot cm$, phosphorus–doped) a stronger screening effect is produced by the Cottrell ambients encircling the microdefects and hampering their interactions with unbalanced vacancies [Perevostchikov and Skoupov (1992)].

If the pressure of the elastic waves arising in the thermal peak is estimated as $p = \dfrac{0.24E}{\pi b^2}$, then for the above irradiation by boron ions we get $p = 2.7 \times 10^4\,Pa$, and the energy of the impulse per a single vacancy will be only $1.2 \cdot 10^{-6}\,eV$. This magnitude is much less than the thermal fluctuations in the lattice and this implies a small probability of cluster dissolution during an impact-acoustic action in the vicinity of the implanted layer. However, this action is not excluded completely, especially near the crystal side reversal to the irradiated side, where a flow of vacancies onto microdefects from the specimen's surface may be initiated by the elastic waves, as a result of their superposition and amplification [Pavlov and Syemin, et al. (1986)].

The experimental data suggested in this subsection manifest the capabilities of ionic irradiation as a way of low-temperature activation of the gettering process. It should be noted here that these processes may run not only during irradiation by ions of average mass and energy but when irradiating the materials by protons [Kozlovscky and Kozlov, et al. (2000)], alpha–particles [Skoupov and Smolin (2000)] and by low-energy ions of chemically active gases used in the reactive ionic–plasmatic etching of crystals [Bogatch and Starkov (1991)]. As was proved in [Bogatch and Starkov (1991)], the plasmic gettering technique used at room temperatures is able to clean a presurface region in the crystal from fast–diffusing impurities as deep as $10 - 30$ mcm.

4.4.2
Effects of irradiation on semiconducting structures

The kinetics of nonstationary processes, for example, diffusion, in homo– and heterostructures is known to be different from that specific for homogeneous monocrystals [Dzafarov (1991)]. The basic reason of this difference lies in that that the defects, internal mechanical stresses and electrical fields in the structures are conditioned by the crystalographical misfit and contact difference of potentials between the conjugative layers. These factors may also affect the defect formation processes during irradiations that result in the long–range effect in the irradiated materials. It is natural to suppose here that depending upon the degree of structural perfection and thermal equilibrium of the irradiated layer the long-range effect is possible to be suppressed or strengthened. The last case is of interest in developing a low–temperature gettering technology.

That the long–rage effect can be strengthened and the gettering ability of the ionic technology be increased has been shown in [Koulikov and Perevostchickov, et al. (1997)], where the effects of ionic irradiation on the silicon microdefect structure through the silicon porous layers have been studied.

Subjected to the study were the structures obtained by the anodic processing of the crystals of $p - Si(001) CZ (\rho = 12 \ Ohm \cdot cm$, boron–doped) in the hydrofluoric acid (HF) solution $HF : H_2O : C_2H_5OH = 1 : 1 : 2$ (vol. portions), with the anodic current density being equal to $10 \ mA \cdot cm_{-2}$. There were formed the layers of porous silicon of the porousness about 35 per cent and 1, 3 and 10 mcm in thickness. Then, the structures were irradiated from the side of porous silicon by argon ions of energies 40 and $100 \ keV$ and dose $1 \cdot 10^{16} \ cm^{-2}$. Upon irradiation, porous silicon was removed in the 15 per cent KOH solution and on the wafer sides contacting with this solution there were performed the by-layer measurements of the surface electrical resistance within the thermal range of $77 - 375 \ K$ and those of the microdefect density (according to the pictures of selective chemical etching in the hydrofluoric acid solution $HF (48\%) : CrO_3 = 1 : 2$ (vol. portions).

The anodic processing and later irradiation of porous silicon by argon ions were experimentally shown to diminish the presurface electrical resistance in a monocrystalline wafer near its interface with the porous silicon; here, the changes were seen the more vividly, the thicker was the porous silicon and the greater was the energy of the incorporated ions (Fig. 4.38) [Perevostchikov and Skoupov (1997)].

Simultaneously, with a reduction of average resistance across the entire surface there has been also detected a fall in the variance of R_s – this was done after anodic processing and etching off the porous silicon by 9 – 10 per cent; and after irradiating and removing off the porous silicon by 25 – 30 per cent. This betrays a growth of homogeneity in the distribution of the components in the extrinsic defect composition within the presurface layer of the monocrystal. The thermal dependencies of surface resistance in the nonirradiated and irradiated "porous silicon – monocrystalline silicon" structures have revealed – within the measuring limits – no specificities that could differ them from the dependencies for the original wafers. This may imply, first, that an anodization and a later ionic irradiation of the porous silicon brings no crucial changes into the energy levels for the prohibited zone, specific for the original silicon, and, second, a fall in the presurface resistance is mainly caused by the charge carriers becoming more mobile, due to a fall in the total concentration of the scattering centres in the monocrystal near the boundary with the porous silicon. After the anodization and irradiation, the depth down to which a wafer by–layer etching changes the presurface resistance will increase, if the thickness of the porous silicon and the energy of ions increase. Table 4.19 holds the depths at which the electrical

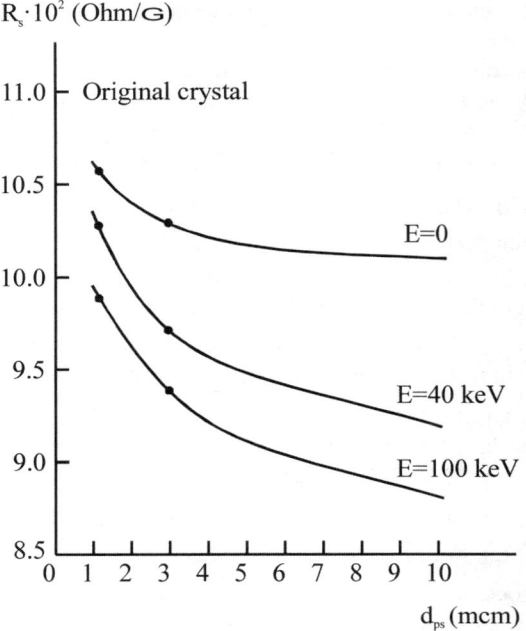

Fig. 4.38. The changes in the presurface electrical resistance of the monocrystalline wafer upon irradiating by argon ions of the dose $1 \cdot 10^{16} \, cm^{-2}$ the porous silicon of various thicknesses.

Table 4.19

Effective depth of gettering in the porous   silicon – monocrystalline silicon structures at various energies of argon ions

Thickness of porous silicon (mcm)	Ionic energy during irradiation (keV)	Gettering depth (mcm)
1	0	4 – 6
1	40	8 – 10
1	100	16 – 18
3	0	8 – 10
3	40	13 – 15
3	100	18 – 20
10	0	17 – 20
10	40	22 – 24
10	100	25 – 28

resistance will coincide, within the confident interval $\pm 3\%$ (with the reliability of 0.95), with R_s for the silicon original wafers, i.e. this is an efficient depth of gettering.

The electrical measurements correlate with the by–layer densities of microdefects in the presurface zone; these densities were found through the pictures of the selective chemical etching in the form of flat–bottomed etching wells. Fig. 4.39 [Perevostchikov and Skoupov (1997)] shows the microdefect

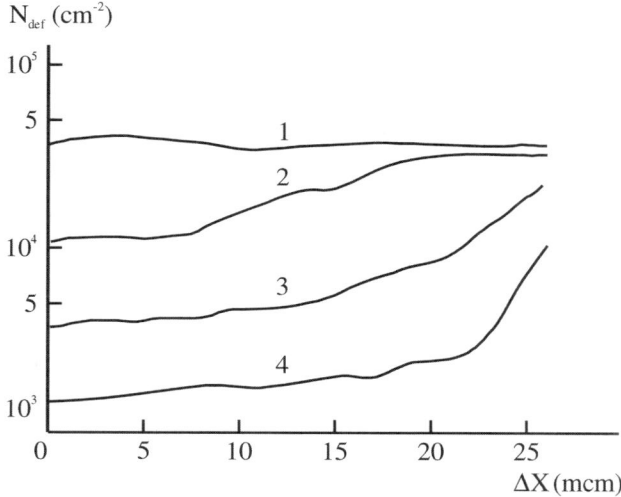

Fig. 4.39. The microdefect densities being dependent **1** on the distance to the "porous silicon– monocrystalline silicon" interface for the original crystals, **2** upon forming the porous silicon and **3** upon irradiating the porous silicon by argon ions of the energies $40\ keV$ and **4** $100\ keV$.

densities being dependent on the depth of the off-etched layer in the structures with 10-mcm porous silicon layers. With the porous silicon becoming thinner, the microdefect low density region comes close to the "porous silicon– monocrystalline silicon" interface.

The results obtained may be explained by the low–temperature gettering of the monocrystalline silicon by the silicon porous layers. When monocrystalline silicon is formed electrically and chemically and later exposed to ionic irradiation, the fluxes of unbalanced IP defects, mostly vacancy-type defects, are being injected into the presurface zone of the wafer. The point defects emerge, first, at the stage of silicon being locally dissolved near the micropores during the anodization process, then, second, during a later active low–temperature oxidation of porous silicon in the air ambient and in the course of irradiation. After anodization and oxidation, ionic irradiation strengthens such relaxation processes as those running in the thermodynamically unbalanced system "porous silicon – monocrystalline silicon", the impurity redistribution and microdefect dissolution. Since the porous silicon layers on the structures under study are far thicker than the runs of argon ions (with a least density of porous silicon accounted, the runs of ions of energies $100 \ keV$ make up $0.43 - 0.04 \ mcm$), it may be supposed that the revealed effect of strengthening the gettering process is determined by the simplest mobile point defects and elastic waves induced within an ion retardation zone in porous silicon.

A strengthening of the long–range effect by ionic irradiation of silicon with a porous layer was detected in [Perevostchikov and Skoupov (1999)].

The experiments were performed on the structures built by anodization of crystals of $p - Si\,(001)\,CZ$ ($\rho = 0.001 \ Ohm \cdot cm$, boron–doped) 360 mcm thick in the hydrofluoric acid solution $HF : H_2O : C_2H_5OH = 1:1:2$ (vol. portions), with the anodic current density being equal to $10 \ mA \cdot cm^{-2}$. The surfaces in the original crystals were prepared by standard technologies. Dislocation density in the original wafers did not exceed $4.1 \cdot 10^2 \ cm^{-2}$. A layer thickness in Si_{por} of the porousness $C = 30 - 33\%$ was $d = 2$, 10 and 20 mcm. The structures were irradiated from the Si_{por} sides by argon ions of energies $40 \ keV$ and dose $6.25 \cdot 10^{16} \ cm^{-2}$. Prior to being measured, the porous silicon layers were removed in the 15 % solution of KOH. The in-depth density distribution profiles for microdefects were registered through the pictures of the selective chemical etching in the hydrofluoric acid solution $HF : CrO_3 = 1:2$ (vol. portions), with the specimens being etched off in a by-layer fashion at the rate of $1\,mcm \cdot \min^{-1}$. At each etching step, the average microdefect density N_{def} was calculated according to the number of etching wells fixed in not less than $15 - 18$ fields of the Neophot –32 microscope, of the error ± 8 per cent and with the confident probability 0.90.

The experimental profiles of the microdefect in-depth density distribution for the structures prior to and upon forming the Si_{por} layers of various thicknesses are given in Fig. 4.40 [Perevostchikov and Skoupov (1997)]. In it, the electrochemical treatment is observed to diminish the microdefect densities both near the $Si_m - Si_{por}$ interface and near the back side of the wafers, i.e. here a long–range gettering occurs. An increase in the thickness of a porous silicon layer increases both the number of dissolved defects and the efficient depths of the cleaned regions on both sides of the structures.

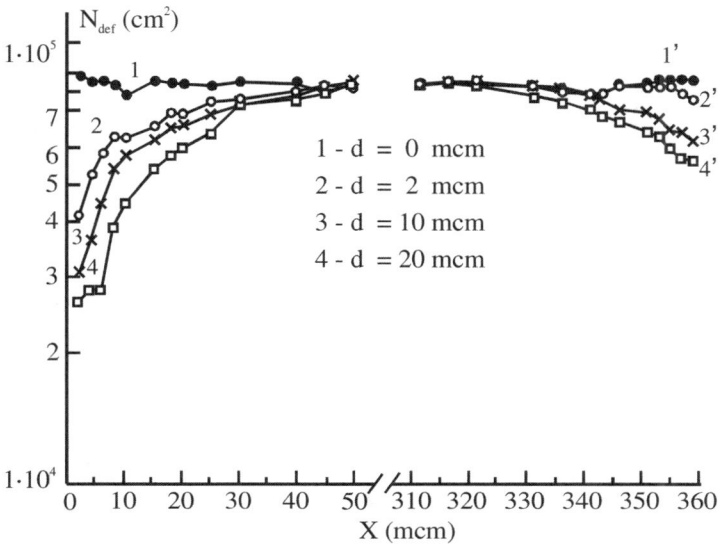

Fig. 4.40. The microdefect density in-depth distribution profiles prior to $(1.1')$ and upon $(\;2.2'\;;\;3.3'\;;\;4.4'\;)$ forming the porous silicon of various thicknesses; 2, 3, 4 – under the porous silicon layer; $2'$, $3'$, $4'$ – near the back side of specimens.

The effects of ionic irradiation are demonstrated in Fig. 4.41 [Perevostchikov and Skoupov (1997)] where the distribution profiles of the relative changes in the defect densities are calculated as

$$\delta N_{def} = N^0_{\;def} - \frac{N_{def}}{N^0_{\;def}}\;,$$

where $N^0_{\;def}$, N_{def} stand for microdefect densities prior to and after irradiation, respectively. They vividly exhibit, on the one hand, that irradiation strengthens the dissolution process and, on the other hand, the efficiency of this process depends on the Si_{por} porous layer in a nonmonotonous fashion. Near the $Si_m - Si_{por}$ interface, an irradiation–stimulated dissolution of defects drastically falls as the Si_{por} layer is becoming more than 10 mcm thick, and this dissolution falls near the back side of the structures at the 2-mcm thickness.

That a transient region is formed in the $Si_m - Si_{por}$ structures and microdefects in the dislocation–free silicon are dissolved is determined by the unbalanced vacancies getting into cooperation with the components of the original

δN_{def} (conv. u.)

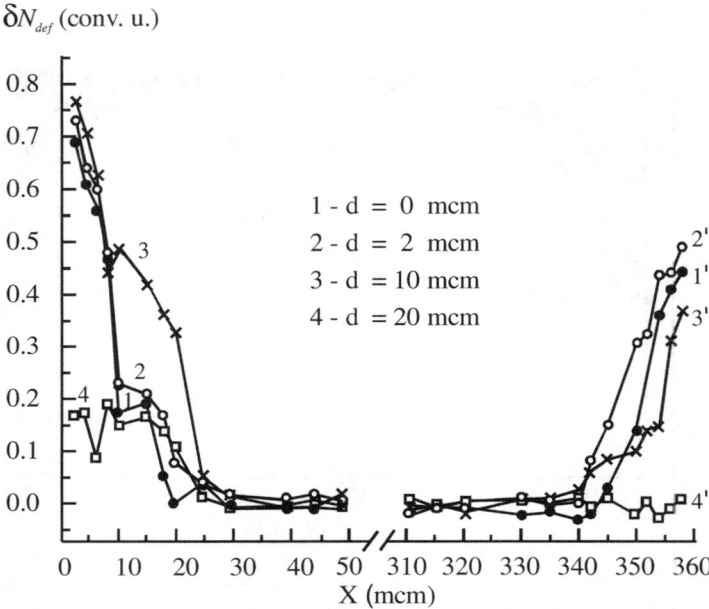

Fig. 4.41. The distribution profiles for relative in-depth changes in the microdefect densities of the structures irradiated from the crystal surface of various thickness (curves 1, 2, 3 and 4) and from the back sides of the specimens (curves 1´, 2´, 3´ and 4´).

extrinsic defect composition; these vacancies arise near the reaction surface when Si_m is being dissolved during the formation of Si_{por} and its later irradiation. Vacancies may be also generated by the impurity ambients that encircle the microdefects and get unbalanced by the alternate–sign mechanical stresses of the elastic waves being born within the zones of electrochemical reactions and retardation of the implanted ions. That vacancies appear in the course of formation of Si_{por} may be naturally related to local active centres lying on the surface of Si_m; on these centres there occur electrochemical reactions, firstly, the reactions of generating SiF_2 and Si being restored from it. Let $U_{act\ c}$ be an energy of the active centre being close to a total activation energy for generating the Si_{por} layer and let $T_{act\ c}$ be a local temperature of the crystalline lattice in the place of the active centre (this temperature being found as $T_{act\ c} = \dfrac{U_{act\ c}}{k}$, where k is the Boltzmann constant), then a magnitude of the vacancy flux will be expressed as

$$I_V = \pi\, n_0\, c\, v_0 \exp\left(-\frac{U_{act\,c}}{kT}\right)\exp\left(-\frac{E^s_{fV}}{kT_{act\,c}}\right),$$

where

n_0 indicates a silicon atom density on the reaction surface (the reticulation density);

c is a spatial porousness in the Si_{por} layer;

v_0 stands for the Debye frequency;

E^s_{fV} is an energy to generate vacancies near the surface, this energy being 2.5 times less than that in the bulk of the crystal [Alekhin (1983)].

Taking $n_0 \approx 1.4 \cdot 10^{15}\ cm^{-2}$, $c \approx 0.3$, $v_0 \approx 10^{13}\ s^{-1}$, $U_{act\,c} \approx 0.041\,eV$

and E^s_{fV} we get $I_V \approx 6.4 \cdot 10^{16}\ cm^{-2}\ s^{-1}$. A reduction of defect density near the back side of the structures occurs, as well as in the event of ionic irradiation, due to elastic waves. The amplitude of elastic waves with the oscillating period τ travelling in crystals at the distances from the $Si_m - Si_{por}$ interface satisfying the condition $z > (2\pi\,\tau\,I_V)^{-0.5}$ may be calculated from the radio [Pavlov and Syemin, et al. (1986)]

$$P = 4G\varepsilon a\,\frac{(2\pi\tau I_V)^{0.5}}{\alpha}\ ,$$

where
ε is a lattice deformation during a formation of vacancies;
α is a coordination number;
$a \approx (\alpha\Omega)^{1/3}$ is an efficient radius of the elastic wave source (Ω stands for atomic volume);
G is a shift modulus.

Taking $\tau = v_0^{-1} \cdot \exp\left(\dfrac{E_{mV}}{kT}\right)$, where kE_{mV} is the energy of vacancy migration, and using the known numerical values of parameters for silicon ($\varepsilon = 0.2$ and $\alpha = 4$ and $E_{mV} = 0.33\,eV$) we obtain $P \approx 4.3\,MPa$ for the wave amplitude on the back side in the structures. The waves of such amplitude initiate a microdefect dissolution and induce the unbalanced vacancies during excitation of impurity ambients and surface sources [Alekhin (1983); Pavlov and Syemin, et al. (1986)]. The dissolution process embraces not only the vacancies and waves generated by the incorporated ions but also the vacancy aggregates. That such aggregates are induced by electrochemical treatment and are later etched off partially by irradiation is proved by the changes in the Si_{por} refraction indicator; they are measured by the ellipsometer ($\lambda = 0.63\,mcm$). On original structures, the average refraction indicator across the Si_{por} surface was 2.435 ± 0.004 , and upon irradiation it was 2.709 ± 0.001, i.e. the density in a porous layer increased [Palatnik and Tcheremskoy, et al. (1982)]. With the Si_{por} layer getting thicker, the significance of point defect drains will increase, as well as there will grow a damping effect of pores on the elastic wave propagation [Palatnik and Tcheremskoy, et al. (1982)], and hence the efficiency of the long–range gettering will fall (Fig. 4.41).

As research results indicate [Koulikov and Perevostchickov, et al. (1997); Perevostchickov and Skoupov (1999)], porous silicon when used as a technological layer to be removed after ionic irradiation prior to next technological operations makes it possible to increase the gettering efficiency. An increase in the sensibility of the silicon microdefect structure to ionic irradiation, when there are layers of amorphous silicon in crystals, was detected in [Kiselyev and Levshounova, et al. (2003)].

Subjected to study in [Kiselyev and Levshounova, et al. (2003)] were the crystals of $n - Si\,(001)\,CZ$ ($\rho = 15\ Ohm \cdot cm$, phosphorus–doped) 630 mcm thick. The original wafers underwent a finishing chemicodynamical polishing of the front side and chemicomechanical polishing of the back side. Prior to ionic irradiation, one part of the surface in the original specimens was coated with amorphous silicon 0.7 mcm thick, this deposition was done by electronic evaporation in the vacuum unit, with the wafer being at 420 K.

The evaporation process was performed by immediately heating the weights of $p - Si\,(001)\,CZ\,(\rho = 40\ Ohm \cdot cm$,   boron–doped). The crystals were irradiated by argon ions of energies $40\,keV$, by the dose of $2.8 \cdot 10^{16}\ cm^{-2}$, with the current density being less than $3\,mkA \cdot cm^{-2}$. The implantation regime was chosen on the basis of experimental results demonstrating for such conditions the greatest efficiency in gettering through a porous silicon layer. Except the irradiated specimens there were also used the original crystals, as controlled, and also the specimens with amorphous silicon layers – in order to detect the effects of increased temperatures (T=420 K), in the course of depositing the amorphous silicon onto a monocrystalline wafer.

A change in the microdefect structure of crystals was detected by a by–layer selective chemical etching in the hydrofluoric acid solution $HF : CrO_3 = 1:1$, together   with   a   by-layer   chemicodynamical   polishing   in $HNO_3 : HF : CH_3COOH = 40:1:1$ at the $1\,mcm/\min$ rate. Upon chemical etching, the micromorphology in the crystal surfaces was recorded at the scanning TMX – 2100 «Accurex» microscope in an noncontact regime. Subjected to studying were both the irradiated (front) and the back side in the specimens.

Fig. 4.42 [Perevostchikov and Skoupov (1999)] shows the density in-depth distribution profiles obtained by chemicodynamical etching of the A – type defects (N_A) after performing various types of actions on the crystals. Here, a low–temperature treatment (an annealing and sedimentation of the amorphous silicon ($a - Si$)) is seen to increase the microdefect densities, especially near the surface. In our opinion, this is caused by a partial dissolution of impurity ambients encircling the microdefects, i.e. an increase in the etching selectivity occurs. In the time when a free surface is being ionically irradiated, simultaneously with this process the microdefects themselves are being dissolved.

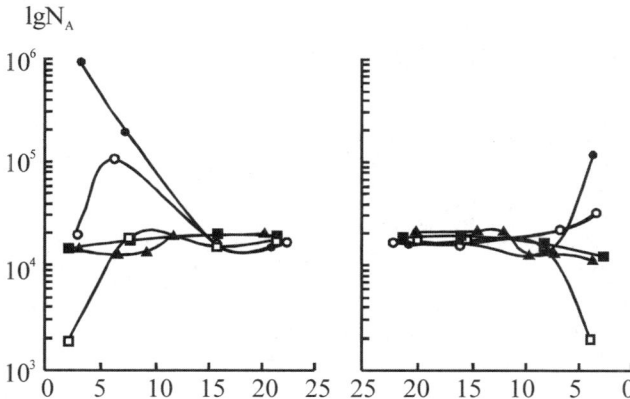

Fig. 4.42. The profiles of microdefect in-depth distributions in silicon monocrystals in the original state (1 – □), upon amorphous silicon deposition (2 – ▲), upon annealing at 420 K (3 –■), upon irradiating the free surface by argon ions (4 –o) and through an amorphous layer (5 – •).

In contrast to this, irradiating through the $a - Si$ film will drastically increase the microdefect concentration on the front sides of the crystals – this is possibly determined by a more intensive interaction of clusters with elastic waves whose amplitude may grow during their passage through a structurally unbalanced amorphous layer. Qualitatively this is also proved by the microdefect densities measured near the back side where the irradiation through the $\alpha - Si$ layer has brought a sharp growth of N_A .

The regularities shown above correlate with the results of the metallographic studies. Fig. 4.43 [Perevostchikov and Skoupov (1997)] suggests the micropictures of silicon surfaces, the original microdefect structure in the specimens and its later transformation during the deposition of the $\alpha - Si$ film, the low–temperature annealing (at 420 K) in the course of irradiation through the amorphous silicon film and without this film. It is seen here that a low–temperature treatment increases the sizes of microdefects and their concentration, and irradiating the surface unprotected by the $a - Si$ film is accompanied not only with an in-depth gettering but with a lateral gettering as well, at the distances of $8 - 20$ mcm from the interface with the implanted region.

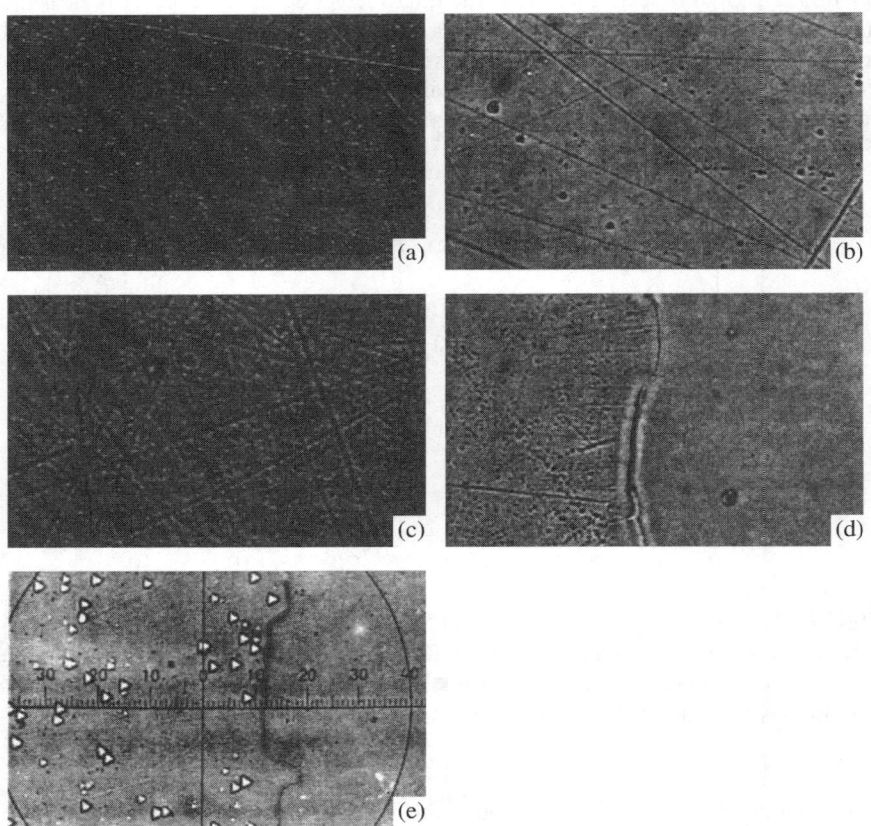

Fig. 4.43. The silicon surface etched at the depth 2 mcm from the irradiated side: **a** in the original state, **b** upon amorphous silicon deposition, **c** upon annealing at 420 K, **d** upon irradiating the free surface (the "irradiated region – nonirradiated region" interface), and **e** upon irradiating through an amorphous layer (the interface "the region irradiated through the $a - Si$ – layer – the region irradiated without it").

Typical microtopograms and histograms for the distribution of lateral sizes of microrelief in silicon crystals etched off to the depth of 2 mcm are presented in Fig. 4.44 [Kiselyev and Levshounova, et al. (2003).

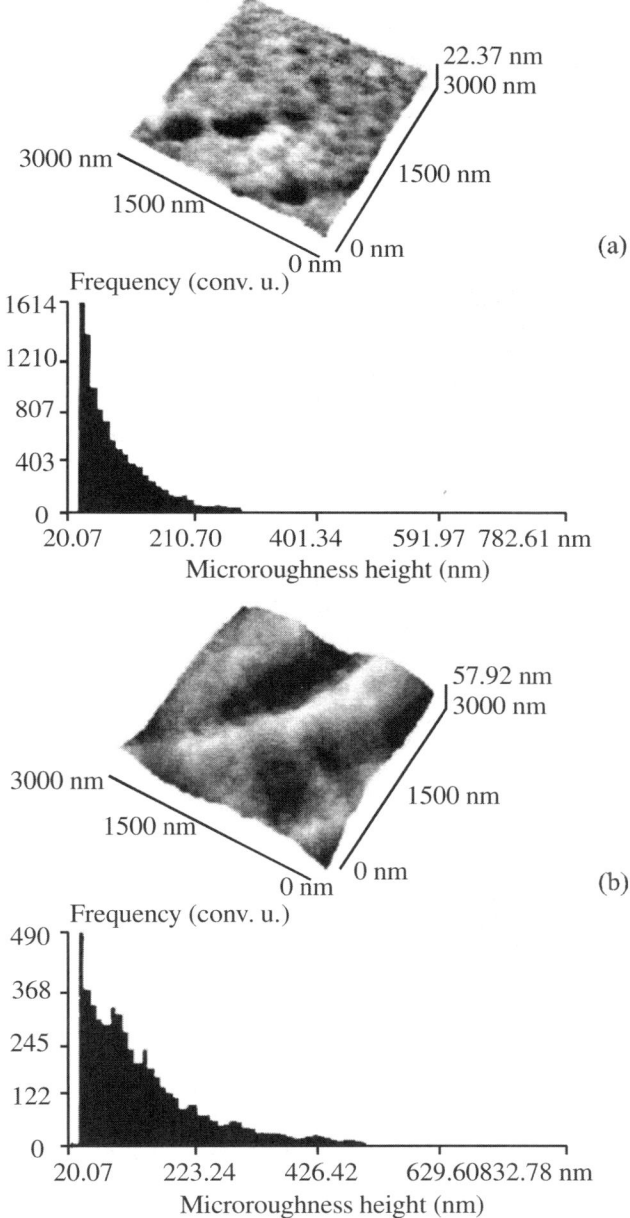

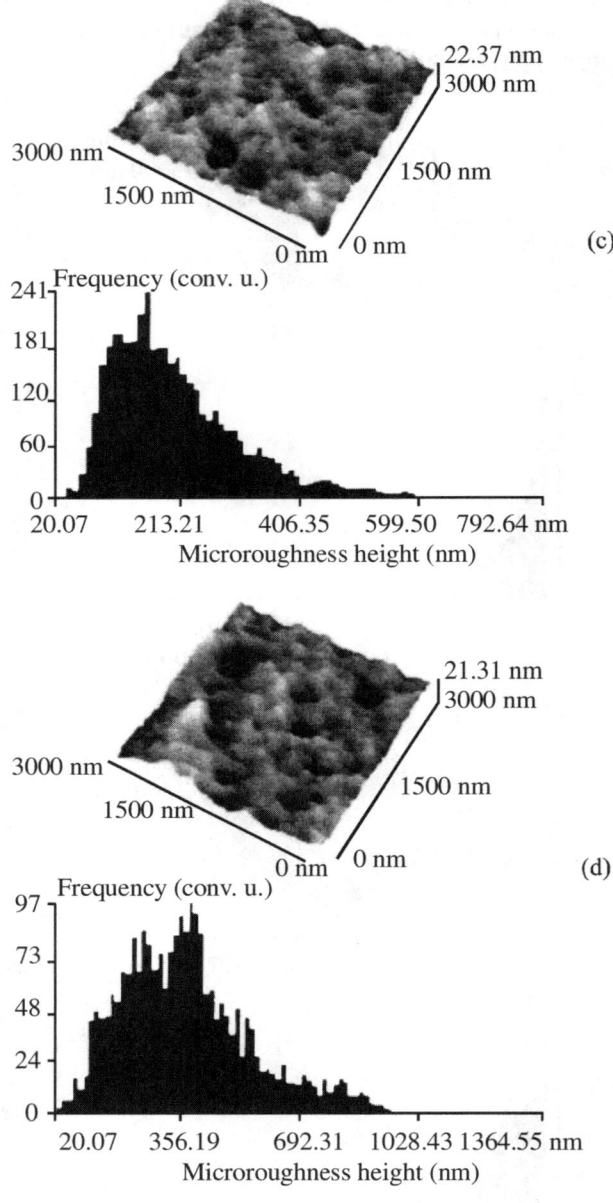

Fig. 4.44. The microrelief of silicon crystals upon etching to the depth of 2 mcm from the irradiated side: **a** in the original state, **b** upon amorphous silicon deposition, **c** upon irradiating the free surface by argon ions and **d** through an amorphous layer.

On the surface of the original silicon one sees nanoinclusions in the form of wells of 40 – 160 nm in diameter and larger ones 250 – 300 nm in diameter, up to 22 nm in depth. Evidently, these wells were formed by the growth microdefects of the $B-$, $C-$ and $D-$types during a selective etching. Upon amorphous silicon being deposited, instead of vanishing small wells there have been detected the clusters up to $\approx 650\ nm$ in diameter and up to $58\ nm$ in depth, they were similar to the $A-$type microdefects. Ionically irradiating a silicon surface free of the amorphous films partially dissolves small $C-$ and $D-$type clusters (Fig. 4.44), and this makes the lateral sizes shifted into the area of great values and leads to a drop of microroughness heights. Ionically irradiating through an amorphous film makes cluster larger (up to $\approx 1200\ nm$), reduces the heights of microroughness and widens the values of lateral sizes of microinhomogeneities. From the back side the effects of ionic irradiation are more clearly seen when treating through the amorphous layer (Fig. 4.45) [Kiselyev and Levshounova, et al. (2003)].

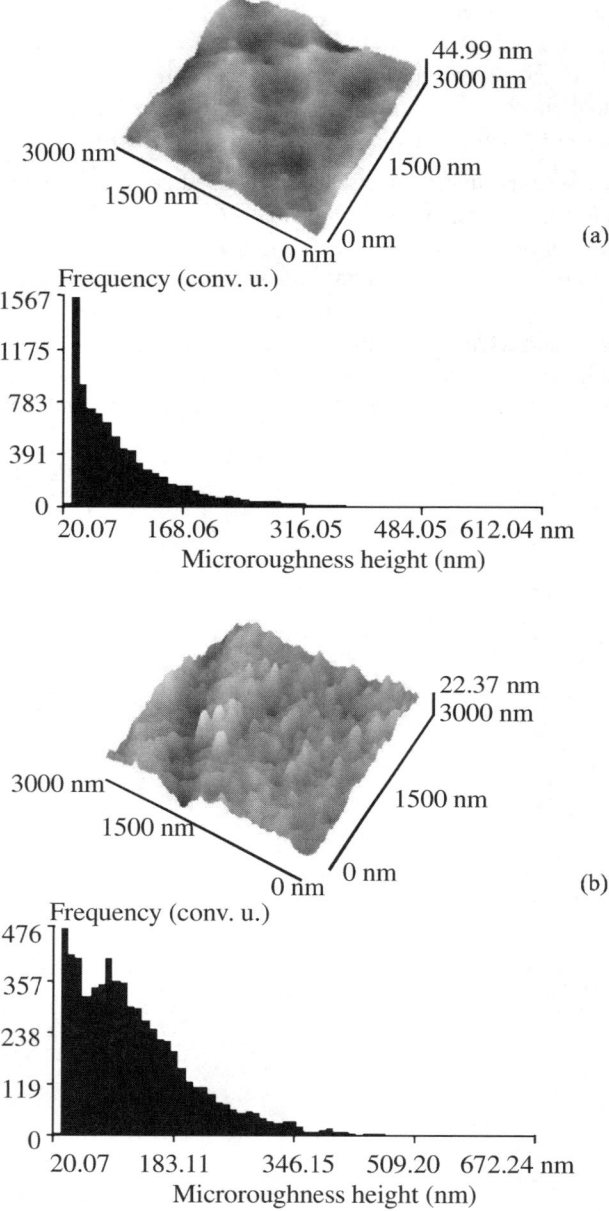

(a)

(b)

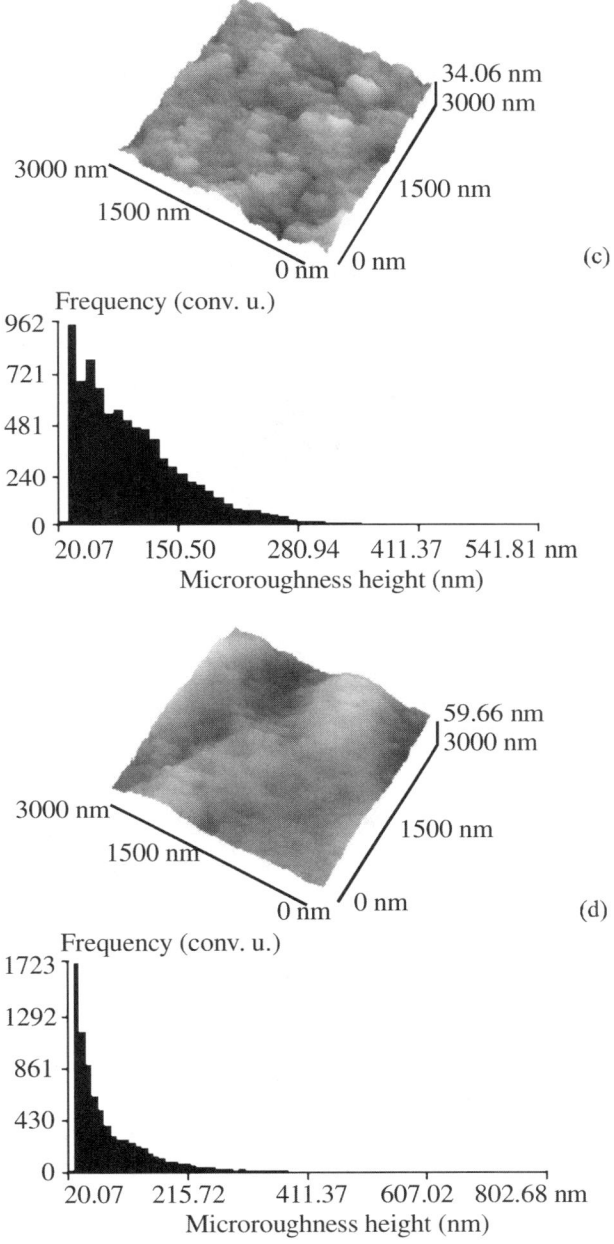

Fig. 4.45. The surface microrelief in silicon crystals at the depth of 2 mcm from the back side: **a** in the original state, **b** upon amorphous silicon deposition, **c** upon irradiating the free surface by argon ions and **d** through an amorphous layer.

With the etching depth increasing, it is observed that the microroughness in the surface of specimens not protected by amorphous silicon during irradiation starts to significantly grow in height, whereas the irradiation through amorphous silicon does not change this parameter practically (Fig. 4.46).

I a

$N_A = 2.9 \cdot 10^4$ (cm^{-2})

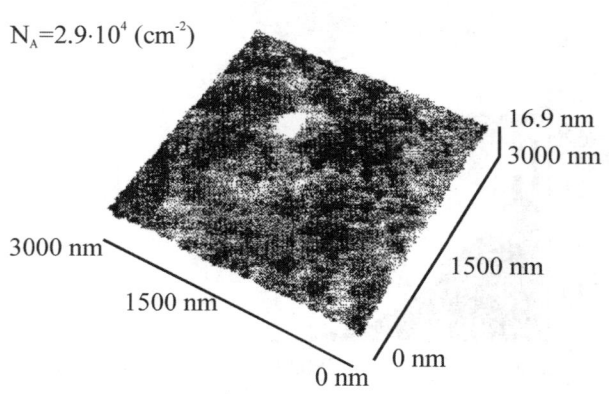

16.9 nm
3000 nm

3000 nm

1500 nm

1500 nm

0 nm

0 nm

II a

$N_A = 4.9 \cdot 10^4$ (cm^{-2})

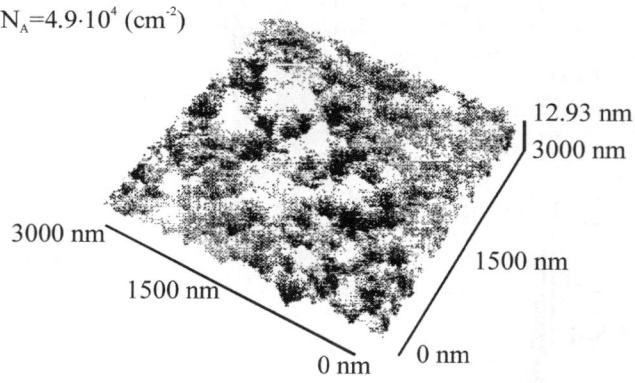

12.93 nm
3000 nm

3000 nm

1500 nm

1500 nm

0 nm

0 nm

I b

$N_A = 7.7 \cdot 10^4 \ (cm^{-2})$

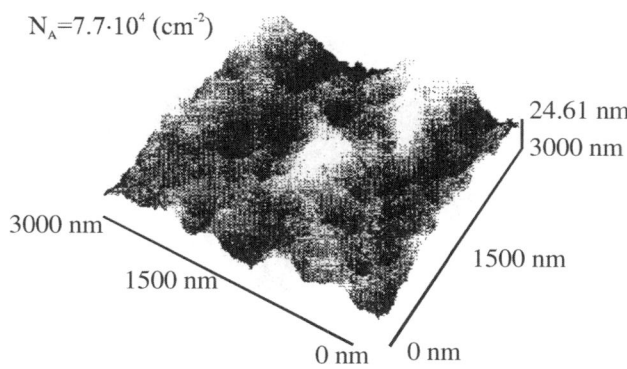

24.61 nm
3000 nm
3000 nm
1500 nm
1500 nm
0 nm   0 nm

II b

$N_A = 7.2 \cdot 10^4 \ (cm^{-2})$

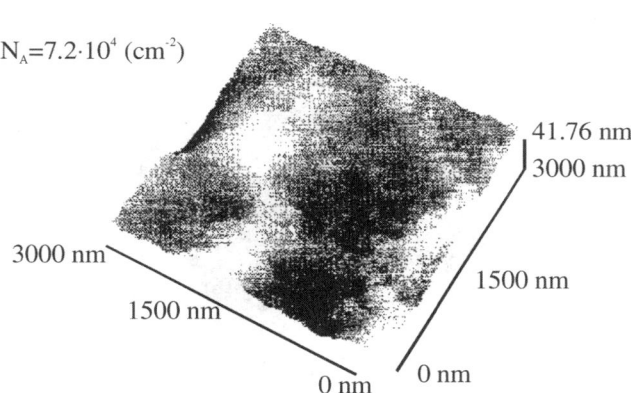

41.76 nm
3000 nm
3000 nm
1500 nm
1500 nm
0 nm   0 nm

I c

$N_A = 4.4 \cdot 10^4$ (cm^{-2})

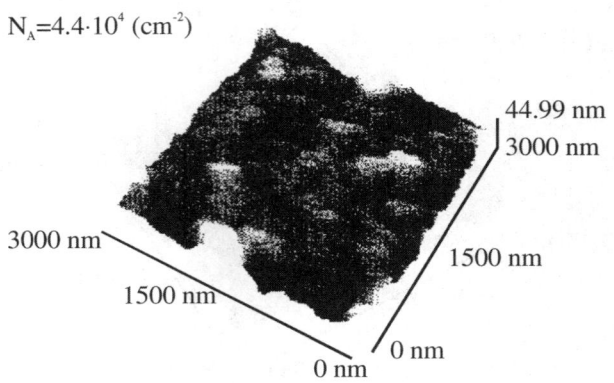

II c

$N_A = 4.0 \cdot 10^4$ (cm^{-2})

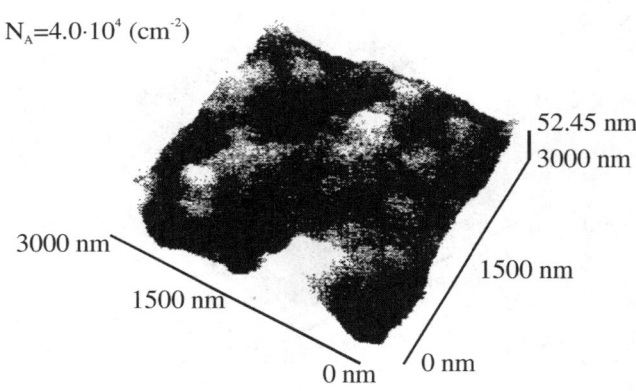

I d

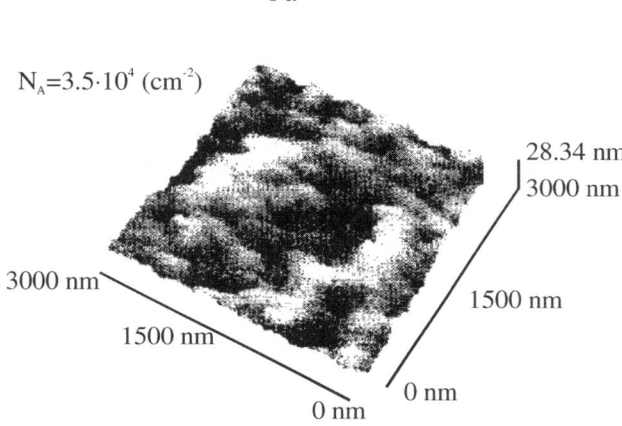

$N_A = 3.5 \cdot 10^4$ (cm^{-2})

28.34 nm
3000 nm

3000 nm
1500 nm

1500 nm

0 nm

0 nm

II d

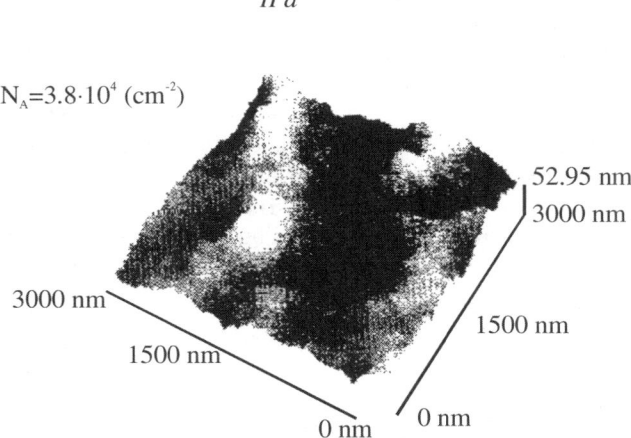

$N_A = 3.8 \cdot 10^4$ (cm^{-2})

52.95 nm
3000 nm

3000 nm
1500 nm

1500 nm

0 nm

0 nm

Fig. 4.46. The surface microrelief of silicon crystal at the 15–mcm depth from the I and II side **a** in the original state, **b** after the amorphous silicon deposition, **c** after irradiating by argon ions the free surface and **d** through an amorphous layer.

The regularities in the changes in the micromorphology of the surface of the by-layer etched crystals and their microdefect structure are explained in [Kiselyev and Levshounova, et al. (2003)] on the basis of the following thinking: in the course of the amorphous silicon deposition, its heating causes a partial dissolution of the impurity ambients screening microdefects; this results in a increased rate of their dissolution in the selective etcher. This very event increases the microroughness in the surface. Irradiating an unprotected surface not only increases a dispersal of ambients but also decreases the concentrations of the microdefects themselves, due to the unbalanced IP defects and elastic waves generated within an ion retardation zone. In irradiating ionically through an amorphous layer, a flux of point defects responsible for the dissolution of the original clusters turns out to be less and, hence, the concentration of the microdefects detected by selective chemical etching "increases". An increase of microdefect density near the nonirradiated side is evidently caused by a flux of IP defects, chiefly vacancies, which are induced by elastic waves travelling from the implanted layer and favour the dissolution of impurity ambients. That N_A increases near the back sides of the $a - Si -$ irradiated specimens, in contrast to the silicon microdefect density upon irradiating a free surface, speaks about a strengthening of elastic waves when they pass through the amorphous layer – this agrees with the conclusions made in [Skoupov and Tetel'baum, et al. (1989)].

Thus, the results obtained show that a presence of the amorphous layer increases a sensitivity of silicon microdefects to ionic irradiation, and this creates a principal possibility to exploit such layers in developing the low-temperature gettering techniques.

The long-range effect and a possibility of regulating the extrinsic defect composition via low–temperature irradiation are specific not only for silicon–based structures but also for semiconducting compounds; this has been proved by the results in [Obolensky and Skoupov (2000)].

In this research, subjected to study were the field Schottky transistors ($n^+ - n - n^- GaAs$) being ionically irradiated from the side of the wafer.

The field Schottky transistors of the 0.5-mcm gate length were formed on the structures obtained by chlorization and consisted of a semiisolating wafer 60 mm in diameter and 400 mcm thick, an undoped buffer layer 1 mcm thick, a silicon–doped active layer 0.1 mcm thick having the electron concentration $3 \cdot 10^{17} \ cm^{-3}$, and a contact 0.15-mcm layer of $n^+ - GaAs$ with the carrier concentration equal to $10^{19} \ cm^{-3}$. The gate and ohmic contact pads represented themselves the vacuum-deposited layers, respectively, of Au with a sublayer of Ti and eutectic $Au - Ge$. The distance between the nearest contact platforms of the neibouring transistors made up $300 \ mcm$. The regions between field Schottky transistors were isolated through proton implantation; proton energies were changed in the sequence $90, \ 60, \ 30 \ keV$; for all energies the dose retained equal to $3.8 \cdot 10^{13} \ cm^{-2}$. Upon production of field Schottky transistors, the structures were polished off chemicodynamically from the side of the wafer to the thickness of $100 \ mcm$. Further, a half of the surface in the wafers was screened from the wafer's side by the metallic mask and the other half was irradiated by argon ions of the energy $E = 90 \ keV$, with the ionic current being of various densities. During irradiation, the temperature of structures did not exceed $310 \pm 5\,K$. Upon irradiation, volt–ampere characteristics and volt– farad characteristics of transistors were measured under normal conditions and with a use of additional illumination by visible light lamp of the power $25\,W$; the potential of the gate in the transistor under measurement being controlled from the contact pads of the neighbouring transistor.

It has been experimentally shown that the long–range effect of irradiation on the parameters of field Schottky transistors became visible at the densities of ionic current being not higher than $0.5\,mkA/cm^2$ and of doses higher than $5 \cdot 10^{15}\ cm^{-2}$. This is manifested by an increase in the mobility of electrons and by a sharper profile of the in-depth distribution of their concentrations within the active layer near the interface with the buffer (Fig. 4.47) [Obolensky and Skoupov (2000)].

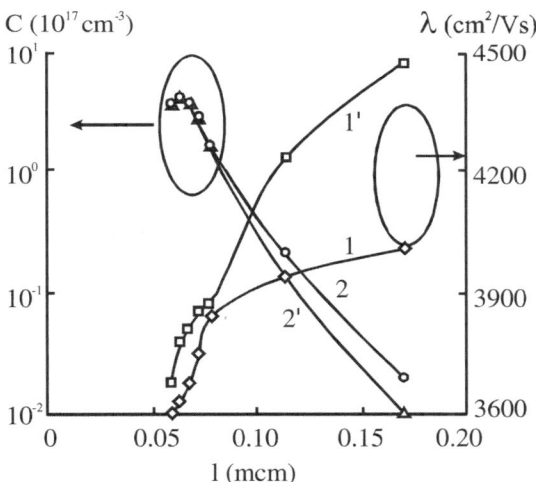

Fig. 4.47. The profiles of $l-$depth distribution of the electronic mobility $\lambda(1,1')$ and concentration $C(2,2')$ prior to $(1,2)$ and upon $(1',2')$ ionic irradiation of the dose $\Phi = 10^{16}\ cm^{-2}$.

During the additional illumination, the volt–ampere characteristics of such transistors are indifferent to whether the current in the transistor is controlled by its own or outside gate (Fig. 4.48) [Obolensky and Skoupov (2000)]. The observable parameters retain stable when the irradiated structures are held

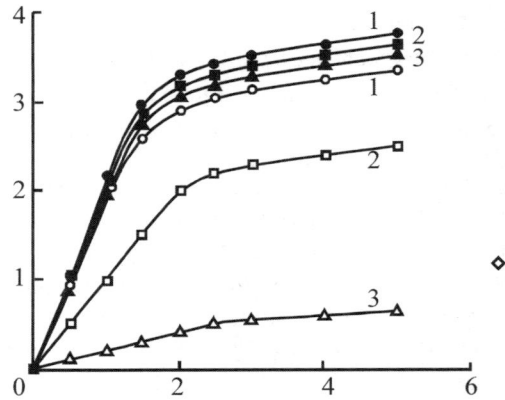

Fig. 4.48. The volt–ampere characteristics in the transistors when being illuminated and controlled from the outside electrode ($V_c = 0\,(1)$; $-0.4(2)$; $-0.8\,V(3)$) prior to (by black signs) and upon (by light signs) ionic irradiation of the dose $\Phi = 10^{16}\ cm^{-2}$.

in normal conditions during a month period or when being annealed for 1h in the air at 400 K. The mobility and concentration of charge carriers measured across the entire surface of the structure with the step equal to $0.25 - 1.25$ mm turned out to be variable not only above the irradiated but also above the controlled region in the specimens near the planar boundary of the interface (Fig. 4.49) [Obolensky and Skoupov (2000)], i.e. it means that alongside with the long–range in-depth effect in the structures under study there occurs a lateral long–range effect as well. These both effects are betrayed by the changes in the average values of the observed characteristics in the field Schottky transistors and in their variance across the surface. For example, in the batch of 110 irradiated transistors the variance of the Schottky resistance fell as much as $1.5 - 2$ times.

The results obtained demonstrate an irreversible transformation of the components in the extrinsic impurity composition in active layers and in the isolated regions formed in these layers by ionic irradiation from the side of the wafer. The specificity in changes of transistors' parameters makes it possible to suppose that this transformation is primarily caused by a fall of the scattering centres in the active layer and by a recombination of charge carriers. In particular, the point defect aggregates induced by epitaxy may serve as such centres [Mil'vidsky and Osvensky (1984); Mil'vidsky and Osvensky (1985)]. It may follow from the experimental results that the currents in the channels of field Schottky transistors located above the implanted part of the wafer are sensitive to the potentials in the contact platforms in the neighbouring transistors.

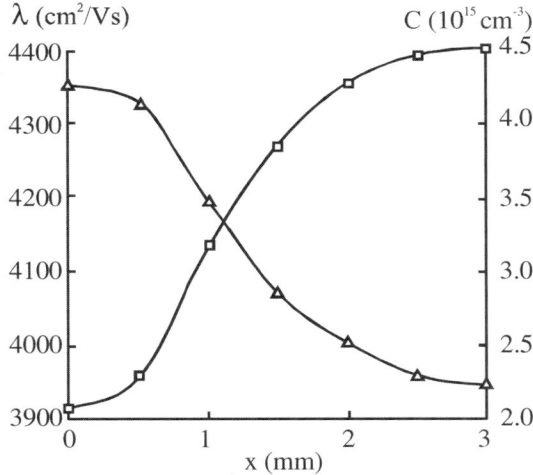

Fig. 4.49. The dependence of electronic mobility and concentration on the distance to the "irradiated – nonirradiated material" interface from the side of the nonirradiated *GaAs* at the depth of 0.15 mcm. The interface boundary is taken to be of zero value.

The irradiation by argon ions brings a partial annealing of the irradiation damages responsible for the isolating properties of the proton–irradiated regions between the transistors (Fig. 4.48).

In the experiments described, the in-depth long–range effect and the lateral long–range effect are, most probably, caused by elastic waves born basically by thermal peaks in the ionic retardation zone [Pavlov and Syemin, et al. (1986); Morozov and Skoupov, et al. (1985)]. Though it should be taken into consideration that when passing through the entire thickness of the structure these elastic waves (let us call them original) drastically reduce their amplitude and, as our estimates indicate, for our case the pressure $P_{max} = 1.44\,GPa$ (the area of a thermal peak) in the transistor's active layer falls to $0.34\,MPa$. The energy of such waves makes up $10^{-4}\,eV/at$ that is not sufficient to immediately activate the dissociation processes in the extrinsic defect aggregates ($\Delta E > 0.1\,eV/at$).

Therefore, they are thought to be basically responsible for exciting the metastable PD aggregates annihilating the Frenkel pairs, the travell of these aggregates to the internal and external drains, a birth of new aggregates and they are also responsible for other processes reducing the total free energy of the defects during irradiation. In other words, a structure as an open thermodynamical system assumes a new state, via recombining the extrinsic defect composition; this state will correspond to its minimum free energy under the varying external conditions, i.e. under the action of the ionic current. Here, each elementary recombination of defects results in a birth of minor elastic waves whose superposition with the original waves strengthens the amplitude in the total field of dynamical stresses diffusing from the irradiated surface of the structure [Skoupov and Tetel'baum, et al. (1989); Syemin and Skoupov, et al. (1988)].

Most intensively these processes run near the external surface and internal boundaries of the interface between the epitaxial layers where the coefficient of static mechanical stresses is maximal and this results in a heavy concentration of spatially–separated vacancies and interstitial atoms. A fall of elastic waves upon such boundary will stimulate an annihilation of alternate–sign IP defects and, as a result of this, will generate minor elastic waves of the amplitude equal to a hydrostatical component in the field of own elastic stresses of defects; this component exceeding P_{max} of original waves by one–two orders [Skoupov and Tetel'baum (1987)]. The numerical estimates for the amplitude and the energy of such waves diffusing through the active layer from the internal boundary of the interface assume on the free surface of the structure the values $46\,MPa$ and $0.013\,eV/at$. Such waves are capable of initiating the surface sources of IP defects which activate the transformation of extrinsic impurity composition in the channels of field Schottky transistors and the annealing of radiation damages incorporated via proton irradiation. The minor elastic waves were diffusing along the interface of the epitaxial layers; they stimulate the reactions between the defects – beyond the area of immediate ionic beam action; this leads to a birth of the lateral long–range effect. This effect is expressed the more, the higher is the concentration of metastable aggregates of IP defects and impurities in the vicinity of the interface, and their spectrum and quantity are directly dependent on the magnitude of the structural misfit of the conjugative layers [Mil'vidsky and Osvensky (1985)]. As some additional stimulation for producing the long–range effect there may evidently be an inhomogeneous field of mechanical stresses, and accordingly the extrinsic defect composition, near the surface structures; this field is determined by homo- and heterophase elements in the topology of the device – like the locally–doped regions, the conducting and dielectric films, the especially built microrelief and the like.

Since the long–range effect is a response of the extrinsic defect composition of the material to external actions and this response minimizes its free energy during the period when the irradiation defects emerge and are accumulated near the irradiated surface and the original extrinsic defect composition is transformed due to interactions with the elastic waves and mobile IP defects – after a cease of irradiation there arise a new integral structural state whose energy may be either less or higher than that prior to irradiation. In the first case, the irradiation has favoured a transfer of the structure to a thermodynamically more balanced state, and in the second case to an unbalanced state, from which the system (i.e. all the components in the extrinsic defect composition and their spatial distribution) will be gradually relaxing into a fresh intermediate metastable state. What specific state the structure will assume during irradiation is determined by numerous factors, including the type and the regime of irradiation and also the specificities in the original extrinsic defect composition of the material. The more unbalanced is the original state of the extrinsic defect composition, the more sensitive will be the material to an external (even slight, so–called, underlimit) action activating and accelerating the relaxation processes. One of the consequences of this process will be the long–range effect.

4.4.3
Models for low–temperature gettering of irradiated structures

As was emphasized in the beginning of subsection 4.4, a physical foundation for the irradiating technologies of low–temperature gettering (and this is one of the manifestations of the long–range effect) is, first, unbalanced IP defects and, second, elastic (striking) waves being induced in the retardation zone or in the zone of initial radiation dissipation. Let us give here a theoretical estimation of the role played by each of these factors in structural transformations of crystals during irradiation.

1. The first sufficiently complete analysis of irradiation defects penetrating to abnormally large depths in crystals irradiated by average–energy ions was seemingly done in [Morozov and Tetel'baum (1983); Morozov and Skoupov, et al. (1985)]. The authors of this research, suggesting a single approach to studying a diffusion of irradiated IP defects, their aggregation and travell to the external and internal drains, have managed to explained the results of other researchers and their own experimental results concerning the long–range effect.

Despite a high value of the shift threshold energy taken in [Morozov N.P. and Tetel'baum D.I. (1983)] equal to $22\,eV$ from [Smirnov (1980a)], it was successfully shown that if the energy in the initial ions increases, then a quantity of deeply penetrating defects will sharply fall, within the energy interval of ions (heavier than gallium) up to $10\,keV$ and in the event of reflecting surface of the crystal. This crucial conclusion is important for understanding the nature of defect formation during ionic–plasmic, ionic–chemical and other types of "dry" etching of crystals.

These studies were further proceeded in [Morozov and Skoupov, et al. (1985)] where the concentration and the distribution of aggregated IP defects diffusing from the generation region have been calculated. In these calculations there was exploited the model for the minor defect formation suggested in [Morozov and Tetel'baum (1983)]. It was supposed that in irradiating by sufficiently high doses the concentration of IP defects of the given type (the calculations were done for vacancies, to make the case definite) approaches the stationary value $N_{vsta}(x)$ at the depth X, for the time much less than the irradiation time t_{irr}. Then, according to [Morozov and Tetel'baum (1980)], the concentration in the largest aggregate N_{vm}, consisting of m vacancies, will be described as

$$N_{vm}(x,t_{irr}) = 0.5k_1 N^2_{vsta}(x)t_{irr} \ , \tag{4.30}$$

where

k_1 stands for the kinetic coefficient of bonding a vacancy into a divacancy;

N_{vstaa} is a stationary concentration of vacancies in the region $x > b$

($b \approx R_r + \Delta R_r$ is a thickness of the generation zone for IP defects);

$t_{reg} = \Phi / j$ (Φ and j stand for a dose and density of the ionic current).

The concentration N_{vsts} is found [Morozov and Tetel'baum (1980)] from the equation

$$D_v \frac{d^2 N_{vsta}}{dx^2} - \left(\frac{D_v}{L^2_{v2}}\right) N_{vsta} - \frac{m}{2} k_1 N^2_{vsta} = 0 \ , \tag{4.31}$$

where D_v is a coefficient for vacancy diffusion; L_{v2} is a diffusion length of vacancies in the region $x > b$.

For $N_{vsta}(\infty) = 0$, $\dfrac{dN_{vsta}}{dx} = 0$ and $x \to \infty$, the equation (4.31) is of the form:

$$N_{vsta}(x) = 4\Omega N_1 \exp\left(\frac{b-x}{L_{v2}}\right)\left[1 - \Omega \exp\left(\frac{b-x}{L_{v2}}\right)\right]^{-2} \ , \tag{4.32}$$

where $N_1 = \dfrac{3D_v}{mk_1 L^2_{v2}}$, $\Omega = \dfrac{N_1}{N_{vsta}(b)}\left(\sqrt{\dfrac{1 - N_{vsta}(b)}{N_1}} - 1\right)$.

The ratios (4.30) and (4.32) determine the function $N_{vm}(x,t_{irr})$. In [Morozov and Skoupov, et al. (1985)], the unknown concentration N_{sta} is found via solving the equation of vacancy concentration balance for the region $x \le b$ and then joining this solution with the value of N_{vsta} for $x > b$:

$$D_v \frac{d^2 N_{vsta}}{dx^2} - \left(\frac{D_v}{L^2_{v1}}\right) N_{vsta} - 0.5mk_1 N^2_{vsta} + j\frac{q}{b} = 0 \ , \tag{4.33}$$

where L_{v1} stands for a vacancy diffusion length within the region $x \le b$.

The vacancy generation rate is supposed to be uniform and equal to $j\dfrac{q}{b}$, where q is the number of vacancies created by a single ion.

If the irradiation is performed by average energy ions, then the condition $L_{v1} < b$ is fulfilled that allows to neglect the diffusion term in (4.33); and then the concentration $N_{vsta}(b)$ will be

$$N_{vsta}(b) = 2\varepsilon \left(\frac{q}{b}\right)\left(\frac{L^2_{v1}}{D_v}\right)\left[\sqrt{1+\frac{j}{\varepsilon}}-1\right],\qquad (4.34)$$

where $\varepsilon = \dfrac{bD^2_{\ v}}{2L^4_{\ v1}k_1mq}$.

The calculating results in [Morozov and Skoupov, et al. (1985)] are represented as an expression for the depth x_i at which N_{vm} is equal to some assigned concentration N_i:

$$x_i = b + L_{v2}Ln(\Omega Z),\qquad (4.35)$$

where $Z = \dfrac{N_1}{N_2}\left(\sqrt{\dfrac{N_2}{N_1}+1}+1\right)^2$, $N_2 = \sqrt{\dfrac{2jN_1}{k_1\Phi}}$.

The values of x_i (in mcm) calculated for the silicon irradiation by argon ions of the energy $40\,keV$ and with $j = 10^{14}\ ions/cm^2\cdot s$ (the target temperatures being different) are given in Table 4.20. In calculations there was used the approximation $x_i \approx \dfrac{1}{j^{0.17}}$ accounting a slight dependence of x_i on the current density and also the numerical values of parameters in the defect formation process suggested in [Morozov and Tetel'baum (1983); Morozov and Tetel'baum (1980)].

Similar calculations are possible to be done for intrinsic interstitial atoms for which a layer with incorporated aggregates must be thicker than for the vacancies, due to their higher mobility. In [Morozov and Skoupov, et al. (1985)] this has been experimentally supported, through the X-ray measurements of the relative complement in the lattice period and through a curvature of silicon specimens irradiated by argon ions, with their by-layer chemical etching.

Table 4.20

The values of depths x_i at which the concentration of vacancy aggregates
is equal to a given value when silicon crystals are irradiated by various doses
of argon ions

Target temperature T (K)	$N_i \cdot 10^{-18} (cm^{-3})$	$\Phi \cdot 10^{-15} (ions / cm^2)$		
		0.6	6.0	60
300	1	0.4	0.7	1.3
	10	0.2	0.4	0.7
400	1	0.5	0.9	1.6
	10	0.3	0.5	0.9
550	1	0.6	1.1	2
	10	0.3	0.6	1.1

It should be emphasized here that the fact that for $x > b$ the aggregates of
irradiated defects accumulate [Morozov and Skoupov, et al. (1985)] in the vicinity
of the irradiated layer aggregates does not contradict the above given results
concerning microdefect dissolution [Demidov and Latysheva, et al. (2000);
Perevostchikov and Skoupov (1999); Kiselyev and Levshounova, et al. (2003)].
First, these aggregates are evidently much less in size than the growth or
technological microdefects. Second, these aggregates lie at small depths;
according to [Morozov and Skoupov, et al. (1985)] this depth makes up
$0.5 - 1.5 mcm$ in the silicon irradiated by argon ions of $E = 40 keV$ and by
doses $(1.2 - 3.7) \cdot 10^{15} cm^{-2}$. Besides, a dissolution process or a growth of
microdefect sizes, as shown in [Italyantsev and Mordkovitch (1983)], are
dependent on the type of defects (whether they are of the vacancy or interstitial
type), and also on the ratios between the concentrations of unbalanced vacancies
and intrinsic interstitial atoms induced by the irradiation within the bulk of the
crystal. The latter depends on the differences in the diffusion lengths of the
Frenkel pairs and on their activity during their interactions with the stable defects
available in the crystal.

When developing the diffusive technologies for irradiation gettering, one
should be aware of the original extrinsic defect composition in crystals and
structures, to control the concentrations and its components during irradiation by a
given type of radiation in some specific (variable) regimes. Here, the theoretical
estimates done in [Dzafarov (1991); Morozov and Tetel'baum (1983); Morozov
and Skoupov, et al. (1985); Italyantsev and Mordkovitch (1983)] make it possible
to preliminarily choose the irradiation regimes for some known initial extrinsic
defect composition. Though it should be noted that a diffusion mechanism alone
does not fully explain the experimental data on the long–range effect, and in
creating the efficient gettering techniques one should take into account one more
factor arising during irradiation – elastic waves.

2. The acoustic waves as one of the factors capable of initiating the structural transformations in the irradiated materials of various interatomic bonds were considered by a number of researchers [Katchanova and Katchanov (1973); Kalinitchenko and Lasourok–El'tzoufin (1973); Davydov and Kalinitchenko (1982)]. This subsection describes a birth of elastic waves whose sources are the thermal peaks [Pavlov and Syemin, et al. (1986)].

Let us first consider the pressure determined by an individual thermal peak.

Let the cascade, upon a cease of ion retardation, approximately look like a cylinder of the length L and radius r. At a sufficiently large distance from the cascade, the cylinder may be replaced with the equivalent semisphere of the radius

$$r_1 = \sqrt{Lr} \, , \tag{4.36}$$

where L is a cylinder length ($L = R_p + \Delta R_p$) and r stands for an efficient radius of the cylinder. A locally heated region in the peak exerts some pressure upon the matrix that can be approximately described through the impulse of the amplitude p_0 and the length τ. Without taking into account the energy dissipation (attenuation) and the reflection of elastic waves from the surface, the amplitude of the pressure p in the point lying from the cascade centre at the distance R is determined, in the acoustic approximation, from the wave equation [Landau and Lifshitz (1953)] having the marginal conditions:

$$p(r_1) = p_0 \quad \text{for } 0 \leq t \leq \tau \; ; \quad p(r_1) = 0 \quad \text{for } t < 0, t > \tau \, ,$$
$$p(\infty) = 0 \, .$$

The solution will be:

$$p(R,t) = p_0 \left\{ \frac{r_1}{R} \exp\left[-\frac{u(t - \tau) - (R - r_1)}{r_1} \right] + \right.$$
$$\left. + \frac{r_1^2}{R^2} \left[1 - \exp\left(-\frac{u(t - \tau) - (R - r)}{r_1} \right] \right\} \tag{4.37}$$

$$\left(\frac{R - r_1}{u} \leq t \leq \frac{R - r_1}{u} + \tau \right)$$

$$p(R,t) = 0 \, , \quad \left(t < \frac{R - r_1}{u}, \; t > \frac{R - r_1}{u} + \tau \right) ,$$

where u is a sound velocity.

For $R \gg r_1$ the second term between the braces is negligibly small as compared to the first. The average value of the time of pressure impulse for $R \gg r_1$ will be equal to

$$P = p_0 \frac{r_1}{R} \left[1 - \exp\left(-\frac{u\tau}{r_1} \right) \right] .$$ (4.38)

Experimentally it was shown that at the distances $R \geq 100 \ mcm$ and for all reasonable p_0 the pressures will be too small to cause noticeable structural changes.

Let us further consider a superposition of waves excited by various ions, leaving the dissipation beyond our consideration. As seen from (4.37), the time an impulse is passing from the given ion through any point is equal to τ. During this time, through the same point there will pass $j\tau ds$ impulses (besides the given) of the ions fallen onto the platform ds. Due to a stochastic nature of ion collisions against the surface, the superposition leads to summing up not amplitudes but their squares. Therefore, an average total pressure in the point R will be described as

$$P^2(z) = \frac{p_0^2 a^2}{z^2} + \int_s \frac{j\tau p^2 va^2 \cos\theta}{R^2(\rho)} ds ,$$ (4.39)

where
z stands for a distance to the surface;
j is a density of the ionic flux;
ρ is a radius–vector of the point on the surface;
s is an irradiation square;
θ is an angle between the normal to the straight line connecting an observation point with the current point ρ ;

$$a = r_1 \left[1 - \exp\left(-\frac{u\tau}{r_1} \right) \right] .$$

For the case of irradiating a round platform of the radius ρ_0 being centred with respect to the observation point, we shall have

$$P = \frac{p_0 a}{z}\left[1 + 2\pi j\tau z^2\left(1 - \frac{1}{\sqrt{1 + \frac{\rho_0^2}{z^2}}}\right)\right]^{0.5} \quad \ldots\ldots\ldots(4.40)$$

From (4.40) it follows that there exists the value $z = z_{osc}$ such that at $z \le z_{osc}$ the basic contribution to the pressure is done by the wave excited by a single ion, and at $z \ge z_{osc}$ the superposition of waves from a large number of ions is important; besides (at $z \ll \rho_0$) we get

$$Z_{osc} = \frac{160}{\sqrt{I\tau}} \, mcm \, , \tag{4.41}$$

where I stands for the ionic current density in mcA/cm^2, $\tau' = 10^{-11} s$. For $\rho_0 \gg z \gg z_{osc}$, the pressure amplitude is not practically dependent on the depth:

$$P = p_0 a\sqrt{2\pi j \tau} \, , \, \rho_0 \gg z \gg z_{osc} \, . \tag{4.42}$$

Until now, no elastic wave dissipation was considered by us. A stochastic nature of wave generation and a lack of exact information concerning the coefficient of hypersound attenuation at the frequencies of the order $\approx 10^{11} Hz$ make it difficult to consider the attenuation strictly. However, approximate estimates are sufficient for our problem. There are known the two extreme cases [Tucker and Rampton (1972)] when hypersound of the frequency $\beta \approx v$ and $\beta \approx v^2$ dies down in a solid body. Let the attenuation coefficient β_0 be known at some frequency. Expanding the rectangular impulse of pressure p_0 into the Fourier integral, taking the smoothing frequency v_{osc} as an upper limit of integration (the frequency is chosen so as to make the monochromatic signal at the given depth z 10 times weaker than the signal for which $\beta = 0$) and using (4.42), we obtain (if $v_{osc}\tau \ll 1$) that the attenuation drops the amplitude of the resulting pressure P as much as n times, where

$$n = \left(\frac{z\beta_0}{v_0 \tau} \right)^{0.5} \text{ for } \beta \approx v , \quad n = \left(\frac{\sqrt{z\beta_0}}{v_0 \tau} \right)^{0.5} \text{ for } \beta \approx v^2 . \quad (4.43)$$

For example, for silicon with $\beta_0 = 0.22 \; dB/cm$ for $v_0 = 9.8 \cdot 10^9 \, s^{-1}$ [Tucker and Rampton (1972)] we have (for $z = 1$) $n \approx 2$ for $\beta \approx v$ and $n \approx 6$ for $\beta \approx v^2$. It is important that the attenuation affects the resulting pressure amplitude less than the amplitude from some separate ion. This happens so because with the attenuation growing with the frequency the basic role will be played by low harmonics for which the duration of passing through the given point (and hence the number of waves participating in the superposition) will be increased, thus partially compensating the attenuation.

For numerical estimations, it is necessary to have an idea concerning the values of p_0. The pressure determined by the thermal peak can be estimated from the ratio [Skoupov and Pashkov (1979)]:

$$p_0 = G \, c_v \, T = \frac{G E_n}{\pi \, r^2 L} , \quad (4.44)$$

where

G is the Grunaisen constant;

c_v is a specific heat capacity of a single unit of volume;

T is a difference of temperatures in the thermal peak and matrix; and

E_n is an ion energy spent for elastic losses.

For estimating purposes, the cascade radius may be accepted to be equal to $5 \; nm$. The choice of the value of r relatively slightly affects the value of the resulting pressure P ($P \approx r^{-0.5}$). From (4.44) for argon ions of $E = 40 \; keV$ in silicon we obtain $p_0 \approx 5 \cdot 10^8 \, Pa$.

In the cascade, higher pressures can arise at the expense of striking waves which in the centre of the cascade can approach the value $\approx 10^{12} \, Pa$. Though at such pressure, according to the energy preservation law, the radius should be reduced as compared against the above given magnitude. Consequently, the heat capacity will drop the cooling time, i.e. the magnitude τ. In the final end, the contribution of this component to the pressure P for $z \gg z_{osc}$ will also fall. For the above given example, at the pressure $p_0 = 10^{12} \, Pa$ the resulting pressure will approximately grow by an order of magnitude. Though, the problem of the adequateness of the wave generation model cannot be thought solved so far.

If attenuation and striking waves at the depths $z \gg z_{osc}$, $z \ll p_0$ are left beyond consideration, then the pressure P in the ionically–irradiated silicon ($E = 40\ keV$, $j = 10^{14}\ ions\,/\,cm^2 \cdot s$) will be $P = 6 \cdot 10^4\ Pa$.

Now we jump to the structural changes to be expected in the case of the above oscillating elastic stresses.

The numerous experimental data on how defects change at large distances from the irradiated surface is suggested in papers listed in [Pavlov and Syemin, et al. (1986)]. On the basis of these experimental results one may conclude that:

a)   there occurs a vanishing of defects, detected through decoration, down to the depths $\approx 200\,mcm$ (supposedly these are the $A-$ clusters whose basic components are interstitial dislocation loops);

b)   removing the dislocations at the depth $\approx 1\ mcm$ from the specimens $\approx 250\ mcm$ thick; the misfit dislocations were induced by impurity diffusion.

In both cases the irradiation was performed by argon ions of $E = 40\ keV$, $j = 10^{14}\ ions\,/\,cm^2 \cdot s$, by doses $3 \cdot 10^{16}$ and $2 \cdot 10^{17}\ ions\,/\,cm^2 \cdot s$, respectively. Here, during irradiation the temperature, according to our measurements, was increasing approximately up to $623\ K$.

Using the above values of pressures P and the rates of the dislocation travell within the field of stresses, it is not difficult to show that in the course of irradiation the conservative travells of dislocations (or loops) are negligibly small and they are unable to cause the observable effects.

However, another effect is possible if the oscillations of the elastic stress are available; this effect will be discussed now.

As known, clusters and dislocations formed at high temperature (for example, during a growth of crystal or diffusion) are encircled with an extrinsic defect ambient [Fridel (1967)]. Let, in the absence of external actions, the quasibalanced concentration of vacancies in an ambient be equal to N_{v0}. When an oscillating elastic stress is applied, the balance will be shifting, synchronously with the stress, so that under the pressure P the vacancy will obtain the additional potential energy ($P\Delta V$), where ΔV is an activating volume. We shall calculate the pressure from the middle level. Then, within the negative pressure phase a portion of vacancies ($\dfrac{N_{v0} P \Delta V}{kT}$ for $|P\Delta V| \ll kT$) will turn out to be excessive and can drain upon a dislocation (or a loop). A vacancy flux per a unit of length of the dislocation during a unit of time will be equal to $\dfrac{\pi N_{v0}\, P\,\Delta V\, D_v\, b}{k\, T\, \lambda}$, where b is the Bürgers vector, D_v is a vacancy diffusion coefficient, and λ is an interatomic distance. Suppose that within the positive pressure phase the vacancy concentration manages to get restored. It is possible, if the extrinsic defect ambient has a sufficiently low energy barrier for vacancy generation. Such situation may really exist: according to computational results [Kiv and Solovjev (1980)], the extrinsic defect aggregates are capable of self-generation of the Frenkel pair components. A complete quantity of vacancies absorbed by l-length dislocations during irradiation time t will be

$$N_u = \frac{\pi N_{u0} P \Delta V D_u blt}{2kT\lambda} .$$ (4.45)

Hence, to have a cluster $1\ mcm$ in diameter dispersed for $t = 300\,s$ it is necessary to have $N_{v0} \approx 10^{19}\ cm^{-3}$. In the above given case, approximately the same value of N_{v0} is needed to remove the dislocations from the specimen.

This mechanism looks like "a vacancy pump" suggested in [Alekhin (1983)]. This mechanism differs from that in [Alekhin (1983)] only by the vacancies being not pumped from the surface but generated inside the extrinsic defect ambients. We suppose that similarly may be explained the numerous results from [Alekhin (1983)] concerning the abnormal behaviour of the actual structure of crystals within the presurface regions at the depths of many dozens of micrometers under variable loads. Direct travell of vacancies at such large distances looks impossible, for their diffusing length does not exceed $10-20\ mcm$. Besides, according to [Alekhin (1983)], the vacancies in the contraction areas are supposed to be of increased energy, whereas; according to generally accepted thinking, their energy must be high in the extension areas.

Now it is necessary to discuss how real are the above suggested concentration of vacancies, dislocation ambients and the $A-$ clusters. For explaining his results, the author [Alekhin (1983)] admits the vacancy concentration at room temperatures in the silicon presurface layers to make up $10^{14} - 10^{17}\ cm^{-3}$. (In accordance with the above said, such concentration sooner refers to extrinsic defect ambients but not to the entire presurface layer). In our case, the temperature was $\approx 573\ K$ and so the concentration $\approx 10^{19}\ cm^{-3}$ obtained by us does not significantly differ from the above given estimate.

Concluding we note here that varying irradiation–induced pressures can structurally change the layers lying closely to the nonirradiated surface of the specimen. This process may be caused by the Frenkel pair components generated in the transient layer "natural oxide – silicon" holding a lot of impurities (oxygen, carbon and the like). During extension and compression phrases, the excessive vacancies or interstitial atoms depositing onto the centres can create extended defects affecting the properties in the presurface layers. Indeed, in silicon irradiated by argon doses $\approx 10^{17}\ ions/cm^2$ we have detected a fall in microhardness on the nonirradiated side, in the specimens up to 1 cm thick. This change vanished upon etching off the layer $\approx 0.5\ mcm$.

Thus, the elastic waves penetrating deeply are able to structurally damage the layers lying as far as hundreds and more micrometers from the ionically–irradiated surface.

The elastic waves we were speaking about earlier should be called original. Later, in [Syemin and Skoupov, et al. (1988)] theoretically and in [Skoupov and Tetel'baum, et al. (1989)] experimentally it was shown that the original waves, while passing through the materials of highly concentrated structural defects (say, microdefects, oxygen-induced stacking faults, low variance crystals, etc.), can give birth to minor elastic waves increasing the amplitude of original waves. This effect can be shown in the crystal whose clusters of the concentration N_{cl} (the cluster has the defect density n and the size R_{cl}) are homogeneously distributed in its entire bulk [Syemin and Skoupov, et al. (1988)]. If a recombination of defects in the cluster under the action of original waves is determined by the annihilation of intrinsic point defects (for example, interstitial atoms composing the cluster and the vacancies from its Maxwell ambient), then in each act of this process there will arise the impulses of minor waves.

Let E be an energy barrier for recombining a single Frenkel pair; then in the field of pressures P, with a corresponding sign of the pressure amplitude, the barrier will drop by the magnitude $P\delta V$, where δV is an activating volume of the annihilation process. The quantity of the reacting defects, additionally to the thermally activated, will be equal to

$$N = n\Delta V \tau_0 v_0 \exp\left(-\frac{E}{kT}\right)\left[\exp\left(\frac{p\partial V}{kT}\right) - 1\right],\qquad (4.46)$$

where

v_0 stands for a thermal oscillation frequency;

ΔV is a cluster volume equal to $\dfrac{4\pi R_{cl}}{3}$;

τ_0 is a period of original wave extremums.

If $N \geq 1$, i.e. this implies at least a single recombination act caused by the original wave, taking $v_0 = 10^{13}\ s^{-1}$, $\tau_0 = 10^{-11}\ s$, $P = 0.1\ GPa$, $\delta V = 10^{-23}\ cm$, $E = 0.4\ eV$ and $T = 300\ K$ we shall obtain $n\Delta V \approx 3\cdot 10^5$.

A recombination, as any structural rearrangement in the crystal, emits an elastic wave of original amplitude $P_{1\max} \approx \varepsilon K$, where ε is a local deformation of lattice in the vicinity of the defect and K is a spatial modulus of elasticity. Taking $\varepsilon \approx 0.1$ and $K = 100 \ GPa$ we get $P_{1\max} \approx 10 \ GPa$, that is two orders more than the original wave amplitude. This minor wave activates the recombination of other defects constituting the given cluster. From (4.46) it follows that for $n > 10^{20} \ cm^{-3}$ there will occur a chain process of defect recombination in the cluster during which the waves generated by separate defects will be overlapped in space and time creating thus a general minor wave. The field of this wave is summed up with the original wave field and compensates an in-depth loss in amplitude and energy of the latter. This effect will be observable at the concentration of clusters equal to $N_{cl} \approx 10^{15} \ cm^{-3}$. Hence, at $N_{cl} > 10^{15} \ cm^{-3}$ the waves will not be dying down during their travell but will be amplified at the expense of minor waves.

Even if some of the amplification criteria listed in [Syemin and Skoupov, et al. (1988)] are not valid for real crystals, say, in sizes and concentration of clusters, one portion of the defects lying deeply from the irradiated surface will be, nevertheless, etched off by the original elastic waves and minor waves, i.e. the defects will be reduced in sizes or dissolved fully. Experimentally this has been found in [Skoupov and Tetel'baum, et al. (1989)], through measuring the lattice deformation, microhardness and charge carrier distribution profiles in the epitaxial structures of silicon where irradiation has brought numerous OSFs.

Therefore, the information in this subsection proves a real possibility of implementing the low–temperature gettering technologies exploiting the irradiation process.

CONCLUSION

The analysis of the present art of gettering the background impurities and extended structural defects in the semiconducting structures reveals that the below problems are urgent for microelectronics:

a) developing the new high– and low–temperature gettering technologies providing reproducible physicochemical properties in the structures. Most crucial is establishing the low–temperature techniques, in return to tendencies in the contemporary microelectronics to manufacture superhigh–speed and ultrasuperhigh–speed ICs and quantum–size structures;

b) developing the gettering technologies utilizing low–energy ("prior to threshold") irradiations of various nature and supplementary statical and dynamical external actions, for example, inhomogeneous elastic or electrical fields;

c) developing the techniques of the combined gettering, with a simultaneous or sequential use of actions, for example, ultrasonic and corpuscular or photonic irradiations, hydrostatic compression and irradiation, etc;

d) developing the technologies for gettering impurities and defects in the structures based on the semiconducting compounds, with a use of the techniques, tested, in particular, on silicon;

e) developing the theoretical models of physical mechanisms for gettering under various external actions, to reliably predict their effects on the extrinsic defect composition in the components of semiconducting structures.

This list of problems is evident to be far from complete; it sooner states basic research directions, within the limits of each direction a complex of problems will have to be attacked.

In this book the authors have tried to show that at present there have been already developed the approaches for attacking many problems. This makes easier to bring them up to the level of commercial technologies.

References

Note: Foreign patents are given in the entry "Patents".

Aglaumov S.N., Skoupov V.D (1993): Gettering impurities in the hydrostatically–compressed silicon monocrystals, Vysokotchistye veshestva (Highly Pure Substances), Moscow, No. 6, pp. 137–140 (In Russian)

Agranat B.A., Bashkirov V.I., Kitaigorodsky Yu.I. (1970): The ultrasonic cleaning. In: Physics and Techniques of Powerful Ultrasound, Vol. 3 "Physical fundamentals for the ultrasound technology", ed. by L.D. Rozenberg, Nauka (Science), Moscow, 688p. (In Russian)

Alekhin V.P. (1983): Physics of Strength and Elasticity of Surface Layers in Materials, Nauka (Science), Moscow, 280p. (In Russian)

Alekhin V.P., Litvinov Yu.M., Moiseenko N.F. (1984): Inducing the point defects by chemicomechanical polishing of silicon and their annihilation by low–temperature annealing, Abstracts of reports at the 4-th All-Russia conference "Structural defects in semiconductors", Novosibirsk State University, Novosibirsk, part 1, pp. 14–15 (In Russian)

Amal'skaya R.M., Bagraev N.T., Klyatchkin L.E. et al. (1992): Gettering in silicon during the process of vacancy generation, Fizika i tekhnika polouprovodnikov (FTP) (Physics and Techniques of Semiconductors), Moscow, Vol. 26, issue 6, pp. 1004–1007 (In Russian)

Bakchadyrkchanov M.K., Zainobodinov S., Teshabaev A.T. et al. (1976): Impact of thermal treatment on interaction of nickel atoms with oxygen in silicon, Fizika Tverdych Poluprovodnikov (FTP) (Physics of Solid Semiconductors), Moscow, Vol. 10, issue 5, pp. 1001–1002 (In Russian)

Baldi L., Cerofolini G. F., Ferla G., Frigerio G. (1978): Gold solubility in silicon and gettering by phosphorus, Physica Status Solidi, (a), Vol. 48, No.2, pp. 523–532

Banishev A.F., Novikova L.V. (1992): Reversable and irreversible structural defects of the silicon surface induced by laser impulses, Fizitcheskaya i Khimitcheskaya Obrabotki Materialov (Physical and Chemical Treatments of Materials), Moscow, No. 4, pp. 55–58 (In Russian)

Baranov Yu. L. (1989): The SOD-technologies: the state of the art and perspectives, Zaroubezhnaya Elektronnaya Tekchnika (Foreign Electronic Engineering), Moscow, No. 11 (342), pp. 19–33 (In Russian)

Baransky P.I., Klotchkov V.P., Potykevitch I.V. (1975): Semiconducting Electronics. Reference book, Naukova doumka (Scientific Thought), Kiev, 704p. (In Russian)

Barch G. R., Chang Z. P. (1967): A diabatic, isothermal and intermediate pressure berivatues of the elastic constants for cubic simmetry // Physica Status Solidi (a), Vol. 19, No. 1, pp. 139–151

Barton V., Kabrera N., Frank F. (1984): Elementary processes of growing crystals, Energoatomizdat, Moscow, 123 p. (In Russian)

Batavin V.V. (1970a): Decomposition of the oversaturated hard solution of oxygen in dislocation – free silicon, Kristallographiya (Crystallography), Moscow, Vol. 15, issue 1, pp. 125–130 (In Russian)

Batavin V.V. (1970): Effects of the SiO_2 particles on the volt-ampere characteristics of the $p - n -$ junctions in silicon, Fizika i Tekhnika Polou–provodnikov (FTP) (Physics and Techniques of Semiconductors), Moscow, Vol. 4, No. 4, pp. 760–763 (In Russian)

Batavin V.V., Drouzhkov A.P., Garnak A.E. (1980): Studying the process–induced defects in silicon epitaxial structures by positron annihilation, Mikroelektronika (Microelectronics), Moscow, Vol. 9, issue 3, pp. 282–284 (In Russian)

Bazhinov A.N., Nickoulov V.V. et al. (1986): Oxygen–and carbon–induced microdefects in silicon epitaxial layers, Elektronnaya Tekhnika (Electronic Engineering), Ser.: SHF Electronics, Moscow, issue 9 (393), pp. 57–60 (In Russian)

Bedny B.I., Yershov S.N., Panteleyev V.A. (1985): Long–range effect during a mechanical treatment of gallium arsenide, Fizika i Tekhnika Polouprovodnikov (FTP) (Physics and Techniques of Semiconductors), Vol. 19, issue 10, pp. 1806–1809 (In Russian)

Beresnev B.I., Martynov E.D., Rodionov K.P. et al (1970): Plasticity and strength of solids under high pressures, Nauka (Science), Moscow, 162p. (In Russian)

Bethe. H.A. (1964): Intermediate Quantum Mechanics, W.A. Benjamin, Inc., New York–Amsterdam

Bevzouck N.S. (1971): A marginal problem of diffusing moisture in the pencilbox–type microcircuits, Voprosy Mikroelektroniky (Problems of microelectronics), Naukova doumka (Scientific Thought), Kiev, pp. 154–160 (In Russian)

Bir G.L., Pikus G.E. (1972): Symmetry and Deformational Effects in Semiconductors, Nauka (Science), Moscow, 167p. (In Russian)

Biryukov V.M., Zotov V.V., Vikhlyantzev O.F. et al. (1976): Effects of carbon on the $p - n -$ junction leakage currents in silicon epitaxial structures, Elekntonnaya Tekhnika (Electronic Engineering), Ser.: Materials, Moscow, issue 5 (98), pp. 23–29 (In Russian)

Bobitsky Ya.V., Bertcha A.I., Demitrouk N.L. et al. (1992): Studying the laser gettering in $GaAs$ by photoluminescence and a disturbed complete internal reflection, Fizika Tverdykh Polouprovodnikov (FTP) (Physics of Solid Semiconductors), Vol. 26, issue 10, pp. 1688–1692 (In Russian)

Bobrova E.A., Galkin G.N., Yenisherlova K.L. (1991): Studying the efficiency of some gettering techniques in silicon with a use of deep layer relaxation spectroscopy, Mikroelektronika (Microelectronics), Moscow, Vol. 20, issue 2,

pp. 124–130 (In Russian)

Bogatch N.V., Gousev V.A., Litovtchenko P.G. (1981): Gettering the stacking faults and the heavy metal impurities in silicon, Polouprovodnikovaya Tekhnika i Mikroelektronika (Semiconducting Engineering and Micro-electronics), Moscow, issue 34, pp. 3–20 (In Russian)

Bogatch N.V., Starkov V.V. (1991): A model for plasmic gettering the fast–diffusing impurities from silicon, Poverkchnost'. Fizika, khimiya, mekhanika (Surface. Physics, chemistry, mechanics), Moscow, No. 6, pp. 100–106 (In Russian)

Boiko V.S., Garber R.I., Krivenko L.F. et al. (1973): Transient sound emission by dislocations, Fizika Tverdogo Tela (FTT) (Physics of a solid body), Vol. 15, issue 1, pp. 321–323 (In Russian)

Bondarenko V.P., Borisenko V.E., Gorskaya L.V. (1984): Redistribution of gold in monosilicon on the interface with porous silicon during a second annealing by incoherent light beams, Zhournal Teoretitcheskoy Fiziki (ZhTF) (Journal of Theoretical Physics), Vol. 54, issue 10, pp. 2021–2026 (In Russian)

Bondarenko V.P., Dorofeev A.M., Taboulina L.V. (1985): Effects of anodization regimes and high–temperature annealing on the specific surface of porous silicon, Poverkchnost'. Fizika, Khimiya, Mekhanika (Surface. Physics, Chemistry, Mechanics), No. 10, pp. 64–69 (In Russian)

Bondarenko V.P., Yakovtzeva V.A., Dolgy L.N. et al. (1994): SOI-structures on the basis of oxidized porous silicon, Mikroelektronika (Microelectronics), Moscow, Vol. 23, issue 6, pp. 61–68 (In Russian)

Bookker G. R., Stickler R. (1965): Two-dimensional defects in silicon after annealing in wet oxygen, Phil. Mag., Vol. 11, pp. 1303–1308

Bourago N.N., Skoupov V.D., Skoupova T.N. et al. (1985): Effects of hydrostatical pressure on structural perfection of epitaxial silicon , Fizika i Tekhnika Vysokych Davleny (Physics and Techniques of High Pressures), Moscow, No. 20, pp. 30–35 (In Russian)

Bourago N.N., Perevostchikov V.A., Skoupov V.D., Tsypkin G.A. (1985): A way of processing the semiconducting slices, USSR patent No. 1299402, IPC: HO1L21/324 (In Russian)

Bourago N.N. Perevostchikov V.A., Skoupov V.D., Tsypkin G.A. (1987): To a nature of irreversible changes in hydrostatically–compressed crystals, Fizika i Tekhnika Vysokych Davleny (Physics and Techniques of High Pressures), Moscow, No. 24, pp. 38–41 (In Russian)

Bourmistrov A.N., Pekarev A.I., Ushakov A.S., Tchistyakov Yu. D. (1978): Effects of mechanical treatment of the silicon substrate back side on the density of defects, Aktivirouemye Protzessy Tekhnologii Mikroelektroniki (Collection:Activated Processes in the Technology of Microelectronics), Taganrog Radio-Engineering Institute after V.D. Kalmykova, Taganrog, issue 3, pp. 91–102 (In Russian)

Boushouev V.A., Petrakov A.P. (1992): Studying by the three–crystal X–ray diffractometry the structure of monocrystalline silicon presurface layers irradiated by millisecond laser impulses, Poverkchnost'. Fizika, khimiya, mekhanika (Surface. Physics, chemistry, mechanics), Moscow, No. 9,

pp. 64–70 (In Russian)

Bruel Michel, Aspar Bernard, Auberton H.A.J. (1997): Smart Cut: A new silicon on insulator material technology based on hydrogen implantation and wafer bonding, Japan, J. Appl. Phys, Part 1, Vol. 36, No. 36, pp. 1636–1641

Buck T. M., Pickar K. A., Poate J.M. et. al. (1972): Gettering rates of various fast-diffusing metal impurities at ion-damaged layers on silicon, Appl. Phys. Lett., Vol. 21, No.10, pp. 485–488

Bullough R., Newman R. (1970): Kinetics of Point Defect Migration to Dislocations, Rep. Progr. Phys., Vol. 33, No. 2, p. 101

Colinge J.P. (1998): Silicon-on-insulator technology: past achievements and future prospects, MRS Bull., Vol. 23, No. 12, pp. 16–19

Craven R. A., Korb H. W. (1981): Internal gettering in silicon, Sol. St. Techno–logies, No.7, pp. 55–61

Damask A.C., Dienes G.J. (1963): Point Defects in Metals, Gorder and Breach Science Publishers, New York, London

Davydov A.A., Kalinitchenko A.I. (1982): Concerning the mechanical effects near the thermal peaks and tracks of pressure fragments, Atomnaya energiya (Atomic Energy), Vol. 53, issue 3, pp. 186–187 (In Russian)

Dornberger E., Ammon W. (1996): J. Electrochem. Soc., Vol. 143, pp. 1636–1641

Demidov E.S., Latysheva N.D., Perevostchickov V.A. et al. (2000): Effects of ionic irradiation on the microdefect structure of silicon crystals, Nonorganic Materials, Moscow, Vol. 36, No. 5, pp. 522–525 (In Russian)

Dougaev V.E., Ivanotchko M.N., Kosyatchenko S.V. (1987): A model for gettering by a damaged layer the fast–diffusing impurities in the semiconducting wafers, deposited in the Res. Inst. for Sci. and Eng. Information (Ukraine), No. 1915–pt 87, 23p. (In Russian)

Dumin D.J., Robinson P.H. (1969): Electrically and optically active defects in the silicon-on-sapphire films, Z. Crystal Growth, Vol. 3, No. 4, pp. 214–218

Dzafarov T.D. (1991): Irradiation–Stimulated Diffusion in Semiconductors, Energoatomizdat (Atomic Energy Publishing House), Moscow, 288p. (In Russian)

Edel'man F.L. (1980): Structure of large IC components, Nauka (Science), Novosibirsk, 225p. (In Russian)

Einspruch Norman G., ed. (1985): GaAs Microelectronics, Vol. 11. In: VLSI Electronics. Microstructure Science, Academic Press, Inc. (Harcourt Brace Jovanovich, Publishers), New York–London

Einspruch N.G., Wisseman W.R. (eds) (1985): Vol 11. GaAs Microelectronics. In: VLSI Electronics. Microstructure Science, Academic Press, Inc. (Harcourt Brace Jovanovich, Publishers), New York, London, Toronto, Sydney, Montreal

Fistoul' V.I. (1977): Decomposition of Oversaturated Semiconducting Hard Solutions, Metallurgia (Mettalurgy), Moscow, 240p. (In Russian)

Fistoul' V.I., Petrovsky V.I., Rytova N.S. et al (1979): Spatial evaporation of crystals, Fizika i Tekhnika Polouprovodnikov (FTP) (Physics and Techniques of Semiconductors), Moscow, No. 7, pp. 1402–1408 (In Russian)

Fistoul' V.I., Tchernova A.I., Yakimova E.E. (1980): Migrating of sodium in silicon, Fizika i Tekhnika Polouprovodnikov (FTP) (Physics and Techniques of Semiconductors), Vol. 14, issue 5, pp. 990–993 (In Russian)

Fistoul' V.I., Sinder M.I. (1981): Vacancy porousness in the presurface region of semiconductors, Fizika i Tekhnika Poluprovodnikov (Physics and Techniques of Semiconductors), Moscow, Vol. 15, issue 6, pp. 1182–1186 (In Russian)

Fistoul' V.I. (1995): New Materials (State of the Art, Problems and Perspectives), textbook for higher education establishments, Institute of Steel and Alloys, Moscow, 142p (In Russian)

Friedel J. (1964): Dislocations, Pergamon Press

Frolov O.S., Zaitzevskaya T.Yu., Kononenko Yu.G. et al. (1986): Studying an efficiency of internal gettering in CZ-grown silicon, Elektronnaya Tekhnika (Electronic Engineering), Ser. 2: Semiconducting Devices, issue 6 (185), pp. 49–54 (In Russian)

Gaidoukov G.N., Kozhevnikov E.A. (1995): Physical model of internal gettering in the silicon technology, Elektronnaya Promyshlennost' (Electronic Industry), Nos. 4–5, pp. 59–61 (In Russian)

Gastev V.V., Mikhailov O.G., Soukhoroukov O.G. et al. (1984): The Effects of the rate of growth in SOS structures on the "silicon–sapphire" interface, Elektronnaya Tekhnika (Electronic Engineering), Moscow, Ser. 6: Materials, issue 7, pp. 31–36 (In Russian)

Gastev V.V., Soukhoroukov O.G., Lotzman A.P. (1985): The Effects of High–Temperature Treatment on the state of the "silicon–sapphire" interface in SOS–structures, Elektronnaya Tekhnika (Electronic Engineering). Moscow, Ser. 3: Microelectronics, issue 1, pp. 95–99 (In Russian)

Gastev V.V., Soukhoroukov O.G., Strizhkov B.V. (1988): Specificity of the transient layer "epitaxial silicon–sapphire", Elektronnaya Tekhnika (Electronic Engineering). Ser. 6: Materials, issue 4(233), pp. 28–31 (In Russian)

Geipl H. S., Tice W. K. (1977): Critical microstructure for ion implantation gettering effect in silicon, Appl. Phys. Lett., Vol. 30, No.7, pp. 325–327

Gegouzin Ya.E. (1962): Macroscopic defects in metals, Moscow, Metallurgizdat (Mettalurgy Publishing House), pp. 151–159 (In Russian)

Gegouzin Ya.E., Kononenko V.G. (1982): Diffusive–dislocational mechanism for curing the isolated pairs, Fizika i khimiya obrabotki materialov (Physics and chemistry of material processing), No. 2, pp. 60–75 (In Russian)

Gegouzin Ya.E., Kaganovsky Yu. S. (1984): Diffusive processes on the silicon surface, Moscow, Energoatomizdat (Atomic Energy Publishing House), 123p. (In Russian)

Gerasimenko N.N., Dvouretchensky A.V., Smirnov L.S. (1972): Surface paramagnetic centres on silicon, Fizika i Tekhnika Polouprovodnikov (FTP) (Physics and Techniques of Semiconductors), Moscow, Vol. 6, pp. 987–989 (In Russian)

Gerasimenko N.N., Dvouretchensky A.V. (1975): Paramagnetic centres on the surface of semiconductors, Elementary physicochemical processes on the surface of monocrystalline semiconductors, Siberian Branch of the USSR Academy of Sciences, Nauka (Science), Novosibirsk, pp. 73–82 (In Russian)

Gerasimenko N.N., Mordkovitch V.N. (1987): The irradiating effects in the semiconductor–dielectric system, Poverkchnost'. Fizika, khimiya, mekhanika (Surface. Physics, chemistry, mechanics), Moscow, No. 6, pp. 5–19 (In Russian)

Gerward L. (1973): X-ray study of lateral strains in ion-implanted silicon, Z. Physik, Vol. 259, No. 4, pp. 313–322

Glaenzer R.H., Yordn A.G. (1968): The effects of contamination on the electrical propertie of edge dislocation in silicon, Phil. Mag., Vol. 18, No. 154, p. 717

Goetsberger A., Shockley W. (1960): Metal Precipitates in Si $p-n-$junctions, J. of Appl. Phys., Vol. 31, No.10, pp. 1821–1824

Gorelyenok A.T., Kryukov V.L., Fourmanov G.P. (1994): Gettering the impurities and defects in Si, $GaAs$ and $InSb$, Pis'ma v Zhournal Teoretitcheskoy Fiziki (ZhTF) (Letters to Journal of Theoretical Physics), Moscow, Vol. 20, issue 13, pp. 60–65 (In Russian)

Gorshkov O.N., Perevostchikov V.A., Skoupov V.D. (1989): Elastic waves in silicon crystals induced by surface chemical etching, Poverkchnost'. Fizika, khimiya, mekhanika (Surface. Physics, Chemistry, Mechanics), Moscow, No. 7, pp. 155–158 (In Russian)

Gousev V.A., Bogatch N.V., Kamensky V.L. (1984): A redistribution of point defects in silicon wafers when gettering by a layer, Dielektriki i polouprovodniki (Dielectrics and semiconductors), Moscow, No. 25, pp. 88–93 (In Russian)

Gousev V.A., Bogatch N.V., Kamensky V.L. (1990): Changing the recombination properties of planar doped regions when gettering by a layer, Optoelektronika i Polouprovodnikovaya Tekhnika (Optoelectronics and Semiconducting Engineering), Moscow, issue 18, pp. 1–8 (In Russian)

Graff K., Fisher G. (1982): Lifetime of carriers in silicon and its effects on the performance of solar elements. In: Transformation of Solar Energy: Problems of Physics of Solids, Energoizdat, Moscow, pp. 151–189 (In Russian)

Grishin V.P., Goulyaeva A.S., Lainer L.V. (1982): Thermal stability of silicon grown by a float–zone melting in the argon atmosphere, The Techiques for Producing and Studying Silicon Monocrystals: Sci. report, Res. Inst. GIREDMET, Moscow, Vol. 110, issue 2, pp. 80–86 (In Russian)

Haller E.E., Hubbard G.S., Hansen W.L., Seeger A. (1977): In: Radiation Effects in Semiconductors, Proc. First Phys. Conf. (1976), Ser. 31, eds. N.B. Urli, Z.W. Corbett, Bristol and London, pp. 309–314.

Hannay N.B. (ed) (1960): Semiconductors, American Chemical Society Monograph Series, Reinhold Publishing Corporation, New York; Chapman & Hall, Lев, London

Heiman A.P. (1967): Donor surface states and bulk acceptor traps in silicon-on-sapphire films, Appl. Phys. Lett. Vol. 11, No. 4, pp. 132–134

Hirth J.P., Lothe J. (1970): Theory of Dislocations, McGraw-Hill Book Co.,
New York–London

Hu S.M. (1973): Diffusion in Silicon and Germanium. In: Atomic Diffusion
in Semiconductors, ed. by D. Shaw, Plenum Press, London and New York,
pp. 248–405

Hu S. M. (1974): Formation of stacking faults and enhanced diffusion
in the oxidation of silicon, J. Appl. Phys, Vol. 45, No. 4, pp. 1567–1573

Indenbom V.L., Orlov A.N. (1977): Longevity of materials under loads
and aggregation of distortions, Fizika Metallov in Metallovedeniye (FMM)
(Physics of Metals and Science of Metals), Moscow, Vol. 43, issue 34,
pp. 469–492 (In Russian)

Indenbom V.L. (1979): A new hypothesis about a mechanism of irradiation–
stimulated processes, Pis'ma v Zhournal Teoretitcheskoy Fiziki (ZhTF) (Letters
to Journal of Theoretical Physics), Moscow, Vol. 5, issue 8, pp. 489–492
(In Russian)

Italyantsev A.G., Mordkovitch V.N. (1983): Transformation of sizes in clusters
of intrinsic point defects in semiconductors, Fizika i Tekhnika
Polouprovodnikov (FTP) (Physics and Techniques of Semiconductors),
Moscow, Vol. 17, issue 2, pp. 217–222 (In Russian)

Italyantsev A.G., Mityukchlyev V.B., Pashenko P.B., Fifer V.N. (1988): Effects
of chemically stimulated injection of vacancies into GaAs on the properties of
layers ionically doped by silicon, Poverkchnost'. Fizika, khimiya, mekhanika
(Surface. Physics, chemistry, mechanics), Moscow, No. 11,
pp. 93–97 (In Russian)

Italyantsev A.G. (1991): Generation of vacancies stimulated by chemical eatching
of crystal surface, Poverkchnost'. Fizika, khimiya, mekhanika (Surface.
Physics, chemistry, mechanics), Moscow, No. 10, pp. 122–127 (In Russian)

Ivanovsky G.F., Petrov V.I. (1986): Ionic–plasmic treatment of materials,
Radio i Svyaz' (Radio and Communication), Moscow, 232p. (In Russian)

Iyer K. R., Kuczynswi G.C. (1971): Effect of the wavelength
of illumination on the photomechanical effect in Germanium.
J. Appl. Physics, Vol. 42, No. 1, pp. 486–487

Jowett C. E. (1979): Failure mechanisms and analysis of procedures
for semiconductor devices, Microelectronics Journal, Vol. 9, No. 3, p. 5

Jung J., Saynovsky W. et al. (1979): High pressure science and technology, Prac.
7-th Int AITRAPT Conf. Le Creusof. No. 1, Oxford, 1980, p. 167

Kaganova I.N., Kaganov M.I. (1973): To the theory of sound generation
by charged particles. Exciting of sound by a Θ – burst, Fizika Tverdogo Tela
(FTT) (Physics of a solid body), Vol. 15, issue 5, pp. 1536–1543 (In Russian)

Kalinin A.A., Kolobova G.A., Mil'vidsky M.G. et al. (1990): Structural
specificities of GaAs silicon–doped monocrystals, Proc. VII-th Intern. Conf.
On Microelectronics, Minsk, pp. 11–12 (In Russian)

Kalinitchenko A.I., Lasourok–El'tzoufin V.T. (1973): Exciting the acoustic
oscillations by a beam of small–density charged particles, Vol. 65, issue 6,
pp. 2364–2368 (In Russia)

Kantchyukovsky O.P., Presnov V.A., Shenkevitch A.L. (1978): Effects of axis pressure on the substance, Naukova doumka (Scientific Thought), Kiev, pp. 58–61 (In Russian)

Kantchyukovsky O.P., Moroz L.V., Presnov V.A. et al. (1980): Structure and properties of silicon compressed by axial pressure, In.: High Pressures and Properties of Materials, Naukova doumka (Scientific Thought), Kiev, pp. 36–38 (In Russian)

Kapoustin Yu. A., Kolokol'nikov B.M., Rovinsky A.P., Chernikov A.M. (1992): The efficient lifetime of unbalanced charge carriers and the electrical properties of the defects induced by mechanical treatment of silicon surface, Poverkchnost'. Fizika, khimiya, mekhanika (Surface. Physics, chemistry, mechanics), Moscow, No. 5, pp. 69–74 (In Russian)

Karban' V.I., Koi P., Rogov V.V. et al. (1982): Processing of Semiconducting Materials, ed. by N.V. Novikov, V. Bertol'di, Naukova Doumka (Scientific Thought), Kiev, 256p. (In Russian)

Karban' V.I., Borzakov Yu.I. (1988): Processing the monocrystals in microelectronics, Radio i Svyaz' (Radio and Communication), Moscow, 104p. (In Russian)

Karzanov V.V., Perevostchikov V.A., Skoupov V.D. (1994): Effects of surface chemical eatching on the Hall parameters of gallium arsenide crystals, Vestnik Nizhegorodskogo Gosudarstvennogo Universiteta: Materiali, prozessi i tekhnologii elektronnoi tekhniki (Reports of Nizhny Novgorod State University: Materials, processes and technologies of electronic devices), Nizhny Novgorod State University after N.I. Lobatchevsky, Nizhny Novgorod, pp. 31–35 (In Russian)

Kasatkin A.P., Perevostchikov V.A., Skoupov V.D., Souslov L.A. (1993): Effects of chemicomechanical and chemicodynamical polishing of the surface on deep centres in n–GaAs , Poverkchnost'. Fizika, khimiya, mekhanika (Surface. Physics, chemistry, mechanics), Moscow, No. 6, pp. 79–84 (In Russian)

Katsutoshi, Izumi (1998): History of SIMOX material, MRS Bulletin, Vol. 23, No. 12

Kazuo, Imai (1981): A new dielectric isolation method using porous silicon, Solid State Electronics, V. 24, No. 2, pp. 159–164

Khokonov Kh.B., Shokarov Kh.B. (1988): Acoustic effect of substance dissolution, Abstracts of reports at 7-th All-Russia symposium on molecular–beam epitaxy, Moscow, Vol. 2, pp. 70–71 (In Russian)

Kikoin I.K. (ed) (1976): Reference. Tables of Physical Magnitudes, Atomizdat, Moscow, 1008p. (In Russian)

Kiselyev V.K., Obolensky S.V., Skoupov V.D. (1999): Effects of internal getter in silicon on the parameters of $Au - Si$ structures, Zhournal Teoretitcheskoy Fiziki (ZhTF) (Journal of Theoretical Physics), Vol. 69, issue 6, pp. 129–131 (In Russian)

Kiselyev A.N., Levshounova V.A., Maximov P.V., et al. (2002): Microdefects in ultrasonically irradiated silicon crystals, Peterburgsky zhournal elektroniky (Petersburg Journal of Electronics), No. 4, pp. 10–15 (In Russian)

Kiselyev A.N., Perevostchikov V.A., Maximov V. et al. (2002): Microdefectness

In silicon crystals upon ultrasonic treatment, Materialy Elektronnoi Tekhniki (Materials of Electronic Engineering), St. Petersburg, No. 4, pp. 10–15 (In Russian)

Kiselyev A.N., Levshounova V.A., Perevostchikov V.A et al. (2003): Specificity of changes in microdefectstructure of silicon when ionically irradiating a free and $a - Si$ −protected surfaces, Poverkchnost'. X-ray, synchrotron and neutron studies, No. 4, pp. 85–91 (In Russian)

Kiv A.E., Solovjev V.N. (1980): Extrinsic aggregates are generators of defects, Fizika Tverdogo Tela (FTT) (Physics of a solid body), Vol. 22, issue 9, pp. 2575–2577 (In Russian)

Kobazeva Z.N., Skvortzov I.M. (1986): Increasing the gettering efficiency by a damaged layer through determining an optimal time for alkali etching, Elektronnaya Tekhnika (Electronic Engineering), Ser. 2: Semiconducting Devices, issue 1 (180), pp. 63–66 (In Russian)

Kock A.J.R. de (1973): Microdefects in dislocation-free silicon crystals, Philips Res. Rept. Suppl., Vol. 26, No. 1, pp. 3–105

Kock A. J. R. de (1977): Microdefects in dislocation–free silicon crystals, Philips Res. Reports Supp., Vol. 26, No. 1, p. 105

Kock A. J. R. de (1981): Defects in Semiconductors, ed. by J. Narayan, T.Y. Tan, Oxford, New York, pp. 309–316

Kőgler R., Peeva A., Lebedev A. et al. (2003): Cu-gettering in ion implanted and annealed silicon in regions before and beyond the mean projected ion range, J. Appl. Phys., Vol. 94, No, 6, pp. 3834–3839

Konakova R.V., Shouman V.B. (1970): Behaviour of recombination centres in silicon during a thermal treatment, Elektronnaya Tekhnika (Electronic Engineering), Ser. 2: Semiconducting Devices, issue 5 (55), pp. 66–69 (In Russian)

Kontzevoy Yu. A. (1970): On the recombination in the semiconductors with macrodefects, Fizika Tverdych Poluprovodnikov (FTP) (Physics of Solid Semiconductors), Vol. 4, No. 6, pp. 1184–1187 (In Russian)

Kontzevoy Yu. A., Litvinov Yu.M., Fattakhov E.A. (1982): Elasticity and Strength of Semiconducting Materials and Structures, Radio i svyaz' (Radio and communication), Moscow, 240 p. (In Russian)

Kontzevoy Yu.A., Filatov D.K. (1987): Reference Sources. Defects in Silicon Structures and Devices. Parts 1 and 2. Central Research Inststitute "Electronika" (Electronics) (In Russian)

Kormishina Zh. A., Perevostchikov V.A., Skoupov V.D. et al. (1997): Structure of the device layers in the SOI–compositions produced by thermocompressive mating, deposited in the All-Russia Institute for Scientific and Engineering Information (VINITI), Moscow, 9-th September 1997, No. 3593–B.97, 11p. (In Russian)

Kosevitch Yu.A. (1981): Step formation and interactions of defects on a crystal surface, Zhournal Eksperimentalnoy i Tekhnitcheskoy Fiziki (ZhETF) (Journal of Experimental and Engineering Physics), Vol. 81, No. 6, pp. 2247–2252 (In Russian)

Koshelev N.I., Yermolaev A.I. (1994): Forming the SOI–structures by mating the silicon wafers through a glass-like dielectric layer, Mikroelektronika (Microelectronics), Vol. 23, issue 6, pp. 55–60 (In Russian)

Kotov A.G., Gromov V.V. (1988): Irradiation Physics of Heterogeneous Systems, Energiya (Energy), Moscow, 232p. (In Russian)

Koulemin A.V. (1978): Ultrasound and Diffusion in Metals, Metallurgia (Mettalurgy), Moscow, 200p. (In Russian)

Koulikov A.V., Perevostchickov V.A., Skoupov V.D., Shengourov V.G. (1997): Low–temperature reaction–stimulated gettering of impurities and defects in silicon by porous silicon layers, Pis'ma v Zhournal Teoretitcheskoy Fiziki (Letters to Journal of Theoretical Physics), Vol. 23, No. 13, pp. 27–31 (In Russian)

Koutchyukov E.G., Shapovalov V.P., Stadnyuk G.A. (1979): Effects of hydrostatic pressure on the parameters of p–n–junctions, Fizika i Tekhnika Polouprovodnikov (FTP) (Physics and Techniques of Semiconductors), Vol. 13, issue 10, pp. 2012–2014 (In Russian)

Kozlovscky V.V., Kozlov V.A., Pomasov V.N. (2000): Semiconductor modification by protonic beams, Fizika i Tekhnika Polouprovodnikov (Physics and Techniques of Semiconductors), Vol. 34, issue 2, pp. 129–147 (In Russian)

Kravtchenko V.M., Boud'ko M.S. (1989): The SOD-technology: the state of the art, Zaroubezhnaya elektronnaya tekchnika (Foreign electronic engineering), No. 9 (340), pp. 3–54 (In Russian)

Krymko M.M., Yenisherlova K.L., Koshelev N.I. et al. (1998): Multilayer silicon structures for powerful electronics and microelectronics, Izvestiya Vouzov. Elektronika. (Reports of Higher Education Institutes. Electronics), No. 5, pp. 45–51 (In Russian)

Kubena J., Hlavka J. (1985): Space correlation of microdefects with recombination of excess carriers in CZ–Si , Phys. State Sol. (a), Vol. 89, No. 1, pp. K 23–25

Kuhl Ch., Schlotterer H., Schwidefsky F. (1976): An optically effective intermediate layer between epitaxial silicon and spinel or sapphire, J. Electrochem. Soc., Vol. 123, No. 1, pp. 97–100

Kveder V., Kittler M., Schröter W. (2001): Recombination activity of contaminated dislocations in silicon: A model describing electron-beam-induced current contrast behavior, Phys. Rev. B 63, paper 115208

Labounov V.A., Baranov I.L., Bondarenko V.P., Dorofeyev A.M. (1983): Contemporary gettering techniques in the semiconducting electronics, Zaroubezhnaya Elektronnaya Tekhnika (Foreign Electronic Engineering), No. 11, pp. 3–49 (In Russian)

Labounov V.A., Bondarenko V.P., Borisenko V.E. (1983): Porous silicon in semiconducting electronics, Zaroubezhnaya Elektronnaya Tekhnika (Foreign Electronic Engineering), No. 15, pp. 3–46 (In Russian)

Lagowski J., Jastrzebski L., Cullen G.W. (1983): Optical probing of silicon–sapphire interface of heteroepitaxial SOS films, J. Electrochem Soc., Vol. 130, No. 8, pp. 1744–1748

Lampert Murray A., Mark Peter (1970): Current Injection in Solids. Academic

Press, New York and London

Landau L.D., Lifshitz E.M. (1953): Mechanics of environments, Moscow, 788 p. (In Russian)

Lannoo M., Bourgoin J. (1981): Point Defects in Semiconductors. I Theoretical Aspects, Springer–Verlag

Laurence J. E. (1969): Stacking faults in annealed silicon surfaces, J. Appl. Phys., Vol. 40, No.1, pp. 360–365

Lefevre H. (1982): Annealing behavior of trap–centers in silicon containing A–swirl defects, Appl. Phys. A, Vol. 29, No. 2, pp. 105–111

Leikin V.N., Zelenov V.I., Mingazin T.A. (1978): Dislocations and their impact on electrophysical parameters of semiconducting devices, Obzory po Elektronnoy Tekhnike (Reviews on Electronic Engineering), Ser. 2: Semiconducting Devices, issue 11 (576), 64p. (In Russian)

Leroy B. (1979): Kinetics of growth of the oxidation stacking faults, J. Appl Phys., Vol. 50, No. 12, pp. 7996–8005

Liblich S., Nassibian A.G. (1973): Effects of diffused nickel on the $Si - SiO_2$ interface, Sol. St. Electron., Vol. 16, No. 12, pp. 1495–1499

Litovtchenko V.G., Romanyuk B.N. (1983): Effect of anisotropic gettering in planar structures, Fizika i Tekhnika Polouprovodnikov (FTP) (Physics and Techniques of Semiconductors), Vol. 17, issue 1, pp. 150–153 (In Russian)

Litovtchenko N.M., Troshin A.L. et al. (1986): The role of swirls in determining the recombination inhomogeneity of the thermally treated silicon, Optoelektronika and Polouprovodnikovaya Tekhnika (Optoelectronics and Semiconducting Engineering), Moscow, No. 10, pp. 69–72 (In Russian)

Litovtchenko V.G., Romanyuk B.N. et al. (1986): Studying the silicon epitaxial structures with buried oxidized and nitrided isolating layers, Optoelektronika and Polouprovodnikovaya Tekhnika (Optoelectronics and Semiconducting engineering), Moscow, No. 10, pp. 58–66 (In Russian)

Litovtchenko V.G., Romanyuk B.N., Shapovalov V.P. et al. (1986a): Effect of planar gettering when growing the silicon films epitaxially, Optoelektronika and Polouprovodnikovaya Tekhnika (Optoelectronics and Semiconducting Engineering), Moscow, issue 10, pp. 84–94 (In Russian)

Litvinenko N.M., Troshin A.L., Sal'nik Z.A. et al. (1986a): The role of swirls in determining a recombinational inhomogeneity in thermally–treated silicon, Optoelektronika i Polouprovodnikovaya Tekhnika (Optoelectronics and Semiconducting Engineering), Moscow, No. 10, pp. 69–72 (In Russian)

Louft B.D., Perevostchickov V.A., et al (1982): Physicochemical Techniques for Processing the Semiconductor Surfaces, ed. by B.D. Louft, Radio i Svyaz' (Radio and communication), Moscow, 136p. (In Russian)

Lourye M.S., Sorokina O.I., Zolotareva T.V. et al. (1991): A gettering effect on the periphery of a slice treated by coherent irradiation, Elektronnaya Tekhnika (Electronic Engineering), Moscow, Ser. 3: Microelectronics, issue 5 (144), pp. 42–43 (In Russian)

Lyubov B.Ya. (1985): Diffusive Changes in the Defective Structure of Solids,

Metallurgia (Mettalurgy), Moscow, 207p. (In Russian)

Lysenko V.S., Nazarov A.N., Roudenko T.E. et al. (1994): Properties
of SOI-structures obtained by laser zone recrystalization of polysilicon
on multiplayer dielectrics, Mikroelektronika (Microelectronics), Moscow,
Vol. 23, issue 6, pp. 32–38 (In Russian)

Mal'tzev P.P., Tchaplygin Yu.A., Timoshenkov S.P. (1998): Perspectives
of the SOI–technology, Izvestiya Vouzov. Elektronika. (Reports of Higher
Education Institutes. Electronics), Moscow, No. 5, pp. 5–9 (In Russian)

Markov A.V., Grishina S.P., Mil'vidsky M.G., Shifrin S.S. (1984): Formation
of aggregates and stability of electrophysical properties of monocrystals
of semiisolating GaAs, Fizika Tverdych Poluprovodnikov (FTP) (Physics
of Solid Semiconductors), Moscow, Vol. 18, issue 3, pp. 465–470
(In Russian)

Marousyak V.I., Kosyatchenko S.V. (1989): Modelling the gettering process
of fast–diffusing impurities in a semiconductor slice, Elektronnaya Tekhnika
(Electronic Engineering), Moscow, Ser. 6: Materials, issue 2 (239), pp. 75–77
(In Russian)

Martynenko Yu. V. (1993): The effects of long-range actions in ionic
implantation, the Collection "Itogi Nauki i Techniki. Pouchki zaryazhennyh
tchastitz i tvergoe telo. Raspolozhenie" (Results of Science and Engineering.
Beams of charged particles and a solid body. Location), Moscow, Russian
Academy of Sciences, Vol. 7, pp. 82–112 (In Russian)

Masters B. J. (1979): Diffusion stimulated by proton irradiation and vacancy
diffusion length in silicon. In: Defects and Radiation Effects in
Semiconductors, ed. J.H. Albany. London, 545p.

Matare G. (1971): Defect Electronics in Semiconductors, Wiley–Interscience.
A Division of John Wiley and Sons, Inc., New York, London, Sydney, Toronto

Matlock J. H. (1977): Semiconductor Silicon. (Eds. H. R. Haff, E. Sirtl),
Electrochem. Soc., New York, pp. 32–52

Maximyuk P.A., Bourbelo M.P, Davidovsky V.M., et al (1977): Effects of natural
anisotropy on the properties of vacancy aggregates in silicon, Ukranian
Physical Journal, Kiev, Vol. 22, No. 5, pp. 717–719 (In Russian)

Mayers M.A., Mur L.E. (eds) (1984): Striking Waves and Phemomena
of High-Speed Deformation of Metals, Metallurgia (Mettalurgy) (Mocow),
512p. (In Russian)

Mc Canghan D.V., Wonsiewicz B.C. (1974): Effects of dislocations
on the properties of metal– SiO_2 –silicon capacitors, J. Appl. Phys., Vol. 45,
No. 11, pp. 4982–4985

Mc Greivy D. (1977): On the origin of leakage currents in silicon-on-sapphire
MOS-transistors, IEEE Trans. Electr. Dev., Vol. ED–24, No. 6, pp. 730–738

Meck R. L., Seidel T. E., Cullis A.G. (1975): Diffusion gettering of Au and Cu
in silicon, J. of Electrochem. Soc., Moscow, Vol. 122, No.6, pp. 786–794
(In Russian)

Milevscky L.S. (1978): Interaction of sliding dislocations with point defects in the crystals of diamand structure. In: Crystalline Structure and Properties of Metallic Alloys, Nauka (Science), Moscow, pp. 161–175 (In Russian)

Milevscky L.S., Vysotzkaya V.V., Sidorov Yu. A. (1980): Dissolving the microdefects in dislocation–free silicon, Fizitcheskaya i Khimitcheskaya Obrabotki Materialov (Physical and Chemical Treatments of Materials), Moscow, No. 1, pp. 153–154 (In Russian)

Milnes A.G. (1973): Deep Impurities in Semiconductors. A Wiley–Interscience Publication. John Willey & Sons, New York, London, Sydney, Toronto

Mil'vidsky M.G., Osvensky V.B. (1984): Structural Defects in Monocrystalline Semiconductors, Metalourgiya (Metallurgy), Moscow, 256 p. (In Russian)

Mil'vidsky M.G., Osvensky V.B. (1985): Structural Defects in Epitaxial Layers of Semiconductors, Metallourgya (Metallurgy), Moscow, 160 p. (In Russian)

Mil'vidsky M.G. (1986): Semiconducting Materials in Contemporary Electronics, Nauka (Science), Moscow, 144p. (In Russian)

Mil'vidsky M.G. (1997): Science of metals, Moscow, issue 5, pp. 36–48 (In Russian)

Mil'vidsky M.G. (1998): Izvestiya Vouzov, Seriya Materialy Elektronnoi Tekhniki (Reports of Higher Education Institutes. Series: Materials of Electronic Devices), Moscow, issue 3, pp. 4–12 (In Russian)

Mil'vidsky M.G. (1999): Semiconducting silicon in the doorway of the 21-st century, Izvestiya Vouzov, Seriya Materialy Elektronnoi Tekhniki (Reports of Higher Education Institutes. Series: Materials of Electronic Devices), Moscow, issue 4, pp. 4–14 (In Russian)

Mineyev V.V. (1982): Static parameters of transistor structures locally doped by gold, Elektronnaya Tekhnika (Electronic Engineering), Moscow, Ser. 3: Microelectronics, issue 1(97), pp. 21–26 (In Russian)

Mishenko K.P., Ravdel' A. (eds) (1974): Short Reference of physical and chemical magnitudes, Khimiya (Chemistry), Leningrad, 200p. (In Russian)

Monkowski J. R. (1981): Gettering processes for defect control, Sol. St. Technology, No.7, pp. 44–51

Morozov N.P., Tetel'baum D.I. (1980): Regularities in defect accumulation in semiconductors irradiated by light ions, Fizika i Tekhnika Polouprovodnikov (FTP)(Physics and Techniques of Semiconductors), Moscow, Vol. 14, issue 5, pp. 934–938 (In Russian)

Morozov N.P., Tetel'baum D.I. (1983):Deep penetration of irradiation–induced defects from the ion–implanted layer into the bulk of the semiconductor, Fizika i Tekhnika Polouprovodnikov (FTP)(Physics and Techniques of Semiconductors), Moscow, Vol. 17, issue 5, pp. 838–842 (In Russian)

Morozov N.P., Skoupov V.D., Tetel'baum D.I. (1985): Defects formation in silicon during the ionic bombardment beyond the region of ion runs, Fizika i Tekhnika Polouprovodnikov (FTP)(Physics and Techniques of Semiconductors), Moscow, Vol. 19, issue 3, pp. 464–468 (In Russian)

Moushnitchenko V.V., Zhourankov L.L. (1990): Studying the process of complete Isolation by porous oxidized silicon, Elektronnaya Tekhnika (Electronic Engineering), Moscow, Ser. 6: Materials, issue 8 (253), pp. 24–28 (In Russian)

Murarka S. P. (1976): A study of phosphorus gettering of gold in silicon by use
of neutron activation analysis, J. Electrochem. Soc.: S-S Sci. and Technologies., Vol. 123, No.5, pp. 765–767

Murarka S. P., Levinstein H. J., Marcus R.B. et al. (1977): Oxidation silicon
no stacking faults generation, J. Appl. Phys., Vol. 48, No.9, pp. 4001–4003

Nagasawa K., Matsushita Y., Kushino S. (1980): A new intrinsic gettering
technique using microdefects in Czochralski silicon crystal: A new double
preannealing technique, Appl. Phys. Lett., Vol. 37, No.7, pp. 622–624

Nemtzev G.Z., Peckarev A.I., Tchistyakov Yu. D., Bourmistrov A.N. (1981):
Geterring the point defects in the manufacture of the semiconducting devices,
Zaroubezhnaya Elektronnaya Tekhnika (Foreign Electronic Engineering),
Moscow, No. 11 (245), pp. 3–63 (In Russian)

Nemtzev G.Z., Pekarev A.I., Tchistyakov Yu. D. (1983): Cleaning silicon
from impurities by the internal getter, Mikroelektronika (Microelectronics),
Moscow, Vol. 12, issue 5, pp. 432–439 (In Russian)

Newman R. C. (1982): Rep. Prog. Phys., Vol. 45, pp. 1163–1210

Nikolaev K.P., Nemirovsky L.N. (1989): Specificity of Production
and Applications of Porous Silicon in Electronic Engineering, Obzory
po Elektronnoy Tekhnike (Reviews on Electronic Engineering), Moscow,
Ser. 2: Semiconducting Devices, issue 9 (1506), 59p. (In Russian)

Obolensky S.V., Skoupov V.D. (2000): The long–range effect in the irradiated
semiconducting structures with internal interfaces, Poverkchnost'. X-ray
structures and neutron studies (Surface. X-ray structures and neutron studies),
No. 5, pp. 75–79 (In Russian)

Oikawa H. (1977): Effect of heat treatment after deposition on internal stress
in molybdenum films on SiO_2/Si substrates, J. Vac. Sci. and Techn. Vol. 14,
No. 5, p. 1153

Okoulitch V.I., Panteleev V.A., Vasin A.S. (1982): Redistribution
of ion–implanted phosphorus in silicon during annealing under all–sided
compression, Fizika i Tekhnika Polouprovodnikov (FTP) (Physics and
Techniques of Semiconductors), Moscow, Vol. 16, issue 8, pp. 1489–1490 (In
Russian)

Osvensky V.B., Proshko G.P., Mil'vidsky M.G. (1967): Effects of dislocations
on the structure of the $p - n -$ junctions in GaAs and on the recombination
irradiation parameters, Fizika i Tekhnika Polouprovodnikov (FTP) (Physics
and Techniques of Semiconductors), Moscow, Vol. 1, No. 6, pp. 911–915
(In Russian)

Palatnik L.S., Tcheremskoy P.G., Foox M.Ya. (1982): Pores in Films,
Energoizdat, Moscow, 216p. (In Russian)

Pall V., Varshower D. (1966): Role of pressure in studying semiconductors,
In: Solids under high pressure, Mir (World), Moscow, pp. 205–283
(In Russian)

Paniotov Yu.N., Toky V.V. (1977): Diffusive mobility of dislocations
in hydrostatically–compressed crystals, In: Effects of High Pressures on the
Substance, Res. Inst. ИСМ of the Ukrainian Academy of Sciences, Kiev,
pp. 23–28 (In Russian)

Papckov V.S., Tzyboul'nickov M.B. (1979): Epitaxial Silicon Layers
  on Dielectric Substrates and the Devices on their Basis, Energya (Energy),
  Moscow, 88p. (In Russian)
Park J.G, Ushio S., Takeno H., et. al. (1994): Electrocem. Soc. Proc., Vol. 94-33,
  pp. 53–57
Park J.G., Rozgonyi G.A. (1996): DRAM wafer qualification issues: oxide
  integrity vs. D-defects, oxygen precipitates and high temperature annealing,
  Solid State Phenomena, Vol. 47-48, pp. 327–353
Patents
      Japan pat. No. 51–34714
      Japan pat. No. 51–35345
      Japan pat. No. 53–1826
      Germany pat. No. 2738195
      Germany pat. No. 2829983
      UK pat. No. 1483888
      UK pat. No. 1501245
      US pat. No. 2573464
      US pat. No. 3905162
      US pat. No. 3923567
      US pat. No. 3929529
      US pat. No. 4018826
      US pat. No. 4029419
      US pat. No. 4053335
      US pat. No. 4131487
      US pat. No. 4144099
      US pat. No. 4177084
      US pat. No. 4244753

Paul W., Warschauer D.M. (1963): The role of pressure in investigation
  of semiconductors. In: Solids under Pressure, McGraw-Hill Book Company
  Inc., New York–London
Pavlov P.V., Pashkov V.I., Genkin V.M. et al. (1973): Changes in the dislocation
  structure during the average energy ion irradiation, Fizika i Tekhnika
  Polouprovodnikov (FTP) (Physics and Techniques of Semiconductors),
  Moscow, Vol. 15, issue 9, pp. 2857–2859 (In Russian)
Pavlov P. V., Syemin Yu. A., Skoupov V. D., Tetel'baum D. I. (1986): Effects
  of elastic ionically–induced waves on the structural perfection of
  semiconducting crystals, Fizika i Tekhnika Polouprovodnikov (FTP) (Physics
  and Techniques of Semiconductors), Moscow, Vol. 20, issue 3, pp. 503–507 (In
  Russian)
Pavlov P. V., Tetel'baum D. I., Skoupov V. D., Syemin Yu. A., Zorina G.V.
  (1986): Abnormally deep structural changes in ion–implanted silicon, Phys.
  Stat. Sol. (A), Moscow, Vol. 94, No. 1, pp. 395–402 (In Russian)
Pavlov P.V., Tetel'baum D.I., Skoupov V.D., Semin Yu.A., Zorina G.V. (1986a):
  Abnormally deep structural changes in ion-implanted silicon, Phys. Stat. Sol.
  A. Vol.94, pp. 395–402 (In Russian)

Pavlov P.V., Skoupov V.D., Tetel'baum D.I. (1987): Mechanical stresses
and elastic waves and their role in structural changes of crystals during the
ionic bombardment and later annealing, Fizitcheskaya i Khimitcheskaya
Obrabotki Materialov (Physical and Chemical Treatments of Materials),
Moscow, No. 6, pp. 19–24 (In Russian)

Penina M.A., Nazarova L.B., Melev V.G. (1988): Effects of mechanical treatment
on the defects in gallium arsenide, Poverkchnost'. Fizika, khimiya, mekhanika
(Surface. Physics, chemistry, mechanics), No. 8, pp. 142–144 (In Russian)

Perevostchikov V.A., Skoupov V.D., Koptelov V.F. (1985): A way
of treating the hard materials, USSR patent No. 1324198, IPC B24B1/00
(In Russian)

Perevostchikov V.A., Skoupov V.D., Koptelov V.F. (1985a): A way
of treating the hard materials, USSR patent No. 1305993, IPC B134B1/00
(In Russian)

Perevostchikov V.A., Skoupov V.D., Koptelov V.F. (1985b): A way
of treating the semiconducting wafers, USSR patent No. 1299401, IPC HO1L
21/324 (In Russian)

Perevostchickov V.A., Skoupov V.D. (1987): Structural changes near the back
side of silicon wafers during a one-sided grinding, Optiko–Mekhanitcheskaya
Promyshlennost' (OMP) (Opticomechanical Industry), Moscow, No. 6,
pp. 35–36 (In Russian)

Perevostchikov V.A., Skoupov V.D (1987a): Structural changes in silicon crystals
during chemical cleaning and annealing of the surface, the Interinstitute
collection "Poloutchenie i analiz tchistyh veshestv (Production and analysis of
pure substances), Nizhny Novgorod State University after N.I. Lobatchevsky,
Nizhny Novgorod, pp. 19–25 (In Russian)

Perevostchikov V.A., Skoupov V.D. (1989): Postoperation accumulation
of defects in mechanically treated silicon wafers, Optiko–Mekhanitcheskaya
Promyshlennost' (OMP) (Opticomechanical Industry), No. 5, pp. 41–44 (In
Russian)

Perevostchikov V.A., Skoupov V.D. (1990): The Long–Range Effect during
Abrasive and Chemical Treatment of Semiconducting Surfaces, Nizhny
Novgorod, the VINITI deposition number 5680–B90, 28p. (In Russian)

Perevostchikov V.A., Skoupov V.D., Tsypkin G.A. (1990): A way
of treating the wafers, USSR patent No. 1736302, IPC H01L21/306
(In Russian)

Perevostchikov V.A., Skoupov V.D. (1992): Specificities in Abrasive
and Chemical Processing of Semiconductor Surfaces, Nizhny Novgorod State
University after N.I. Lobatchevsky, Nizhny Novgorod, 198 p. (In Russian)

Perevostchikov V.A., Skoupov V.D., Shengourov V.G. (1994): Mechanical
and chemical properties of the "porous silicon–monocrystalline silicon"
structures, Elektronnaya Tekhnika (Electronic Engineering), Moscow,
Ser. 7: Technology and arrangements for equipment production, No. 1 (182),
pp. 10–13 (In Russian)

Perevostchikov V.A., Skoupov V.D. (1994): The reversible long–range effect in semiconducting crystals, Pis'ma v Zhournal Teoretitcheskoy Fiziki (ZhTF) (Letters to Journal of Theoretical Physics), Moscow, Vol. 20, issue 23, pp. 12–16 (In Russian)

Perevostchikov V.A., Skoupov V.D. (1994a): A use of long–range effect in preparing a surface in semiconducting wafers, Collection of reports at International conference "Microelectronics–94", Zvenigorod, Russia, part. 2, pp. 281–282

Perevostchikov V.A. (1995): Processes of chemicodynamical polishing of a semiconductor surface, Vysokotchistye veshestva (Highly Pure Substances), Moscow, No. 2, pp. 5–29 (In Russian)

Perevostchikov V.A., Skoupov V.D. (1997): Physical and Chemical Fundamentals of Processing the Semiconductor Surfaces, Nizhny Novgorod State University after N.I. Lobatchevsky, Nizhny Novgorod, 254p. (In Russian)

Perevostchikov V.A., Skoupov V.D., Shengourov V.G (1998): Multilayer construction of structures with porous silicon, Poverkchnost'. Fizika, Khimiya, Mekhanika (Surface. Physics, Chemistry, Mechanics), Moscow, No. 4, pp. 44–46 (In Russian)

Perevostchikov V.A., Skoupov V.D. (1999): Long–range gettering of microdefects in silicon monocrystals when forming on their surface the porous silicon layers during ionic irradiation, Pis'ma v Zhournal Teoretitcheskoy Fiziki (ZhTF) (Letters to Journal of Theoretical Physics), Vol. 25, issue 8, pp. 50–54 (In Russian)

Perevostchikov V.A., Skoupov V.D. (2002): Gettering the Impurities and Defects in Semiconductors, Nizhny Novgorod State University after N.I. Lobatchevsky, Nizhny Novgorod, 220p. (In Russian)

Perevostchikov V.A., Skoupov V.D. (2002a): Nanometric changes in the structure of ionically–irradiated silicon crystals at the depths exceeding the run of the incorporated ions, Pis'ma v Zhournal Teoretitcheskoy Fiziki (ZhTF) (Letters to Journal of Theoretical Physics), Vol. 28, issue 14, pp. 77–82 (In Russian)

Perevostchikov V.A., Skoupov V.D. (2004): Minor effects in semiconductor crystals during abrasive and chemical treatments of the surface, Reports of the A.M. Prokhorov Engineering Academy "Technology of materials and components of electronic engineering", Moscow–Nizhny Novgorod, Vol. 7, pp. 28–34 (In Russian)

Petroff P. M., De Kock A. Y. R. (1975): Characterization of swirl defects in floating–zone Si crystals, J. Crystal Growth, No. 30, pp. 117–124.

Petrova V.Z., Koshelev N.I., Yermakov A.I. et al. (1998): Technology of forming the SOI–structures and multiplayer silicon structures, Izvestiya Vouzov. Elektronika. (Reports of Higher Education Institutes. Electronics), No. 5, pp. 30–33 (In Russian)

Pilyankevitch A.N. (ed) (1987): Effects of high pressures on a substance, Naukova doumka (Scientific Thought), Kiev, Vols. 1-2, 237p. (In Russian)

Polyakova A.L., Shklovskaya–Kordi V.V. (1969): Electrical characteristics of the silicon $p - n -$ junctions subjected to inhomogeneous deformation, In: Physics of Electronic–Hole Junctions and Semiconducting Devices, Nauka (Science), Leningrad, pp. 141–147 (In Russian)

Polyakova A.L. (1979): Deformation in Semiconductors and Semiconducting
Devices, Energiya (Energy), Moscow, 168 p. (In Russian)

Pomerantz D. (1967): A cause and cure of stacking faults in silicon epitaxial
layers, J. Appl. Phys., Vol. 38, No.13, pp. 5020–5026

Popov V.P. (1998): Producing the SOI-structures for ultralarge integrated circuits,
Izvestiya Vouzov. Elektronika. (Reports of Higher Education Institutes.
Electronics), No. 5, pp. 22–29 (In Russian)

Prokofyeva V.K., Sokolov E.B., Sergeeva Zh.M. et al. (1991): Optimizing
the process of internal getter formation in silicon substrates with various
contents of oxygen, Elektronnaya Tekhnika (Electronic Engineering), Moscow,
Ser. 6:Materials, issue 6 (260), pp. 26–29 (In Russian)

Puzanov N.J., Eidenson A.M. (1992): The effect of thermal history during crystal
growth on oxygen precipitation in Czochralski-grown silicon, Semiconduct.
Sci. Technol. Vol. 7, No.3, pp. 406–413 (In Russian)

Puzanov N.J., Eidenson A.M. (1997): Selective interaction of twin boundaries
with vacancies and self-interstitials in dislocation-free Si tetracrystals,
J. Cryst. Growth, Vol. 178, No. 4, pp. 459–467

Radcliffe. S.V. (1970): Pressure induced effects on defect structure and properties.
In: Mechanical Behaviour of Materials under Pressure, ed. H. L. D. Pugh,
Elsevier Publishing Company Ltd, Amsterdam – London – New York,
pp. 638–679

Ravi K. V. (1974): On the annihilation of oxidation induced stacking-faults
in silicon, Phil. Mag., Vol. 30, No. 5, pp. 1081–1090

Ravi K. V. (1981): Imperfections and Impurities in Semiconductor Silicon,
A Wiley–Interscience Publication, John Wiley & Sons, New York, Chichester,
Brislane, Toronto

Roudenko T.E., Roudenko A.N., Nazarov A.N. et al. (1994): Electrophysical
properties of zone–melt recrystallization of SOI–structures: the studying
techniques and experimental results, Mikroelektromika (Microelectronics),
Moscow, Vol. 23, issue 6, pp. 18–31 (In Russian)

Roumack N.V., Soukchanova N.B. (1982): Effects of diffusing nickel, zirconium
and palladium on the characteristics of the $Si - SiO_2$ interface, Elektronnaya
Tekhnika (Electronic Engineering), Moscow, Ser. 3: Microelectronics, issue 1
(97), pp. 33–36 (In Russian)

Rozgonyi G. A., Kyshner R. A. (1976): The elimination of stacking faults
by preoxidation gettering of silicon wafers, J. Electrochem. Soc.: S-S Sci.
and Technologies, Vol. 123, No.4, pp. 570–576

Rozgonyi G. A., Pearce C. W. (1977): Gettering O_2 in monocrystalline silicon,
Appl. Phys. Lett., Vol. 31, No.5, pp. 343–345

Runge H., Wörl H. (1978): Autoradiographic detection of getter effect
of argon-implanted layers in silicon, Phys. Stat. Sol. (a), Vol. 45, pp. 509–512

Rzaev S.G. (1991): Electrical activity of secondary dislocations in silicon,
Elektronnaya Tekhnika (Electronic Engineering), Moscow, Ser. 2:
Semiconducting Devices, issue 7 (261), pp. 23–26 (In Russian)

Sally I.V, Fol'kovitch E.S. (1970): Semiconducting Silicon Production,

Metallurgia (Mettalurgy), Moscow (In Russian)

Schwuttke G. H. (1978): Damaged profiles in silicon and their impact on device reliability, Tech. Rep., No.7, AD-A022756

Seidel T. E., Meck R. L., Cullis A.G. (1975): Direct comparison of ion-damaged and phosphorus diffusion gettering of Au in Si, J. Appl. Phys., Vol. 46, No. 2, pp. 600–601

Shapovalov V.P., Gryadoun V.I., Korolev A.E. (1995): Defect formation in the silicon surface region during thermal oxidation, Fizika i Tekhnika Polouprovodnikov (FTP) (Physics and Techniques of Semiconductors), Moscow, Vol. 29, issue 11, pp. 1995–2001 (In Russia)

Shelpakova I.R., Yudelevitch I.G., Aupov B.M. (1984): The By-layer Analysis of Electronic Materials, Nauka (Science), Novosibirsk, 181p. (In Russian)

Shinyaev A.Ya. (1973): Phase Transformations and Properties of Alloys under High Pressure, Nauka (Science), Moscow, 154p. (In Russian)

Sigson T. W., Cspredi L., Mayer I.W. (1976): Ion-implantation gettering of gold in silicon, J. of Electrochem. Soc., Vol. 123, No.7, pp. 1116–1117

Sinha A.K., Sheng T.T. (1978): The temperature dependence silicon substrates, Thin Solid Films, Vol. 48, No. 1, pp. 117–126

Skoupov V.D., Uspenskaya G.I. (1975): The combined X-ray spectrometer for measuring deformation in monocrystals, Pribory i Teknika Eksperimenta (PTE) (Instruments and Techniques of Experiment), Moscow, No. 2, pp. 210–213 (In Russian)

Skoupov V.D., Pashkov V.I. (1979): The X-ray measuring of small deformations in monocrystalline slices, Pribory i Teknika Eksperimenta (PTE) (Instruments and Techniques of Experiment), No. 1, pp. 212–213 (In Russian)

Skoupov V.D. Sherban' M.Yu. (1986): Relaxing the residual stresses in the semiconducting hydrostatically–compressed structures, Fizika i Teknika Vysokih Davleny (Physics and Techniques of High Pressures), Moscow, No. 21, pp. 24–27 (In Russian)

Skoupov V.D., Tetel'baum D.I. (1987): The effects of elastic stresses on the transformation of defect aggregates in semiconductors, Fizika i Tekhnika Polouprovodnikov (FTP) (Physics and Techniques of Semiconductors), Moscow, Vol. 21, issue 8, pp. 1495–1497 (In Russian)

Skoupov V.D, Tzypkin G.A., Shemtchourov V.G. (1989): The effects of hydrostatic pressure on the Schottky diode characteristics, Fizika i Tekhnika Polouprovodnikov (FTP) (Physics and Techniques of Semiconductors), Moscow, Vol. 23, issue 3, pp. 554–556 (In Russian)

Skoupov V.D., Tetel'baum D.I., Zhengourov V.G. (1989): The impact of extended defects in original crystals on the long–range effect during ionic implantation, Pis'ma v Zhournal Teoretitcheskoy Fiziki (ZhTF) (Letters to Journal of Theoretical Physics), Vol. 15, issue 22, pp. 44–47 (In Russian)

Skoupov V.D. (1994): Mechanical and chemical properties of the "porous silicon–monocrystalline silicon" structures, Elektronnaya Tekhnika (Electronic Engineering), Ser. 7: Technology and Arrangements to Produce the Equipment, Moscow, No. 1 (182), pp. 10–13 (In Russian)

Skoupov V.D. (1996): A use of irradiation to form an internal getter in silicon wafers, An abstract of the report at the 1-st All-Russia conference on the theory of materials and physicochemical fundamentals of manufacturing the doped silicon crystals, Moscow Building Institute (MISI), Moscow, p. 127 (In Russian)

Skoupov V.D., Smolin V.K. (2000): Increasing the sensitivity of the selective chemical etching to microdefects in silicon during the alpha–particle irradiation, Zavodskaya Laboratorya. Diagnostika Materialov (Industrial Laboratory. Diagnostics of Materials), Moscow, Vol. 66, No. 4, pp. 31–33 (In Russian)

Smirnov L.S. (ed) (1980): Physical Processes in Irradiated Semiconductors. Nauka (Science), Novosibirsk, 242p. (In Russian)

Smirnov L.S. (ed) (1980a): Problems of Irradiation Technology for Semiconductors, Nauka (Science), Novosibirsk, 296 p. (In Russian)

Smoul'scky A.S. (1979): Dislocation–Free Silicon and Production of Contemporary Semiconducting Devices, Obzory po Elektronnoy Tekhnike (Reviews on Electronic Engineering), Ser. 2: Semiconducting Devices, Moscow, issue 12, 59p. (In Russian)

Sprokel G. J., Fairfield J. M. (1965): Diffusion of gold into silicon crystals, J. Electrochem. Soc., Vol. 112, No. 2, pp. 200–203

Stengl R., Tan T., Gosele U. (1989): A model for the silicon wafer bonding process, J. Appl. Phys., Japan, Vol. 28, No. 10, pp. 1735–1741 (In English)

Stepanov V.A.,Pestchanskaya N.N., Speizman V.V. (1984): Strength and Relaxational Phenomena in Solids, Nauka (Science), Leningrad, 246p. (In Russian)

Stroukov F.V., Khromov S.S., Astakhov V.P. (1992): Planar gettering during a high-temperature oxidation of silicon, Mikroelektronika (Microelectronics), Moscow, Vol. 21, issue 2, pp. 91–93 (In Russian)

Sugita Y., Shimizu H., Yoshinake A. (1974): Shrinkage and annihilation of stacking faults in silicon, J. Vac. and Technol., Vol. 14, No. 1, pp. 44–46

Syemin Yu.A., Skoupov V.D., Tetel'baum D.I. (1988): Strengthening the ionically–induced elastic waves during their travel in crystals with defect clusters, Pis'ma v Zhournal Teoretitcheskoy Fiziki (ZhTF) (Letters to Journal of Theoretical Physics), Vol. 21, issue 8, pp. 1495–1497 (In Russian)

Takeno H., Kato M., Kitagowara Y. (1996): Proc. 2-nd International Symposium on Advanced Technologies of Silicon Materials. Ed. by M. Umeno. Osaka, Japan, pp. 294–299

Takeshi H., Toshiharu S. (1978): Elimination of stacking fault formation in silicon by preoxidation annealing in N_2:HCl:O_2 mixtures, Appl. Phys. Lett., Vol. 33, No. 4, pp. 347–349

Tarui Ya. (1985): Technological Fundamentals for Superhigh-Speed Integrated Circuits, Translation from Japanese, Radio i Svyaz' (Radio and Communication), Moscow, 480p. (Tr. from Japanese)

Tchernyaev V.N. (1987): Technology for Manufacturing the Integrated Microcircuits and Microprocessors. Textbook for institutes of higher education. Radio i Svyaz' (Radio and Communication), Moscow, 464p. (In Russian)

Tchistyakov Yu. D. and Rainova Yu.P. (1979): Physicochemical Fundamentals of Technologies in Microelectronics, Metallurgia (Metallurgy), Moscow, 408p. (In Russian)

Tikhomirov G.B., Kititchenko T.S., Korovin A.P. et al. (1984): A study of deep levels in thin heteroepitaxial layers of silicon on sapphire, Elektronnaya Tekhnika (Electronic Engineering), Ser. 2: Semiconducting Devices, Moscow, issue 4 (170), pp. 21–25 (In Russian)

Tikhomirov G.B., Korovin A.P., Korotkova N.V. et al. (1984): Regulating the electrophysical parameters of submicronic layers of silicon on sapphire, Elektronnaya Tekhnika (Electronic Engineering), Ser. 2: Semiconducting Devices, Moscow, issue 6 (172), pp. 37–45 (In Russian)

Tucker J.W., Rampton V.W. (1972): Microwave Ultrasonics in Solid State Physics. North-Holland Publishing Company, Amsterdam

Tyapounina N.A., Naimi E.K., Zinenkova G.M. (1999): Ultrasonic Action on the Crystals with Defects, Moscow State University, Moscow, 238p. (In Russian)

Ustinov V.M., Zakharov B.G. (1977): Macrostresses in Epitaxial A_3B_5 − Based Structures, Obzory po Elektronnoy Tekhnike (Reviews on Electronic Engineering), Ser. 6: Materials, Moscow, issue 4 (492), 34p. (In Russian)

Vanhellemont J. (1999): Defect engineering in the development of advanced silicon crystals and wafers, Solid State Phenomena, Vol. 69-70, pp. 111–120

Vasilyeva E.D., Sokolov V.N., Shapiro I.Yu. (1991): Effects of defective structure of silicon wafers on the internal getter formation and the parameters of the $Si - SiO_2$ interface, Mikroelektronika (Microelectronics), Moscow, Vol. 20, issue 4, pp. 392–396 (In Russian)

Vasilyeva E.D., Kolotov M.N., Sokolov V.I. et al. (1992): The properties of the $Si - SiO_2$ interface and the internal gettering process in MOS–and SMOS–structures, Mikroelektronika (Microelectronics), Moscow, Vol. 21, issue 5, pp. 74–80 (In Russian)

Vasin A.S., Okoulitch V.I., Panteleev V.A., Tetel'baum D.I. (1985): The effects of pressure on the recrystalization rate of the amorphous silicon layer during a postimplantation annealing, Fizika Tverdogo Tela (FTT) (Physics of a solid body), Moscow, Vol. 27, issue 1, pp. 274–177 (In Russian)

Vasudev K. (1983): Analysis of leakage currents in CMOS/SOS devices, Mikroelektronika (Microelectronics), Moscow, Vol. 14, No. 6, pp. 45–48 (In Russian)

Vavilov V.S., Kiselev V.F., Moukhashov B.N (1990): Defects in Silicon and on its Surface, Nouka (Science), Moscow, 216p. (In Russian)

Velchev N., Toncheva L., Dimitrov T. (1980): Electrical properties of MOS-structures with process-induced defects, Crystal Lattice Defects, Vol. 8, No. 4, pp. 154–166 (In Russian)

Vendik O.G., Gorin Yu.N., Popov V.F. (1984): Corpuscular–Photonic Technology, Vysshaya Shkola (Higher Education School), Moscow, 240p. (In Russian)

Verkhovsky E.I. (1981): Gettering Techniques in Silicon, Obzory po Elektronnoy

Tekhnike (Reviews on Electronic Engineering), Ser. 2:Semiconducting
   Devices, Moscow, issue 2 (838), 48p. (In Russian)
Vigdorovitch V.N., Kryukov V.L., Fourmanov G.P. (1982): Spacial processes
   of regeneration and recombination of interstitial atoms of silicon as factors of
   impurity gettering, Doklady Akademii Nauk (DAN) (Reports of the Academy
   of Sciences), Moscow, Vol. 325, No. 6, pp. 1181–1185 (In Russian)
Vinetski V.L., Kholodar' G.A. (1979): Irradiational Physics of Semiconductors,
   Naukova doumka (Scientific Thought), Kiev, 336p. (In Russian)
Volkov A.F., Zaitzev N.A., Sourovikov M.B. (1983): Impact of thermal
   operations on silicon performance, Obzory po Elektronnoy Tekhnike (Reviews
   on Electronic Engineering), Ser. 6: Materials, Moscow, issue 10 (992), 49 p.
   (In Russian)
Volle V.M., Voronkov V.B., Grekhov I.V. et al. (1991): Studying a homogeneity
   and electrical characteristics of semiconducting $p^{+}-n$, $n-n$, $p-p-$
   structures formed by direct silicon bonding, Elektronnaya Tekhnika (Electronic
   Engineering), Ser. 6:Materials, Moscow, issue 7 (261), pp. 60–62 (In Russian)
Vorobyev N.N., Ignatyeva L.A., Kleinfeld Yu.S. et al. (1979): Effects
   of crystallographic defects in silicon on the parameters of powerful transistors,
   Elektronnaya Tekhnika (Electronic Engineering). Ser. 3:Microelectronics,
   Moscow, issue 5, pp. 95–99 (In Russian)
Voronkov V.V. (1982): J. Cryst. Growth, Vol. 59, pp. 625–636
Voronkov V.V., Falster R.J. (1998): Vacancy-type microdefect formation
   in Czochralski silicon, J. Cryst. Growth, Vol.194, No.1, pp. 76–88
Voronkova V.V., Zhoukova L.A., Mil'vidsky M.G. (1997: Stability of the regular
   step structure of vicinal surface during gaseous phase epitaxy, Crystallography,
   Vol. 42, No. 6, pp. 1114–1123 (In Russian)
Vyatkin A.F., Italyantsev A.G., Kopetsky Tch.V. et al. (1986): Recombination
   of structural defects in semiconductors stimulated by chemical reactions on the
   crystal surface, Poverkchnost'. Fizika, khimiya, mekhanika (Surface. Physics,
   chemistry, mechanics), Moscow, No. 11, pp. 67–73 (In Russian)
Vysotskaya V.V., Gorin S.N., Menshikova V.A. (1984): Effects of thermal
   processes on forming defects in silicon, Elektronnaya Tekhnika (Electronic
   Engineering). Ser. 6:Materials, Moscow, issue 1 (186), pp. 26–27 (In Russian)
Weber S. (1987): A better way to protect VLSI circuits from radiation,
   Electronics, A McGrow-Hill publication, November 26, 1987. Vol. 60, No. 24,
   pp 127–129
Westwood A. (1972): The effects of adsorption on the hardness and mobility
   of dislocations near the surface of nonmetals, In: Micropasticity, Metallurgia
   (Mettalurgy), Moscow, pp. 301–344 (In Russian)
Wittkower A. (1999): New technology boots chip performance, Vacuum
   Solutions, 1999 March / April, p. 1641
Yackovencko A.G., Gvelesiani A.A. (1975): Oxygen in silicon, Zarubezhnaya
   Elektronnaya Tekhnika (Foreign Electronic Engineering), Moscow, No. 14,
   pp. 23–43 (In Russian)
Yefimov I.E., Kozyr' I.Ya., Gorbounov Yu.I. (1986): Microelectronics. Physical

and Technological Foundations, Reliability. Textbook for special instrumentation constructing institutes, Vysshaya Shkola, Moscow, 464p. (In Russian)

Yemtzev V.V., Mashovetz T.V. (1981): Impurities and Point Defects in Semiconductors, Radio and Svyaz' (Radio and Communication), Moscow, 248p. (In Russian)

Yenisherlova K.L., Marounina N.I. (1984): Uncontrolled impurities in silicon wafers and their redistribution during a gettering process, Elektronnaya Tekhnika (Electronic Engineering), Ser. 2: Semiconducting Devices, Moscow, issue 2 (168), pp. 10–15 (In Russian)

Yenisherlova K.L., Alekhin A.N., Mordkovitch V.N. et al. (1991): The ion-doped layers as a getter for epitaxial structures, Elektronnaya Tekhnika (Electronic Engineering), Ser. 6: Materials, Moscow, issue 6 (260), pp. 17–19 (In Russian)

Yenisherlova K.L., Batchourin V.V., Kontzevoy Yu. A. et al. (1991): Composite silicon structures for high-volt MDS–transistors, Elektronnaya Tekhnika (Electronic Engineering), Ser. 6:Materials, Moscow, issue 7 (261), pp. 74–79 (In Russian)

Yenisherlova K.L., Mil'vidsky M.G., Reznik V.Ya. et al. (1991): Studying the specificities of internal getter formation in silicon structures, Elektronnaya Tekhnika (Electronic Engineering), Ser. 6:Materials, Moscow, issue 6 (260), pp. 29–32 (In Russian)

Yenisherlova K.L., Rousak T.F., Shmelev G.G. et al. (1994): Studying the processes of directly mating silicon wafers when forming the SOI–structures, Mikroelektronika (Microelectronics), Moscow, Vol. 23, issue 6, pp. 46–54 (In Russian)

Yershov S.N. (1978): Migration of Intrinsic Point Defects in Various Charge States in Elementary Semiconductors, PhD dissertation, Nizhny Novgorod State University after N.I. Lobatchevsky (In Russian)

Zaitzev N.A., Pavlov E.A., Potapov E.V. et al. (1981): Impact of the cooling rate on structural and electrophysical properties of the $Si - SiO_2$ interface, Elektronnaya Tekhnika (Electronic Engineering), Ser. 3: Microelectronics, Moscow, issue 6 (99), pp. 13–17 (In Russian)

Zaitzev V.I. (1983): The physics of plasticity in the hydrostatically–compressed crystals, Naukova doumka (Scientific Thought), Kiev, 183p. (In Russian)

Zaitzev V.I., Toky V.V. (1987): The specificity of dislocational mechanism in plastic deformation during the гидроэкструзии and other types of pressures, In: The effects of High Pressures on the Substance, Naukova doumka (Scientific Thought), Kiev, Vol. 2, pp. 5–25 (In Russian)

Zeeger A., Föll H., Frank W. (1977): Intrinsic interstitial atoms, vacancies and their accumulations in silicon and germanium. In: Radiation Effects in Semiconductors, The Institute of Physics, Bristol and London

Zeeger A., Fell X., Frank W. (1979): Intrinsic interstitial atoms, vacancies and their aggregates in silicon and germanium, In: Point Defects in Solids, Mir (World), Moscow, pp. 163–186 (In Russian)

Zorina G.V., Popov Yu. S., Tetel'baum D.I., Shargel' V.L. (1983): The effects
of long-range actions in the ionically-irradiated silicon, Abstracts
for the reports at the International Conference on Ionic Implantation
in Semiconductors and Other Materials, Vilnius, Lithuania, pp. 193–194
(In Russian)

Zouev V.A., Larionova T.P., Minaev N.S. (1988): Defects of structures
in heteroepitaxial films of silicon on sapphire irradiated by high–energy
particles, Elektronnaya Tekhnika (Electronic Engineering), Ser. 6: Materials,
Moscow, issue 4 (233), pp. 32–36 (In Russian)

Subject Index

Authors' Index

Aglaumov S.N.
Agranat B.A.
Alekhin V.P.
Amal'skaya R.M.
Ammon W.
Aspar Bernard
Astakhov V.P.
Auberton H.A.J.
Ayupov B.M.

Bagraev N.T.
Bakchadyrkchanov M.K.
Baldi L.
Banishev A.F.
Baranov I.L.
Baranov Yu.L.
Baransky P.I.
Barch G. R.
Barton V.
Bashkirov V.I.
Batavin V.V.
Batchourin V.V.
Bazhinov A.N.
Bedny B.I.
Beresnev B.I.
Bertcha A.I.
Bethe H.A.
Bevzouck N.S.
Bir G.L.
Biryukov V.M
Bobitsky Ya.V.
Bobrova E.A.
Bogatch N.V.
Boiko V.S.
Bondarenko V.P.
Bookker G. R.
Borisenko V.E.

Borzakov Yu.I.
Boud'ko M.S.
Bourago N.N.
Bourbelo M.P.
Bourgoin J.
Bourmistrov A.N.
Boushouev V.A.
Bruel Michel
Buck T.M.
Bullough R.

Cerofolini G.F.
Chang Z. P.
Chernikov A.M.
Colinge J.P.
Corbett Z.W.
Craven R.A.
Cullen G.W.
Cullis A.G.
Cspredi L.

Damask A.C.
Davidovsky V.M.
Davydov A.A.
Demitrouk N.L.
Dimitrov T.
Demidov E.S.,
Dienes G.J.
Dolgy L.N.
Dornberger E.
Dorofeyev A.M.
Dougaev V.E.
Drouzhkov A.P.
Dumin D.J.
Dvouretchensky A.V.
Dzafarov T.D.

Edel'man F.L.
Eidenson A.M.
Einspruch N. G.

Fairfield J.M.
Falster R.J.
Fattakhov E.A.
Ferla G.
Fifer V.N.
Filatov D.K.
Fisher G.
Fistoul' V.I.
Főll H.
Fol'kovitch E.S.
Foox M.Ya.
Fourmanov G.P.
Frank F.
Frank W.
Friedel J.
Frigerio G.
Frolov O.S.

Gaidoukov G.N.
Galkin G.N.
Garber R.I.
Garnak A.E.
Gastev V.V.
Gegouzin Ya.E.
Geipl H. S.
Genkin V.M.
Gerasimenko N.N.
Gerward L.
Glaenzer R.H.
Goetsberger A.I.
Gorbounov Yu.I.
Gorelyenok A.T.
Gorin S.N.

Springer Series in
ADVANCED MICROELECTRONICS

Printing: Strauss GmbH, Mörlenbach
Binding: Schäffer, Grünstadt